CLIMATE AND HUMAN CHANGE

DISASTER OR OPPORTUNITY?

CLIMATE AND HUMAN CHANGE

DISASTER OR OPPORTUNITY?

JONATHAN COWIE
Northumberland Heath, Kent, UK

The Parthenon Publishing Group
International Publishers in Medicine, Science & Technology

NEW YORK LONDON

Published in the USA by
The Parthenon Publishing Group Inc.
One Blue Hill Plaza
PO Box 1564, Pearl River
New York 10965, USA

Published in the UK by
The Parthenon Publishing Group Ltd
Casterton Hall, Carnforth
Lancs. LA6 2LA, UK

First published 1998

Library of Congress Cataloging-in-Publication Data

Cowie, Jonathan
Climate and human change : disaster or opportunity / by Jonathan Cowie
p. cm.
Includes bibliographical references and index.
ISBN 1-85070-971-8
1. Environmental policy. 2. Climatic changes—Social aspects.
I. Title.
GE170.C69 1997
363. 736'74—dc21 97–24533
CIP

British Library Cataloguing in Publication Data

Cowie, Jonathan
Climate and human change : disaster or opportunity
1. Climatic changes 2. Global warming 3. Man – Influence of climate
4. Environmental policy
I. Title
551.6

ISBN 1-85070-971-8

Cover picture: Land, sea, atmosphere interactions by Peter Tyres

Typeset by Books Unlimited, Nottingham NG16 4FS

Printed and bound by J. W. Arrowsmith Ltd., Bristol, UK

CONTENTS

ACKNOWLEDGEMENTS

Left to myself this work, certainly nothing like it, would probably never have seen the light of day. As can be discerned from the following introduction, this is not a specialist text other than its specialism is highly interdisciplinary. To cover such a broad expanse, indeed to attempt such an exercise, requires help and encouragement. It is impossible to mention everyone who has assisted for a variety of reasons. For instance I should individually thank *all* those who replied to my appeal for information on perception and the greenhouse effect in *The Psychologist*, but trust that thanking the editor of that journal will largely suffice. However, of those who replied I found the contributions, or lines of enquiry identified, by Joanne Borril and Dr Ken Manktelow particularly useful. Then there were those encountered before (in the organising of), during, and as a consequence of the 1995 International Ecology Congress, and those who in passing through The Institute of Biology in Queensberry Place provided comment, snippets and the occasional offprint. There were also those whose help (in lifting a corner of the curtain to see back stage) was given in confidence.

Not least I am indebted to a number of librarians dotted about the country, but in particular to those at Salford University during three short periods of study over six years. It has to be said that these folk provided an excellent, friendly and helpful service and are a credit to their profession.

I must, of course, thank those who have spent some considerable time either on the text itself and/or for encouragement: my two principal readers Simon Geikie and Tony Chester; the Concatenation Science team and Louise Tyres for help with some of the diagrams and further encouragement; Elaine Sparkes for not being too bored those evenings Simon and I were involved with dishwashers' energy ratings, biogeocycles and such. Meanwhile Helen Lee has done wonders subbing the manuscript and taking the publishing forward.

However I really would like to give special thanks to Hertfordshire University's Dr Brian Perry (for his 'global overviews') and Dr Alan Cheshire (who [unknowingly?] sparked me off on palaeoclimatology); they fanned my original interest in energy and climate change issues and their guidance, back when I was an undergraduate, set the chain of events in motion that ultimately led to this work. In addition, for a similar role that enabled me to build on this interest, my appreciation goes to Dr Mike Pugh-Thomas and Rev Dr Stan Frost of Salford University's Environmental Resources Unit.

Naturally it goes without saying that any errors or wayward thoughts you, the reader, may detect herein are my very own, and not the good folk above.

Jonathan Cowie
Northumberland Heath
Kent

INTRODUCTION

This book has been written as an exercise in the public understanding of science. At this point I should state that one of my extra-curricular activities consists of an exercise that many refer to as the 'public understanding of science'. There is a fine line between the *understanding* of science and its *appreciation* to which many public understanding of science exercises actually pander. However in this instance, this text is very much one that will hopefully improve understanding first, if not stimulate the appreciation of science second.

You may, or may not, be surprised at the range of opinions within the scientific community on climate change issues: I certainly have been. This surprise has been compounded by the numerous specialist perspectives from which as disparate a range of views are expressed. While many have been stimulating, sadly not all of these views have been rigorously formed and all too often relevant key grounding has been missing: the example of the peat bog specialist on methane fluxes being unaware of the Greenland ice core record of methane is all too illustrative. If scientists have problems how then are policy makers, let alone the public, expected to obtain a handle on climate change issues? What are the questions to ask? When asking such questions, is it important to recognise that current global warming may or may not be absolute evidence of human impact on climate? Above all, what ever may (or may not) be happening, should we do anything about it or will human life continue unabated? These are the bottom-line questions that policy-makers need to answer.

Then again, why yet another book on climate change? There certainly have been many works on the subject. Yet, if one engages in a computer-assisted literature search, one soon finds that, apart from a handful of texts, nearly all 'greenhouse-type' books look at global climate principally from just one or two specialist dimensions. There are works that relate climate change to sea level rise, or alternatively energy policy, or agricultural impact, or economics, or deforestation, or renewable energy, or low energy strategies, or environmental perception. But if you key in 'climate change' *or* 'greenhouse effect' *and* just three of the above key-word subject areas you will be surprised at the paucity of response: in fact key in four or five of the above and most commonly used bibliographies in universities will give you no citations at all! This text then is to provide an *introductory overview* in *all* these areas. Hopefully the sociologist will find the ecology of interest, the energy policy analyst the palaeoclimatology, the psychologist the sea-level rise ... whatever. From such an introduction this book's references can begin to illuminate paths that can be followed in an academic library.

What this text will *not* do is provide (for instance) an ecology undergraduate with a sound grounding in the ecology of climate change: there are many specialist ecology texts devoted to that subject (run a computer-assisted literature search at any university library and see). Instead, this book provides an introductory interdisciplinary overview, albeit with an opinion or two at its very end – a finale with which you can either agree or disagree. It *is* a whirlwind tour of the issues and concerns, the science and the social science, of climate change and the possible human impact on the Earth's climate system. It is a starting point, a beginning, and as such has a number of simplifications and generalizations. For the detail, the full range of complications, the minutiae, undergraduates must engage in further study, but there are plenty of leads here from which to start.

Alternatively, if you are not a student but have an interest in things environmental, you will find that this book will present you with many of the numerous facets that connect climate change to our species. You need not be studying at university, for just as the ecology covered herein has to be understood by climatologists, and the economics and energy production aspects have to be comprehended by conservationists, you should not need any particular embryonic expertise to digest this book. For this reason, one of my readers has no formal post-16 year science education, but is a regular viewer of popular science documentaries on TV: his ability to cope with the concepts presented herein has been central to this text's writing. This book should therefore be of use to policy-makers who have no formal science training.

Turning now to how this book might be read and used, you may want to take on the following points. Because both the undergraduate student and the interested lay-environmentalist probably will only be interested in some (not all) of the ground covered, each chapter is relatively self-contained and a number of chapters may be skipped without unduly spoiling the reading of this book's conclusions. In addition to the chapters, the book is further divided into three parts. The first section briefly examines the global problems we face as a species, introduces international environmental policy and thinking from the 1970s to 1990s, and summarises the way the economics (in driving our planetary culture) relates to the environment. The second, and main, section reviews past climate change, how we may be affecting climate change today, and outlines the various existing relevant issues and policies. The final part looks at the problems of adapting our economics and policies to climate issues, comments on how we perceive and relate to climate (and other environmental) concerns, before discussing whether or not policy-makers need to adopt greenhouse policies today.

Through examining climate issues in an interdisciplinary way (not just with individual sciences but also the social 'sciences'), readers from many walks of life will hopefully obtain a feel for the issues at stake. From there on in understanding might percolate through to others in the public sector, from school pupils through to readers who are biology or geography teachers, or those neighbours sharing a beer at the end of a working day. This is the goal that makes this book an exercise in the public understanding of science. One can but hope.

A final important note. Given that some climatologists may read this work (as an introduction to energy policy, ecological, social issues and such (but not for the climatology)) they may well pick up on one simplification referred continually to herein. In considering whether or not climate change is taking place, and in looking at the climate-change-*is*-taking-place case, the 1990 Intergovernmental Panel on Climate Change (IPCC) Business-as-Usual forecast is used and not the later 1992 IPCC equivalent which was again used in the 1995 IPCC Assessment published in 1996 (albeit with a minor cooling due to regional short-term sulphur aerosol pollution). There are two reasons for this emphasis on the earlier 1990 forecast. While the latter forecast is mentioned there was a considerable wave of literature published after 1990 relating to the 1990 Business-as-Usual (B-a-U) projection, and where possible I have kept as close as possible to the references cited and have not modified them so as not to confuse should the student wish to follow up material for his or her self. Secondly a forecast or projection is only a guestimate (hopefully an educated one), it is not a reality. In terms of the broad argument, and the general overview, when considering whether human-induced climate change is taking place or not, there is little extra benefit to be gained from lowering a warming estimate over a century by 20% or 30%: the principal question is whether or not any significant human-induced climate change *is* taking place? In any case the IPCC forecasts are gentle curves over a century, whereas the reality is that the global climate warms and cools in fits and starts, so from a policy-making perspective a slightly different guestimate is of little consequence. (It is virtually inevitable that more rapid shifts in climate than the IPCC predicts may occur, even if the extent of the overall change is, as suggested by the IPCC in 1992 and 1996, less than the IPCC 1990 forecast). Readers, as is pointed out in the text (but which might be missed by those who skip-read) need not be worried by this use of the earlier B-a-U scenario. Indeed neither should climatologists (who if nervous in their proximity to the subject) should remember that *only* the IPCC's '*best estimates*' for the respective 1990 and 1992 Business-as-Usual forecasts differ. A close reading of the IPCC assessments reveals that the 'High estimate' for the 1992 forecast for the year 2100 is higher than the 'Low estimate' for the 1990 forecast. In short when looking at the detail of these forecasts (and not just the IPCC 'best estimate' I have used and others frequently quote) there is a considerable overlap. Furthermore, these estimates were arrived at by a committee of the IPCC after lengthy and considerable wrangling. They are meant only as a very general guide. So climatologists should not be worried that I have decided to make greater use of the less up-to-date, but more frequently quoted in the early

1990s, IPCC forecast. I would remind those still with concerns that future surprises (such as the possible collapse of the Broecker oceanic salt conveyor) will kick the IPCC forecasts right out of the window: indeed the IPCC made this very point in its 1995 assessment. Notwithstanding all of this, such detail as to whether warming over a certain time might be exactly *x*, *y* or *z* degrees is not a bottom-line question. The bottom-line question is whether 'human-induced climate change is taking place'? Fundamentally that question can be answered with a 'yes' or a 'no'. Indeed, as will be shown, this and other details, such as the precise costs incurred through climatic change, are not needed when deciding the other bottom-line question as to whether or not to instigate a low-energy greenhouse policy: but let me not spoil things for you.

If the above all seems a little too obvious, I should point out that in informally discussing the public understanding of science with colleagues some (fortunately a small minority) have opined that such exercises are frustrated by science's rapid pace of advancement. Indeed one referee for this work (fortunately just one) was of the opinion that unless the latest work – defined as published within the last three months – was included then books such as this would be hopelessly out of date. This might well be true for writing that is a detailed examination of a highly specialist climatological area, but not – I contend – when contemplating a broad-brush introductory, interdisciplinary overview. Furthermore, such short time horizons preclude the publishing of books whose peak sell lifetime is expected to be in the order of two to four years. While postgraduate researchers live off a diet rich in academic journal papers, undergraduates rely more on books, and the general public rely solely on the latter source. What hope for the public understanding or appreciation of science if overviews such as this are considered hopelessly out of date on publication? Furthermore, and paradoxically, as the piles of unused offprints littering my study testify, the vast majority of journal papers in being so specialised do date quickly. The trick, with the benefit of hindsight, is to see which material has stood the test of time and to include it.

Of course the above does pose the question as to why some (albeit a minority) of experts hold the view that understanding of science exercises must include the latest minutiae. My opinion, for what it is worth, is that academics over the past decade have become increasingly pressurised. Over this period the general trend in many OECD (developed) countries has been for their respective Governments to devote a smaller proportion of their Gross Domestic Product to state-funded research. Scientists on both sides of the Atlantic have felt the pinch and invariably some more so than others. In the UK the trend has increasingly been to funding more research through short-term contracts, typically three years long, which means that by the end of year two researchers start worrying about where their next funding is to come from. This drives researchers further into ever more specialist areas, gives them little opportunity to stand back, look around, and see the overall context (as opposed to their funding context) in which the work stands. While there are undoubted benefits to accrue from short-term contracts, I am far from sure that we currently have the balance right. Nor am I convinced that the current vogue for all UK university lecturers to engage in research, or for that matter for some researchers to lecture can be justified. This supposes that all good researchers are good communicators and *vice versa*, which is patently not true as anybody who has been a university inmate will testify! I make these points because science is the most powerful tool modern humans have developed. (Just try living for one week without using *any* of the technology arising from science over the past 150 years: no mains electricity, no filtered chlorinated mains water, no modern processed and packaged foods, not electric fridge, no automatic machines, manufactured clothes, etc, etc. In this light the lack of commercial radio, CDs, TV, video, etc becomes a minor inconvenience). If we are to sustain six billion (and counting) humans on the planet (let alone increasingly maintain them in the lifestyles to which they aspire) then we are going to have to use resources effectively and more sustainably; we are going to have to invest in science! Such acceptance of public funds morally requires that academics explain to the public what they are doing. Additionally, as our planet becomes more crowded and its problems more complex (which incidentally are not a result of science *per se*), we will need digestible appraisals of such issues. As part of this, specialists need to know how their work relates to others without becoming as expert in these other disciplines, as well as provide a window for interested non-experts. This book is just one such attempt.

All that remains for me to say is that, because climate change concerns potentially affect much of our species' activities, possibly greatly affecting our global culture's future, and in having their basis in the Earth's past climate change over many millennia (if not far longer), the whole subject makes for an intricately rich tapestry. It is one of the most fascinating detective stories our species has and is facing. It takes us orbiting the Sun through the millennia. It takes us from ice cores near the Earth's poles to tropical forests on the equator. It takes us to the heights of the Himalayas, to the oceans' abyssal currents. Indeed because we are right in the middle of it all – we have yet to encounter what our children and their descendants will face – numerous options are still open. What we have yet to do will determine its ending centuries away, so making this on-going detective story somewhat enigmatic. I hope you come to share my fascination. Now, on with the show ...

PART 1

THE GLOBAL COMMONS, POLICY AND ECONOMICS

CHAPTER 1

THE GLOBAL PROBLEM

Our planet is changing, and it is currently changing as it has never done before in its entire history. As our planet changes, we change, but we – our species, ourselves – *are* the primary causal agents. Humans have brought much beauty to, and demonstrated wondrous abilities in, the World. Yet we are altering our planet as we live and it, in turn, is altering our lives. This reciprocal arrangement arises from our being an intrinsic part of the planetary biosphere. But is part of our planetary impact an effect on the Earth's climate? If so, what effects are we talking about and what are the consequences? Indeed, to carry out an analysis of such grand and complex relationships, where does one begin?

We could start with the Earth itself. Our world is bountiful and, together with energy from the Sun, it nurtures all life on its surface. Yet today, while we still have just the one finite planet – all 5.976×10^{21} tonnes of it – there are over 6.0 billion of us humans with more on the way: this is more than enough to make a noticeable, a measurable, impact on the planetary biosphere, the very system that sustains our 6.0 billion fellows, you and me. As we shall see later, since the industrial revolution, human numbers have grown eightfold. Today's human environmental impacts are discernible in a number of ways: levels of pollution, changes in land use, amount of soil erosion, and loss of species, to cite just a few. Specific examples include the fact that in two centuries it is estimated that the planet has lost over 6 000 000 km^2 of forest, and that over 60 000 km^2 of agricultural land is made unproductive by erosion each year. Furthermore, over this time we have become more dependant on resources, as exemplified by the human use of freshwater which has grown from 100 to 3600 km^3 a year.

But does this matter? We have had soil erosion, pollution, and loss of species throughout human history, so is there any real cause for concern now? If there is, where does the balance of seriousness, or priority, lie? And if the concern warranted were great enough, can we do anything, or if we cannot, what then?

Cushioned by an international food chain, and warmed at the flick of a switch, it is easy for the developed nation readers of this book to forget the true environmental cost of these self-same systems of diet and warmth, or the scale of biosphere change they incur. Today, human environmental impacts are on a level that has not been seen since civilisation began, though of course this in it self does not imply terminal damage to the biosphere. Yet in terms of species loss, the last time there was a greater period of extinction was when the dinosaurs died out: that was 65 million years ago! That extinction was thought to be caused by an asteroid colliding with the Earth (which, incidently, scattered iridium planet-wide, and so provided us with the key clue to the event's passing). Today, while not as dramatically cataclysmic as an asteroid impact, the current wave of changes are still pronounced; they affect the biosphere's fundamental components and processes (though it must again be remembered currently not terminally). Using the criteria of atmospheric mix, carbon dioxide and methane concentrations are currently higher than at any time since the end of the last glacial some 12 000 years ago: we have to go back over 160 000 years to find higher concentrations. As for the CFCs (chlorofluorocarbons), our species has the dubious honour of being the first and only agent to introduce these into the biosphere.

All of these atmospheric alterations have received considerable attention recently. They have spurred scientists into studying our planet's climate; for changes in the atmospheric constituents affect the planet's energy budget as sure as night follows day (or, more appropriately, interglacial intercedes glacial). Similarly, the lay environmental movement together with politicians, especially over the past two decades, have become more concerned with the way our species is exploiting resources and changing our world. Oil consumption, for instance, has been steadily increasing for over a century since the development of the internal combustion engine; by 1979 World consumption had reached a peak of 65 780 thousand barrels a day[1], a pinnacle not to be topped for over half a decade (it has just been passed at the time of writing (1997)). The year following this peak in consumption, 1980, saw the publication of two international reports on the state of the planet. The first, the *World Conservation Strategy*[2] drew attention to the threat from unrestrained development on the natural systems that uniquely provide our species with the very biological resources which feed and clothe us, as well as help to keep us housed and warm. The second report focused on the unequal distribution and consumption of resources between the rich 'northern' and the poorer 'southern' countries, as well as the north-south interdependence – this publication is known as the *Brandt Report*[3]. (We will return to these reports later.) The subject matter of both documents could well be described as our 'Global Problem' – that is the uncoupling of the policies ensuring future resource supply from those of meeting current resource demand, while ignoring environmental impacts. Given that the present moves into the future, this is an inequitable situation that may (if not *will*) come to a head (given post-World War II and current trends prevailing) in just a few decades time. That is to say that matters could come to a climax somewhere in the middle of the twenty-first century.

Allegedly human-induced (or anthropogenic) global warming (resulting from the so-called 'greenhouse effect' induced by gases such as carbon dioxide and methane mentioned earlier) embodies virtually all the dimensions relating to this 'Global Problem'. These include:

- demography (human population changes);
- the manner and efficiency of resources use (the way we live);
- environmental perception;
- environmental economics and policy.

Importantly, it follows that the connection between alleged global warming and human activity cannot be broken. Carbon dioxide from the burning of fossil fuels (the 'anthropogenic', that is 'human-generated' contribution of carbon dioxide) is central to the global warming question. All of which presents us with quite a dilemma since we, at the tail-end of the twentieth century, are all but physically *and* psychologically addicted to burning tremendous quantities of coal and oil. (Currently, we rely on fossil fuels to meet over 85%[3] of World demand for primary energy.) Yet if (given for a moment that global warming is a problem) we could arrive at an equitable solution to such climate change concerns, then we would inevitably find that we have also gone a long way towards solving many of the other energy-related environmental difficulties currently confronting us, and indeed more mundane economic ones.

So where does one begin to look at the global problem? Well, the one thing that all these threads have in common is that they have their roots in what economists call 'the fundamental economic problem' ...

THE FUNDAMENTAL ECONOMIC PROBLEM

The fundamental economic problem is at the heart of the World's developmental problems and is unique to *Homo sapiens*, it being a product of our sentience. Simply expressed, the fundamental economic problem is the discontinuity - between human wants (as opposed to 'needs') and the ability to utilize resources to satisfy them (see Figure 1.1). As such the fundamental economic problem can never be completely resolved, though it can be, in part, reduced. Since human wants are essentially unlimited (people tend to want more than they have), so wants grow, whereas resources are finite.

For example, leisure, being an area of human activity in which the key decisions are made by individuals themselves, provides an excellent example of the way 'wants' have grown as our society has developed. For instance, whereas Edwardian Londoners would aspire to an annual holiday in the nearby south coast resort of Brighton, their modern counterparts have set their sights further afield on the Eastern Mediterranean and North Africa; these last incur a far greater resource-use cost per capita than the Londoners going to Brighton.

As society strives to fulfil its members increasing 'wants', standards of living rise with many former 'wants' becoming more easily gratified (so ceasing to be

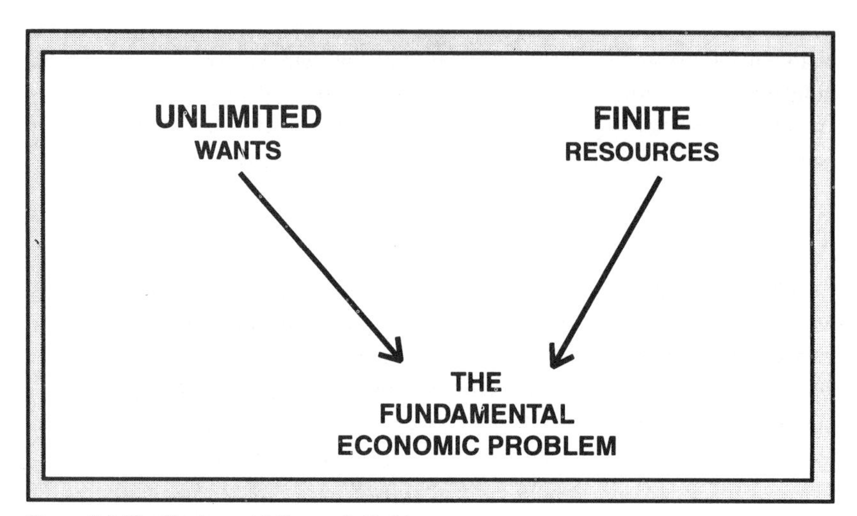

Figure 1.1 The Fundamental Economic Problem

'most-wanted wants') and with new 'wants' created. It is not difficult to see that the result is a never-ending self-reinforcing positive feedback system whereby ever more resources are consumed at an ever increasing rate. Furthermore, the fundamental economic problem is compounded on the global scale by the growing World population. The World is in the midst of an unprecedented expansion of human numbers. It took hundreds of thousands of years for the population to reach 10 million, 10 000 years ago. 8000 years later this number increased ten fold to 100 million at the time of the Roman Empire. Between then and 1950 World population grew to 2.5 billion, and doubled again by 1987. In 1990 the World population stood at 5.3 billion with an annual increase of 90 million. The United Nations estimate for 1990 was that low-income countries' populations would grow at between 1.4–3.9% per annum, and high-income industrial nations by 1% or less[4]. Halting this growth should be governments' top priority according to a report from the World's top 58 scientific academies compiled in October 1993[5], prior to the 1994 United Nations Conference on Population and Development in Cairo.

Given the finite nature of our world, the limitations of *such* growth will inevitably be manifest. Nonetheless, outside of economic theory, little is being done to manage population growth despite increased wealth. Economically the combined gross domestic product (GDP) of the World's nations is of the order of tens of thousands of billions of dollars and, in real terms, global GDP has doubled to this magnitude in just two decades[4].

Yet while it is largely human economically-driven activity causing environmental impact, it is important to note that dollars 'produced' by a nation do not *necessarily* relate to the *degree* of environmental impact being caused (or for that matter of human well-being, nor standard of life be synonymous with material resource consumption). Economic growth is symptomatic of the fundamental economic problem both individually and (aggregating up over 6 billion times) on the global scale: the fundamental economic problem is, if you will, the driving force behind the Global Economic Problem.

THE GLOBAL PERSPECTIVE

On a planetary scale, though connected to the fundamental economic problem, the environmental view of the World's economic problems can be divided into two key components. The first is that our economic activities are beginning to pollute at a global level, and that this not only threatens the Earth's life-support systems, but the ability to economically exploit its biological resources[2].

The Earth's global commons, of oceans and atmosphere, are increasingly being polluted, as exemplified by rises in pollutant background levels and the increasing impact of transfrontier pollutants such as the sulphurous and nitrous gases that result in acid rain (and, in the case of sulphates, short-term regional cooling). Oil and chemical pollution incidents have in the past been both local and regional in nature, but increasingly there are now signs of a global impact. That the atmosphere has measurable quantities of anthropogenic carbon and CFCs affecting the global climate is now firmly established (though to what degree in the future remains to be determined), and the results of human economic activity can now be seen on the oceans worldwide. (One telling instance is plastics. The chemical industry relies on small, 3–4 mm pellets of a variety of plastics as an industrial feedstock. On a regional scale these have been found off industrial coastlines and shipping routes since the 1960s. Now they are being found in the South Atlantic and the Pacific, several thousand kilometres from any industrial source[6]. Larger still, non-biodegradable plastic litter (bottles, crates, wrappings, etc) has a surface ocean residence time of at least three years, it too affects the oceans planet-wide – for example see Table 1.1). Such 'global common' pollution arising from our economic activities can be considered the thin end of a wedge that ultimately leads to major long-term environmental degradation, and this includes anthropogenic (human-induced) climate change.

The second component to this global perspective is the unequal way that a few people/nations exploit the World intensively compared to the majority. Though the entire planet shares the global commons of oceans and atmosphere, so that the cost of any pollution is in theory shared equally, the benefit from the economic activity that caused this degradation only accrues to a minority. For example, those hunter-gatherers who live on the edge of a desert that is expanding are unlikely to have benefited from the energy harnessed by the burning of the fossil fuels that released carbon dioxide that caused the climate to change, and induced the loss of their marginal lands. Where impacts do affect the wealthy, the affluent can afford to buy their way out of the problem, the poor

Table 1.1 Rubbish reported by Robert France on Ellesmere Island (the northernmost land in North America, the size of Great Britain but inhabited by only 100 permanent residents) illustrating that though some parts of the world are technically classified as 'wilderness', there is nowhere today really free from human waste. The rubbish was collected while on an 800 km journey through the non-glaciated part of the island[7]

Category	*Number*
Gasoline/petrol drums (45 gallon)	47
Metal food containers	41
Rifle cartridges	15
Gasoline/petrol drums	14
Snowmobile parts	10
Self-aggrandizing cairns made by army personnel	10
Food crates (wooden and metal each containing about 50 further items)	7
Recent news magazines	6
Metal cans and cylinders	6
Cigarette butts	6
Angle iron (3 m lengths for cartography)	6
Pieces of clothing	5
Bales of wire	4
Scientific crates	3
Snow gauges	3
Tarpaulins (plastic and cotton)	3
Tent heaters	2
Cotton and plastic cloth (15 m lengths)	2
Wooden boxes	2
Styrofoam drinking cups	2
Flare cannisters	2
Wooden tower (for radio antenna)	1
Graffiti (3 m letters for incoming helicopters spelling out sexual slang)	1
Surveying steel cable	1

cannot. Worldwide the gap between the rich and the poor is large. In 1986 the average person in, what The World Bank calls the low income countries (such as Niger and Benin) earned only US$270, middle income countries (such as Colombia and Chile) US$1270, and industrialised countries US$12 960. On average, those in the wealthy industrialised countries earn some 48-fold more than the low-earners. This last represents two ends of the spectrum. Overall, the industrialised countries represent only 15% of the World's population yet in 1986 they earned over 16 times the amount of the low-income countries containing half the World's population[4]. The dollar disparity is clearly related to a difference between the two extremes in resource use. In the late 1980s (around 1986/7) the 15% of the World's population in the developed nations used 2.9 times the amount of cereals per capita than those in the developing countries,

81 times more milk, 5.7 times more meat, 11.1 times more sawn wood, 13.9 times more paper and paperboard, 4.7 times more fertilizers, 20.6 times more copper, 13 times steel, 19.1 times aluminium, 20.3 times more inorganic chemicals, and 16.7 times more organic chemicals[8]. In short, from a global perspective, the world's commons of ocean and air are being polluted planet-wide, so affecting all more or less equally, but the polluters are not distributed evenly across its surface.

With specific regard to the chief greenhouse gas, carbon dioxide, energy use powers economic activity, so it is not surprising that the world distribution of wealth mirrors that of energy consumption. Indeed when the World Conservation Strategy 1991 follow-up report (more of which later) was being drafted[9], the authors used national carbon dioxide (CO_2) generation figures as an indicator of environmental demand in that CO_2 production relates to the consumption of a key economic resource which is also finite. The problems of, and solutions to, environmental degradation (and including human-induced global warming should it be happening) really are of central importance with respect to the way we manage our planet. Indeed, it would not be difficult to argue for minimising environmental demand (in other words maximising environmental sustainability) and that this should be somewhere on our top agenda both nationally and internationally. We are currently faced with a choice. Our world can either be managed in a sustainable way so that future generations can continue to harvest resources, or we can continue to over-exploit natural systems, so degrading the biosphere. Which path we take will determine the rapidity and extent of environmental degradation (hence human-induced climate change should it exist) and in turn the changes imposed on us as a species. As will be shown, we have a certain amount of choice regarding the former, and less choice with the latter. Yet the choice of policy we make will in no small part be influenced by past and existing environmental policies. By looking at these policies we can better understand the likely thinking behind future policy.

CHAPTER 2

ENVIRONMENTAL POLICY, ECONOMICS AND THE GLOBAL COMMONS

INTERNATIONAL REPORTS AND CONFERENCES

Concern over the increasing depletion of planetary resources and the associated increasing threat to, and impact on, the environment is not new. As the problems have grown, so too has the need for environmental issues to be discussed at an international level, at conferences and in reports. The first of these – that is the first of which much of today's environmentalism owes its roots – was the 1972 UN Conference on the Human Environment in Stockholm.

Stockholm Conference on the Human Environment

During the 1960s the state of the environment was clearly recognised as becoming a major global problem. It was being addressed by a number of bodies (well known examples include the World Wildlife Fund (now the Worldwide Fund for Nature) and the International Union for the Conservation of Nature). Such efforts while scoring notable, but isolated, successes, were overall uncoordinated and ineffective (the World's environmental problems are, after all, still with us). Yet even back in the 1960s it was recognised that more needed to be done planet-wide. By the late 1960s the United Nations saw itself as the body that could focus efforts on an international scale, and so it began to plan a major conference to be held in Stockholm to address environmental issues – the Conference on the Human Environment.

Work on the conference agenda was begun in 1970 by a preparatory committee, with representatives from 27 nations. Working groups were set up to explore specific areas of concern, and national reports were submitted by 80 countries. At the time it became apparent that the developing countries were reluctant to participate in the debate. They saw such discussion as a diversion from their need to develop. Furthermore, they viewed pollution as a problem associated with the rich, and so were worried that concerns over environmental issues would put a brake on their own development. It was to allay such fears that the General Secretary to the Conference on the Human Environment held a preliminary meeting on 'Development and Environment' in June 1971. The meeting primarily concentrated on trade agreements in an attempt to reassure the less developed nations. For instance, it declared that if goods were to become more expensive as a result of new environmental policies, then the developing countries should receive subsidies to offset the increases. Though the preliminary meeting failed to deliver any policy agreements, by claiming that the main Conference on the Human Environment would point the way to development without detrimental environmental side effects, it encouraged representatives of the less developed nations to attend. And so it was that a year later in 1972 from June 5th to 16th the UN Conference on the Human Environment met in Stockholm under the slogan 'Only one Earth': this was later to be the title of the conference's unofficial, but widely read, report[10].

The conference concluded by producing its *official* report which contained the Stockholm Proclamation: a list of 'Principles', and an 'Action Plan'. The Proclamation is perhaps most famous for its statement that, 'We hold that of all things in the World, people are most precious'. Of the 26 'Principles' many are still of relevance today, and they are still frequently referred to when formulating international policy. Those that are especially applicable to the 'greenhouse' issue include: number 21 States may exploit their resources as they wish but must not endanger others; and number 22-Compensation is due to states thus endangered. The conference Action Plan on the other hand contained 109 recommendations of which energy was covered by recommendations numbers 57 to 59. These received little attention as the Energy Crisis of 1973 had yet to take place and though the Conference recognised the 'greenhouse effect', it was more concerned with another energy-related transfrontier pollution issue – that of acid rain. Recommendation 59, though, was of relevance to today's energy issues. It called for a comprehensive and prompt study of energy sources, new technology, and consumption in order to provide the basis for global policy. This was to have been completed by 1975 but was never undertaken. However, the conference did establish the United Nations Environment Programme (UNEP): a body that subsequently proved to be an important source of information for environmental scientists, and which favourably contrasted with other environmental activities from the UN proper[11]. It is also worth noting that, though the Conference succeeded in focusing attention on environmental issues and in raising awareness, subsequent to Stockholm most countries planned to consume *more* fossil fuels per annum, and indeed over two decades they have fulfilled this goal.

Club of Rome's Limits to Growth Report

1972 also saw the publication *for* the Club of Rome (by a commissioned team at the Massachusetts Institute of Technology) of a report called *The Limits To Growth*[12]. The Club of Rome, which is often called 'the invisible college', was founded by the industrialist Dr Aurelio Pecci, and invited experts to address particular problems. The 1972 *Limits to Growth* was an attempt to look, using a computer model, at the future prospects for the world and its economy. The model's designers were heavily influenced by Malthusian thinking[13]. (The eighteenth century Rev Thomas Malthus proposed that while natural resources (such as food supply) could be increasingly developed such production was additive: that is to say they grew arithmetically: 2, 4, 6, 8, 10, 12, ... On the other hand human population, Malthus contended, tended to grow exponentially, 2, 4, 8, 16, 32. What kept human population down were a variety of population checks such as war, famine, and moral restraint (see also the beginning of Chapter 6).

The *Limits to Growth* study was criticised for not clearly identifying the limitations of its computer model. Whatever a computer programmer puts into a computer will determine what comes out at the other end: hence the programmers' adage 'garbage in, garbage out'. As such, the report was seen by many as somewhat self-fulfilling. Essentially, the report said that many of the resources we rely on are finite, and that this limits economic growth. Nonetheless, the report had a wide appeal, especially as the following year saw the oil crisis which demonstrated the reliance of the world economy on finite resources, and seemed to aptly illustrate the nature of the type of limits to growth of which the report warned. At the time this seemed to corroborate the report's theme.

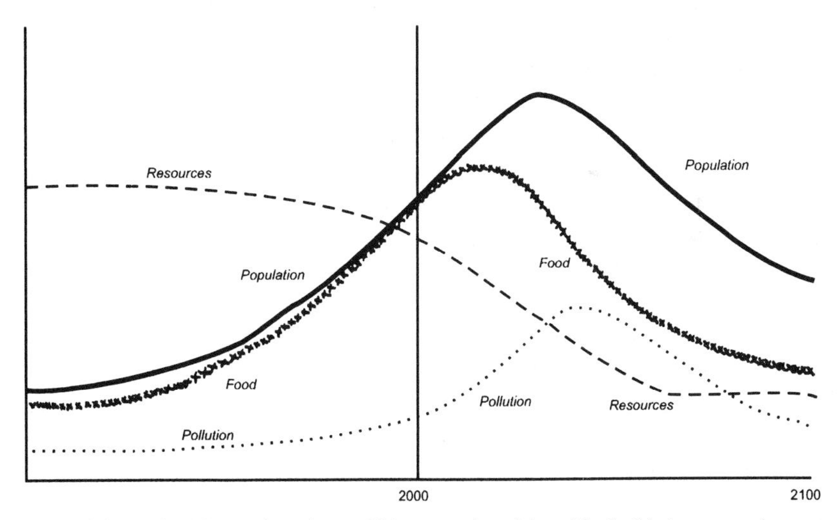

Figure 2.1 Standard Output from the world (computer) model used in the *Limits to growth* report for the Club of Rome. The output was standardised to 1900 and 1970 figures and plotted against respective scales for each of the four parameters shown here. The model (programmed with Malthusian constraints in mind) reflects a possible response to Malthusian parameters. This is the standard (business-as-usual) output. Different forecasts result from programming different assumptions about the future. However, the standard output (if it is to be believed) reveals continued growth through to the first quarter of the next century followed by a period of stress through the middle of the next century. How accurate this output is cannot be ascertained, however in the two decades since the report was published the World population has grown as the model predicts

However, though the report was criticised, it is important to note that just as there is an element of truth in the Malthusian approach to population and resource use, so *Limits to Growth* should not be dismissed totally out of hand. The key prediction the model made (and one that showed through despite making several computer runs while changing a number of the model's parameters) was that, given Malthusian trends, there would be a period of global stress as finite resources diminished and World population grew. This was forecasted to take place sometime between the year 2020 and 2080 (Figure 2.1). It is interesting to note that nearly two decades later the UN estimates for world population indicate that this period will see the World's population grow rapidly and approach what is thought to be a peak. This also is the period of time in which oil and natural gas will cease to dominate the world's primary energy market, as supplies run out.

Brandt Report (1980)

At the end of the 1970s, eighteen leading World statesmen formed an independent commission (mentioned above) to provide an analysis on the World economy. Called the Brandt Commission, after its chairman the former West German Vice Chancellor and Foreign Minister, Willy Brandt, it consisted of renowned politicians such as the US Presidential Assistant Peter Peterson, and the former UK Prime Minister Edward Heath. With such a line-up the Commission could hardly be ignored. It was not. Over a million copies were sold in editions published in over twenty languages.

The Commission particularly wanted to tackle the question of the gulf between the World's rich 'northern' and the poorer 'southern' nations. This gulf they saw as the major impediment to the future stability, and development, of the World Economy. The Commission's report[3] spelled out the extent of the mutual interests between North and South, and appealed for an emergency programme to avert disaster for the worst hit countries, for a longer-term reorganisation of the global economic system, and for a summit of world leaders.

In addition to the principal recommendations above, the report included proposals on energy that have implications for the 'greenhouse effect' (a term not mentioned in the report itself). The report wanted to see an energy agency (that, it said, would preferably be affiliated to The World Bank) established to increase energy production in developing countries. It called for support for increased energy research, and the results of this to be disseminated by the new energy agency. Finally, it saw that far closer dialogue between energy consuming and producing nations was required.

The report was welcomed by governments worldwide, though with some reservations by non-governmental organisations (NGOs) who had little confidence in the established international machinery which the report's proposals relied upon. The big question, though, was would anyone take enough notice to do anything? If the governments of the rich northern nations were hoping that they need not, and that Brandt and his commission would go away, then they were wrong.

Three years later, in 1983, the Brandt Commission published a follow-up report.[14]. By then the World Summit the original report called for had taken place in Cancun, Mexico in 1981, but as far as the Brandt Commission was concerned this summit: 'fell far short of [its] expectations. It produced no new guidelines nor any clear impetus for future negotiations. It did not even come close to launching the idea of a world economic recovery programme'.

Finally in 1985, though not from the Brandt Commission itself, but from its Chairman, Willy Brandt, came a further 'call for action'[15]. He saw that there was already tremendous slack in the World economy as represented by huge arms expenditure. He pointed out that only a fraction of this expenditure would be required to boost the World economy and reduce the differences between the developed and less-developed nations. Though this plea did not receive the considerable public attention of the original Brandt Report, the latter half of the 1980s did see leaders in the developed nations make serious attempts to reduce military budgets as traditional East-West tensions thawed. This continued throughout the early 1990s with the dissolution of the Soviet Union. However, economic problems afflicted the former Soviet nations, and combined with a general World depression, the desire for politicians to provide tax breaks etc hindered the freed capital from funding sustainable economies.

World Conservation Strategy

In 1980, on the heels of the Brandt Report, another major document was published – *The World Conservation Strategy*[2]. This was prepared by the International Union for the Conservation of Nature (IUCN), with the advice, cooperation and financial assistance of UNEP and the World Wildlife Fund. It also drew upon the expertise of bodies such as the Food and Agricultural Organisation (FAO) of the UN, as well as the UN's Educational, Scientific and Cultural Organisation (UNESCO): many of these bodies had simultaneously provided information to the Brandt Commission.

The World Conservation Strategy had three main objectives:

(1) To maintain essential ecological processes and life-support systems (such as soil regeneration and protection, the recycling of nutrients, and the cleansing of waters) on which human survival and development depend.
(2) To preserve genetic diversity (the range of genetic material found in the World's organisms), on which depend the functioning of many of the above processes and life-support systems, the breeding programmes necessary for the protection and improvement of cultivated plants,

domesticated animals and microorganisms, as well as much scientific and medical advances, technical innovation, and the security of the many industries that use living resources.

(3) To ensure the sustainable utilization of species and ecosystems (notably fish and other wildlife, forest and grazing lands) which support millions of rural communities as well as major industries.

It established an overall framework of policy goals for nations, but recognised that each had to tackle its own problems individually and so recommended their formulating national conservation programmes. Importantly, in terms of the 'greenhouse effect', *The World Conservation Strategy* considered the threat to the global commons (primarily the oceans and atmosphere). Here it highlighted the threats imposed by chlorofluorocarbons and carbon dioxide, and urged that international support be given to the World Meteorological Organization (WMO) for its World Climate Programme and the importance of the Convention on Long-range Transboundary Air Pollution.

As with the Brandt Report, *The World Conservation Strategy* was warmly received by political leaders many of whom – once the immediate glare of the publicity-of-the-day had passed – then failed to initiate any action. Worse, some nations (including the UK) did produce their own responses (their own national strategy) without due regard as to how these would be implemented. However, a handful of nations, such as Norway and Holland, were more successful in formulating and implementing environmental policy. Importantly, regarding the 'greenhouse' issues, *The World Conservation Strategy's* fears have subsequently proved to have sufficient grounds for the strategy's climate-related recommendations to be implemented with the creation of the Montreal Protocol and the formation of the Intergovernmental Panel on Climate Change (IPCC).

Again, as with the Brandt Report, the authors have not rested with the publication of their original report, and have since published their World Conservation Strategy for the 1990s called *Caring for the Earth*[16].

Caring for the Earth (aka The World Conservation Strategy II)

In 1992 the report *Caring for the Earth*[16] (from the IUCN, WWF, and UNEP) referred to the climate change issue. The report's preliminary draft[9] drew on the concept of carbon dioxide release as an indicator of environmental demand through unsustainable development. (This last is an important point for it could be argued that both the development-focussed Brandt Report and the conservation-oriented 1980 *World Conservation Strategy* would have benefited from closer cooperation in their writing and launch. As it was, the authoritative summary of development issues and of environmental conservation were to wait a further seven years, for the Brundtland Commission (see next section), before being addressed by a major international report for the first time since Stockholm.)

The goal put forward by *Caring for the Earth* was to help improve the condition of the World's people, by defining two requirements. One was to define a new ethic for sustainable living, and to translate its principles into practice. As such it outlined a broad programme, including timetables for action.

With regard to anthropogenic climate change, it set energy efficiency targets and called for the use of energy standards, taxes and charges to encourage this. It also called for national energy strategies that were not so dependent on fossil fuels, and international technological, and financial, transference from the wealthier countries to poor nations to improve energy efficiency. Finally, it proposed that governments encourage energy efficiency through granting awards, and providing energy information.

Brundtland Report

Following the Brandt Report, in December 1983 the Secretary General of the United Nations called upon Norway's Prime Minister, Gro Harlem Brundtland, to establish and chair an independent commission. Its aims were:

(1) To propose long-term environmental strategies for achieving sustainable development by the year 2000 and beyond;
(2) to recommend ways in which environmental concern could be translated into greater cooperation among developing countries and between countries at different stages of economic development;
(3) to consider ways by which the international community could deal effectively with environmental concerns; and to help define shared perceptions of long-term environmental issues.

The commission, called the 'World Commission on Environment and Development', consisted of members from 21 countries (including the Canadian Maurice Strong, the Secretary General of the Stockholm Conference). It published its report in 1987 under the title *Our Common Future*[17].

The Brundtland Report in effect consolidated and affirmed the principles arising from Stockholm in 1973. It summarised the current World situation emphasising the urgent need that development be sustainable, bringing together much of the policy discussion behind the Brandt Report and *The World Conservation Strategy*. (Though, importantly, for the first time space (as in 'outer space') was to be considered as part of the global commons.) However, in its foreword Brundtland referred to three 'problems bearing on our very survival' that were symptomatic of our current non-sustainable manner of development – those of: desertification, ozone depletion, and global warming.

Regarding the latter, an entire section of the report was devoted to energy. Here the report listed four key elements of sustainability that it suggested should be reconciled when considering the question of an equitable World energy policy:

(1) sufficient growth of energy supplies to meet human needs (which meant accommodating a minimum of 3% per capita income growth in developing countries);
(2) energy efficiency and conservation measures, such that waste of primary resources is minimised;
(3) public health, recognising the problems of risks to safety inherent in energy issues; and
(4) protection of the biosphere and the prevention of more localized forms of pollution.

It also highlighted four areas of uncertainty and risk:

(1) global warming,
(2) urban-industrial air pollution,
(3) acid rain,
(4) risk of nuclear accidents, dangers of nuclear proliferation, and the problems of nuclear waste disposal.

The report called for the period ahead to be regarded as a transitional one from an era in which energy has been used in an unsustainable manner. However, the report could not identify a generally acceptable way forward (though it favoured high energy-efficient/low energy-use strategies) but pointed out that the international community had yet to address this problem with a sufficient sense of urgency, and to adopt a global perspective. As we shall see in the chapter on perceptions, at the time of the report's publication political leaders in both the USA and the UK either did not accept that such problems existed, or they claimed that sustainable development was an unachievable concept[18]. This attitude was to begin to change as the decade ended.

UNCED (1992)

Finally, in 1992 there was the United Nations Conference on the Environment and Development in Rio de Janeiro. It was convened on much the same basis as the Stockholm Conference. However it focused on two areas: biodiversity and climate change. At the conference two conventions were signed relating to these areas. Of these the Framework Convention on Climate (FCC) called upon the World's nations to formulate their own strategies to address global warming. We will return to the FCC again in Chapter 10 (on energy policy). In addition to the major international reports and conferences, there have been other national and regional initiatives. Overall, throughout the 1970s and 1980s, these have exhibited an increasing environmental awareness even if they have not been completely successful in having their conclusions implemented. Translating environmental concern into action in the real world has proven difficult, not least because it is not easy to connect environmental values with the everyday standards, which would make it easier for governments, industry and the average person to relate. What value, for example, has an acre of ancient woodland in crude dollars or pounds, or, if we are to begin to tackle the thorny question of possible human-induced climate change, what is the cost of pollution?

THE ECONOMICS OF POLLUTION

One of the keys to success of any policy, whether germinating from an international report or originating from a national government, is that of society's willingness and ability to pay for it. If a policy cannot be afforded then it simply will not survive in an effective form. Consequently, when looking at the question of possible human-induced (anthropogenic), or pollution-generated, climate change, the way we economically value pollution becomes of central importance.

At the end of the day, pollution causes damage. This is an unarguable fact almost by virtue of the definition of the word. It might be damage in a concrete way, such as by causing disease (*cf.* the former London smogs), or in a more ephemeral way (for example the noise pollution from a party disturbing neighbours' sleep). Either way pollution causes damage, but human-generated pollution is also a by-product of economic activity, and so cannot be considered independently of that activity (even a party is an economic activity: the activity being one of consumption).

For instance, if someone engages in an industrial activity producing, say widgets, then any pollution arising from the widget production has to be considered in the light of the benefits that widgets provide, both directly to the individual widget consumer and overall to society. The costs of widgets, including those of pollution, has somehow to be compared to those of its benefits if we are to appreciate the overall net benefit (or cost) to society. (Of course, as frequently happens, it is possible to ignore some of the benefits and costs, - particularly those of pollution – until recently fridge manufacturers did not think of the cost to society of releasing CFCs to the atmosphere.)

From here on, for a little while until the end of this chapter, it is going to become a bit technical while we explore the world of environmental economics. But it is worth it from the understanding economics gives us.

With regard to energy, the benefits of energy production accrue to the end use to which the energy is put: food production, warmth, entertainment, etc. In most cases, the amount of pollution generated in product manufacture is nearly directly proportional to the amount of product produced. For example, the production of x units of fossil fuel will ultimately produce y units of a pollutant (such as carbon dioxide) when it is consumed so that, approximately, the production of 2x units of fuel will ultimately generate about 2y units of carbon dioxide. This perspective simply relates the amount of pollution to production so that it is possible to see that the factors that govern production of our widgets also govern those of pollution. Here the principal factor determining production in a free market economy is the law of supply and demand. This is the 'law' that in the free enterprise culture determines the optimum quantity (Q_o) of a product produced, hence the pollution.

It works like this. When the supply of the product is low, the demand is high so that the price of the product, our widgets or whatever, is also high. Conversely, when the supply is great, so that most consumers can have enough of the product, the demand is low (everyone can easily have access to a widget) and so the price is also low. In practice the supply is never so high so that the demand, hence the price (p), reaches zero, otherwise the widget would not be worth producing: a balance has to be struck between the ability to supply and the satiation of demand (Figure 2.2).

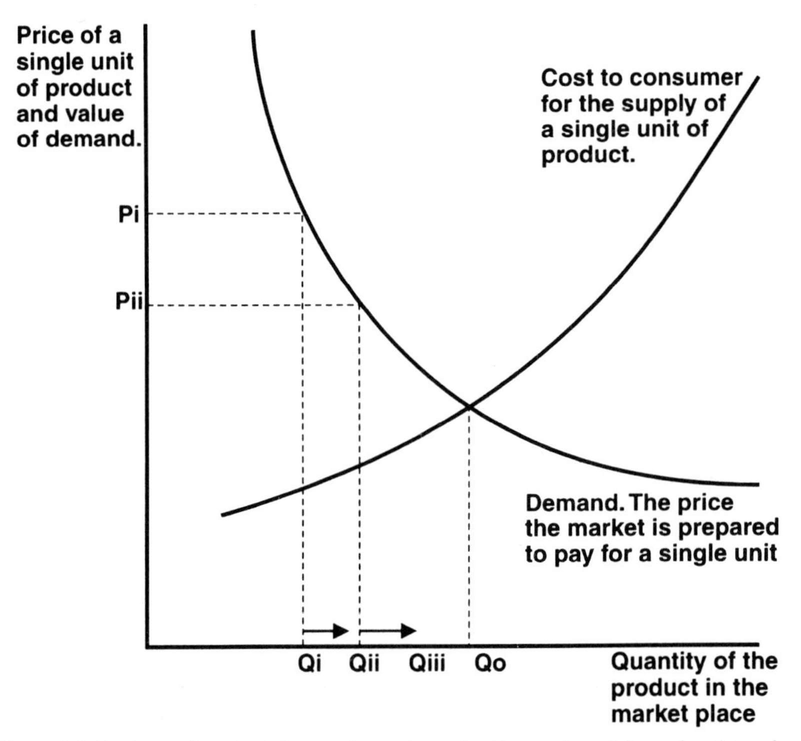

Figure 2.2 'Optimum Quantity' of production as determined by supply and demand ... Assuming the demand for a produce is seen by business as being as Q_i then the market will bear a price of P_i without leaving a surplus. At this price the business is able to supply more produce (Q_{ii}) but at this level of supply the market will only pay the lower price of P_{ii} if all the product is to be sold. At this price the business can still afford to make more of the produce Q_{iii}. The process goes on until supply equals demand and the optimum quantity (Q_o) is sold on the market at an optimal price (P_o). This also gives the level of pollution that the industry is likely to generate and/or control

The optimum quantity (Q_o) produced is reached when supply equates with demand[19]. It is the result of the two opposing trends. As the price of the product increases so business (wishing to maximise profits) increases production. Conversely as supply increases so the price the market will bear decreases.

As has already been mentioned, typically, the level of pollution generated by an industry is often nearly proportional to the amount of product produced, so that the law of supply and demand will determine the amount of pollution that industry is likely to produce. Of course, not all the pollution produced by an industrial process will necessarily be released into the environment – some of it might be trapped by the industry itself using filters or whatever – nonetheless production level of product is closely related to production level of pollution, be it released or not.

If some or all of the pollution is released into the environment beyond the environment's ability to assimilate it, then there will be a cost associated in clearing it up not directly borne by the industry. The industry is said to impose

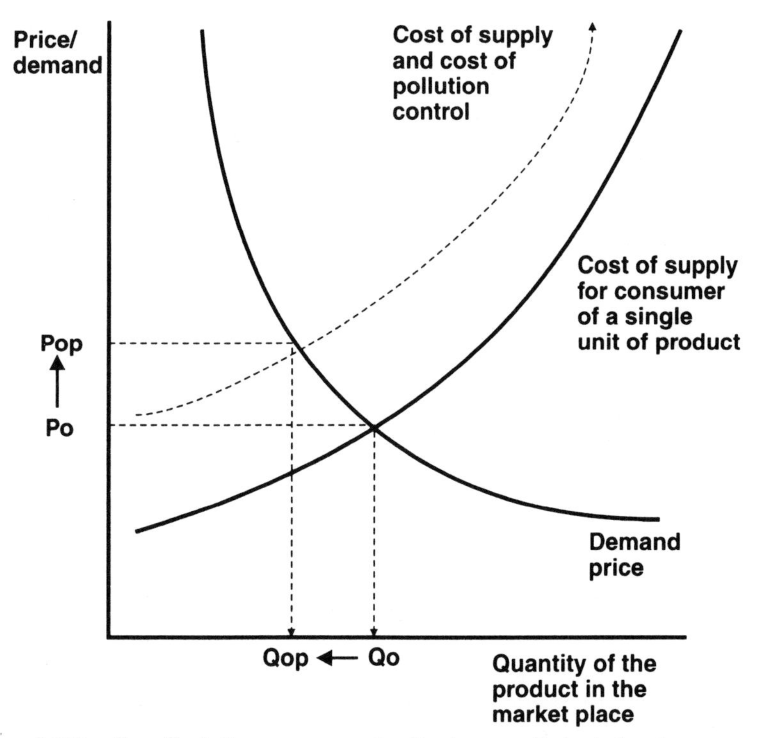

Figure 2.3 The effect of including costs to control and/or clean up pollution in the price consumers pay and the optimum quantity of product in the market. The optimum price for competing producers to sell at is forced up from $\mathbf{P_o}$ to $\mathbf{P_{op}}$ so lowering demand and raising the demand price those who still wish to purchase the product are prepared to pay. Additionally, the optimum quantity of goods in the market is lowered from $\mathbf{Q_o}$ to $\mathbf{Q_{op}}$

an *externality* cost on society: that is to say a cost external to the market system. For example, a food take-away business might pay for the provision of a waste bin outside of its premises for used food containers, but if it does not then the cost of cleaning up the food containers (by, say, the local authorities) is an externality. Even if the food containers are not cleared up, this externality cost still exists as the 'quality of life' of others is lowered in their refuse-strewn street.

Supposing our business decides to increase its level of pollution control (or has controls imposed upon it), then the cost of supplying the product increases. This is passed on to the consumer which in turn lowers consumer demand, and so the optimum quantity (Q_o) of the product produced in the market[19] (see Figure 2.3). Consequently the concept of an ideal economic level of production is not an absolute, but a variable depending on many market vagaries including those of society-imposed pollution constraints. If there are many firms supplying the market, and especially if they are made to cover the costs of their principal externalities, then demand will be lowered in the face of an increased product price so forcing the firms to become more competitive (spending more, say, on promotion and marketing) so lowering profits. Human nature being what it is, firms do not favour this state of affairs, firms' investors like large profits, and businesses may fight off having to pay such costs. (However as we shall see in the Chapter on perceptions, there are exceptions).

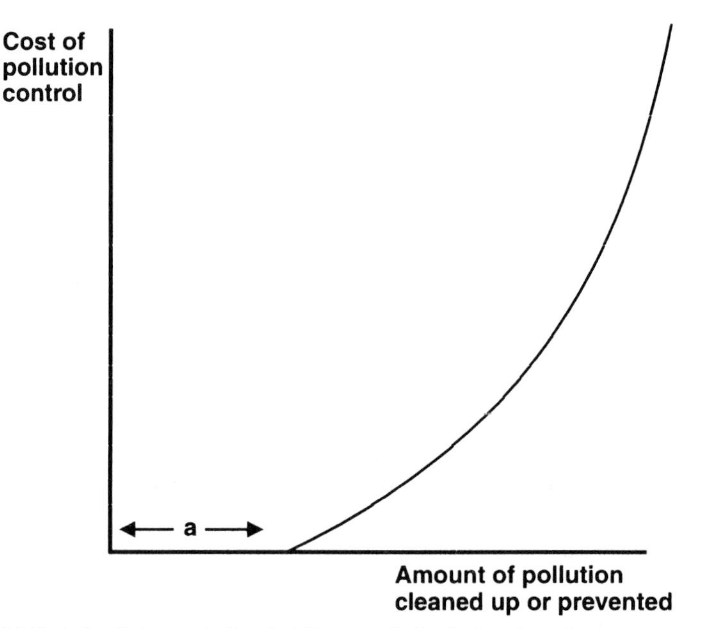

Figure 2.4 Cost of pollution control against the amount of pollution cleaned up or prevented. In many instances some pollution can be assimilated naturally by the environment (**a**) so the producer may not pay to clean this low level of pollution up. However, determining the environment's assimilative capacity varies in most instances for various pollutants and sites of release

At this point one might be forgiven for wondering whether the optimum level of pollution released could ever be zero (a view often held by members of the lay environmental movement); after all if there was no pollution released whatsoever, then there would be no costs imposed upon society or the environment. Unfortunately this is not true. To begin with, the cost to clear up pollution is not directly proportional to the amount of pollution cleaned up (see Figure 2.4). It is easier to lower the first quarter of pollution released than the second quarter, just as it is easier to quickly run a duster over exposed surfaces in one's house than it is to be more thorough by lifting objects to dust beneath before replacing them. In short, there is no straight line relationship between the cost of pollution control and the level of pollution cleaned up; rather an upward curve. Again, just as it is impossible to get one's house absolutely clean, so it is impossible to ensure that absolutely all of a pollutant is removed from an effluent discharge.

There are two reasons why it is not desirable to remove 'all' pollutants from an effluent discharge. First, the aforementioned cost of cleaning up pollution does not have a straight line relationship with the level of pollution. Invariably the easiest, hence cheapest, pollution is cleaned up first: consider the costs involved in the dusting example above, quickly dusting around would be cheaper than having to lift up objects. Second, the environment is often capable of assimilating some pollution (<-a-> (see Figure 2.4)). A small paper bag blown down the street will hardly be noticed and will, quite naturally, rot away in time. 1000 paper bags blowing down the street will be highly visible before rotting away, during which time they may incur other costs, such as clogging drains.

Given that it is neither possible to remove all pollutants from effluent discharges, nor desirable, and since too much pollution is certainly not wanted, then the question remains as to what is the optimum pollution level?

At the end of the day, whether or not pollution is cleaned up, and to what extent (be it level of pollution tolerated by society or the market), will depend on the costs the pollution is *seen* to cause set against those of abatement or control[19]. Clearly, if the damage pollution causes is seen to be less than the cost of control or abatement, then the control will not be cost-effective – it would be cheaper to pay to have the damage put right. The 'optimum' level of pollution (O_p) calculated this way, is determined by the minimum cost of the combined abatement and damage costs (Figure 2.5). On the other hand, selfish commerce would not want to pay any costs, so selfish (as opposed to less-selfish) commerce's optimum pollution level would be O_c. Either way, in terms of releasing greenhouse gases, there are big problems due to the extent of alleged climate change and the costs associated with it: we need to know the (climate) costs to calculate optimum pollution levels (unless you adopt the 'selfish commerce' perspective).

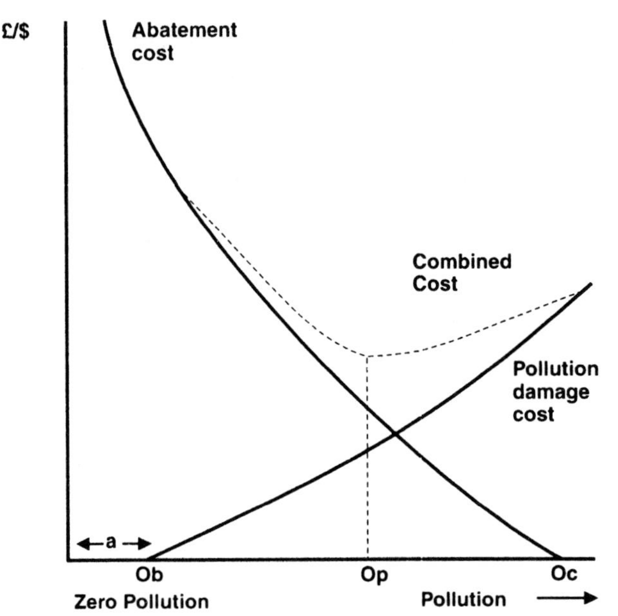

Figure 2.5 The optimum level of pollution as determined by minimising pollution damage and abatement costs, O_p. On the other hand O_c is the optimum level of pollution for a selfish commercial operation not wanting to pay for pollution abatement. O_b is the biological optimum where the pollution level is just within the environmental assimilative capacity, **a**. The optimum level of pollution depends on your perspective: biological, social or selfish commercial

THE COST–BENEFIT APPROACH

Another way of looking at 'optimum' pollution levels is to take into account *all* the costs associated with the product (not just those of pollution damage and abatement discussed above, but also of the product itself (which does damage and incurs costs right through to its disposal at the end of its lifetime)), together with *all* the benefits arising from its production *and* use. Provided that the total benefits outweigh the total costs, then there is a 'profit' or net gain for society. Here, while pollution's costs may well be, at least superficially, appreciated, it is more difficult to see how pollution benefits anyone. This is because the - benefits that are generated from pollution arise indirectly, they are in fact chiefly provided by the benefits from providing the product to the market (not the pollution itself). It is the economic activity that generates the pollution which provides the benefits.

Here, the 'benefits per unit of product produced' (what economists call the marginal benefit) becomes a useful way of looking at the value of a product – they tell us the value of an individual item of product. These marginal benefits are maximised if fewer units of the product are made – in a world with only one car the benefit that could be obtained from that car would be more than in a world overflowing with cars (see Figure 2.6a). Similarly, the marginal benefits from pollution (which in actuality arise from the creation of a product, be it widgets or whatever) decreases with the amount of product, hence pollution, produced. Conversely, the marginal costs of pollution increase the more pollution there is. (That is to say, the cost of each individual unit of pollution increases

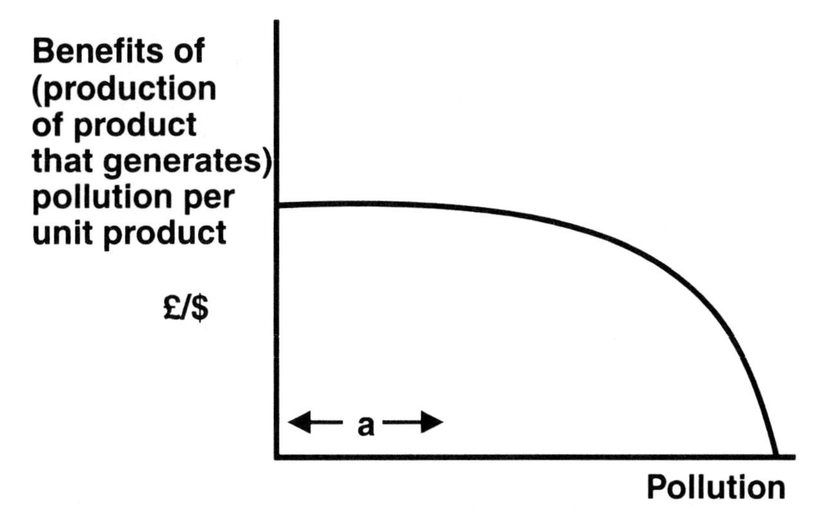

Figure 2.6a The benefit of pollution per unit cost of product against pollution. With few products produced the pollution is assimilated by the enviroment, so the net benefit is positive. As more products are produced so there is more pollution and so the overall benefit of production against

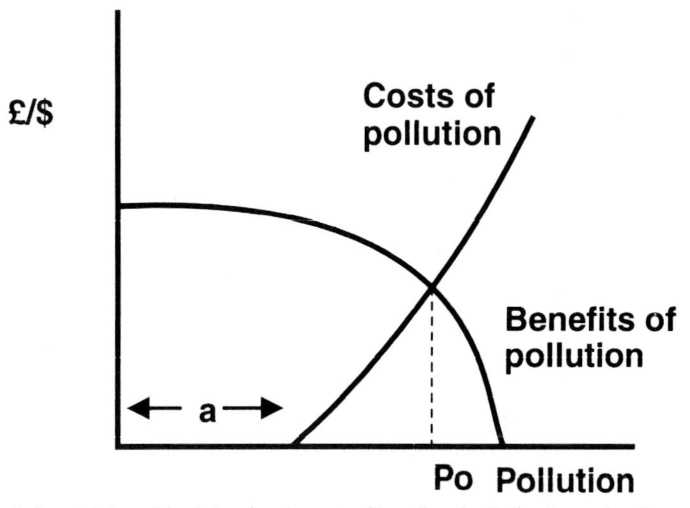

Figure 2.6b Figs 2.6a and 2.4 combined showing the costs of benefits of pollution (assuming that either the producer pays to prevent pollution and/or society and/or the polluter pays to clean up the pollution). $\mathbf{P_0}$ is the optimal level of pollution from this cost-benefit perspective

the more units there are.) At low levels of environmental pollution, within the assimilative capacity of the environment, the pollution costs are zero: the environment will (by definition) assimilate the pollution fairly quickly. At high levels of pollution, any extra pollution will not be assimilated by the environment (the assimilative system of which would be saturated) and so the pollution's full impact will be felt. The optimum level of pollution, from a cost–benefit perspective, will not be so low that additional produce (hence pollution) is prevented, which itself could provide extra net benefit. Neither will it be so high that costs outweigh the benefits. Rather the optimum level will occur when the marginal costs equal the marginal benefits (see Figure 2.6b).

The problem with the cost–benefit approach is that it is difficult to identify and quantify *all* the costs and benefits. Traditionally, business assigns costs to each operation. Subsequent decisions also have to be taken (such as scale and type of operation) and a cash value is assigned to these too. One ends up with a branch flow chart with probabilities as to the likely financial outcome calculated at each branch end. By assigning probabilities at each branch, it is possible to quantify the likely costs and benefits of adopting the particular strategy associated with that branch.

In the 1980s another economic tool became increasingly important, that of environmental impact analysis. It has increasingly been employed to determine environmental costs of strategies or projects, especially where there are many dimensions to the problem involved. There are five principal ways of carrying out an environmental impact analysis:

(1) *Check Lists.* The various factors that are to be considered in assessing the project are listed and dealt with in turn. These might include: ecological effects, pollution, aesthetics, human interest, etc. At the very least this provides standardization when comparing options.
(2) *Matrix Methods.* Here a large matrix is generated where each kind of impacting activity is given a column, and each row a particular form of impact that can occur. For example a mining operation could have heading a column 'generation of slurry' and a row could contain river pollution. One of the most commonly used matrices is the Leopold matrix.
(3) *Map Overlays.* This technique is particularly useful when considering projects that use large areas of land. A map of the area affected by the project is divided up into small units. In each unit the environmental or human resource value can be identified enabling planners to decide where development should take place.
(4) *Computer Modelling.* This helps to enable multi-dimensional problems to be solved by harnessing the tremendous number-crunching abilities of computers. The advantages are that it is possible to discern patterns in what would otherwise be an impenetrable mass of data. The disad-

vantages are that the value of the system is only as good as the model programmed into the computer.

(5) *Delphi Poll.* With this method various experts from different disciplines are called upon to give their views on a project. In the light of their peers' reports the experts provide a second modified opinion. This cycle is repeated a number of times until a consensus is reached.

With regard to the greenhouse issue, the last two techniques have been used extensively to identify the scope and scale of the problem. Computer models, primarily global climatic models (GCMs) have been used to see how the Earth's weather will probably be changed under different, warmer conditions. A Delphi poll approach was taken by the Intergovernmental Panel on Climate Change (IPCC) that reported in the Autumn of 1990 as to the likely scale and impacts of the greenhouse effect. Here some 700 experts were involved in providing a consensus opinion for politicians. We will be continually referring to the IPCC's first working group report[20] throughout much of the remainder of this text.

Both environmental impact and decision analysis are treated as separate types of economic appraisal to cost–benefit analysis, however it is worth noting that they both have their uses within the cost–benefit approach. The value of attempting to account for costs and benefits – with regard to issues such as the pollution of the global commons – is that it broadens the appraisal beyond the immediate market: beyond the producer–consumer considerations. The individual's interests even at a national level, be s/he a fossil fuel producer or consumer, does not always coincide with those of the general (society's) interest. Taking a small-scale example which we shall return to later, an area of common grazing could be considered analogous to that of a global common. Here it is in every individual's personal interest in using the common for grazing to put out more of their animals to pasture, but after the grazing carrying capacity of the common is reached then every new animal's presence is contrary to the general interest[21].

As we have seen in this section, society has to make decisions about pollution. Either we can forget about paying to minimise or clean up pollution, and so live in a polluted world, or we can pay for some, or indeed all, of the costs associated with pollution. In by far the majority of industrial cases it is the middle road that is taken, but as has been indicated, the level of pollution that should be tolerated in society's 'best interests' is debatable and depends on how one looks at the problem (see Chapter 12 on perceptions and responses). The techniques and approaches outlined above do, though, provide a means for assessing the policy options open to us. Certainly they are not as simplistic as portrayed by the hard-nosed caricature of an industrialist's philosophy of 'pollute and be damned provided there is a profit in it'. Equally, the idealistic lay-green view of a pollution-free planet, whilst tempting, is neither pragmatic nor realistic. Just as nobody wants to walk with the devil, neither is it possible to fly with the angels. For better or worse human beings want to live in a world with warm (or cool) houses, to have the ability to travel, to have a technical base to provide communications, medicines, education, and a host of other energy-requiring services. However, if we are going to arrive at some workable policy, using the above techniques, we must first appraise the ideas and evidence behind so-called global warming.

PART 2

CLIMATIC AND HUMAN CHANGE

CHAPTER 3

HISTORY OF THE GLOBAL CLIMATE

THE EARLY ATMOSPHERE AND CLIMATE

Our solar system, of sun and planets, is some 4.6 billion years old. It began with a dust cloud floating in space. The planets themselves, including the Earth, are thought to have rapidly condensed out of this cloud within a few hundred million years or so (i.e. within a couple of percent of the solar system's life to date). The Earth's atmosphere that subsequently formed was very different to that of today's: it had little oxygen, a small fraction of the amount found today; indeed perhaps in only trace amounts, far from today's atmospheric concentration of 21% (it was originally less than 1% of today's percentage). At that time carbon dioxide was the major atmospheric constituent. Yet it was in this atmosphere that life arose somewhere between 3.5 and 4.2 billion years ago, albeit in the simple, prokaryotic, forms of bacteria and algae. Then two billion years ago life as we commonly know it, with the evolution of photosynthetic organisms, caused the atmosphere to begin to change by consuming the carbon dioxide and generating vast amounts of oxygen, increasing the oxygen content[22, 23]. Over the next billion years, until one billion years ago, the oxygen in the atmosphere continued to increase to close to today's value. The Earth had experienced its first pollution catastrophe: by comparison the current human generation of greenhouse gases is a mere hiccough in the biosphere's cycling processes. Referring to the creation of an oxygen atmosphere as 'the worst atmospheric pollution incident that this planet has ever known'[24], may seem to be somewhat of an odd statement, but at the time it foreshadowed great change for the then living species. The availability of oxygen, albeit at first in small concentrations, enabled life to begin to exploit the new thermodynamically advantageous metabolic pathways and this further enhanced the rate of oxygen production. Fossils of one of the first aerobic species have been found. It was similar to a current species of alga, - *Grypania spiralis*, that grows in long, spirally coiled filaments, about 1–2 mm wide[25]. From then on the anaerobic species were to lose their domination over the planet to these new aerobic species that generated and used oxygen.

The more complex organisms that evolved two billion years ago were markedly different, not just in their metabolic pathways but in structure. These eukaryotic organisms had cells (like yours and mine) with a distinct membrane-bound nucleus and other intracellular membranes. By 600 million years ago, in an atmosphere close to that of today's, life evolved to a multicellular level, though creatures with a backbone had yet to evolve. These non-backbone, or invertebrate, species included trilobites, clams and snails.

Prior to the massive input of oxygen into the atmosphere, there was far more carbon dioxide in the air than today. We do not know what the early, primordial climate of the Earth was exactly like, except that it was probably warmer than today due to carbon dioxide's ability to trap heat: then there must have been a massive greenhouse effect warming the Earth. Even so, the Earth could not have been too warm for life to flourish. It is thought that the Sun was generating less heat than today, so offsetting the primordial greenhouse effect, and that climate feedback systems (such as cloud cover reflecting heat) were either different then, and/or operating at a different magnitude to those of today.

However, just as oxygen was put into the atmosphere by plant life, so the carbon dioxide was removed. The carbon dioxide in the atmosphere was chemically transformed into carbonates that were laid down as great beds of chalk, which we see today as banks of rock such as the cliffs of Dover in England.

To give a very greatly simplified description between 600 million and 2.5 million years ago the Earth's climate, though for most of the time warmer than today's, was fairly (or, more accurately, *comparatively*) stable (albeit with a few notable, but quite separate, exceptions)*. However from 2.5 million years ago, the Earth cooled and entered an ice age from which it has yet to leave. It may at first seem strange to think that we are currently living in an ice age, but importantly this current ice age consists of both cool episodes, or glacials, and warm interludes, or interglacials such as the one we are currently in. As we shall see, the study of these past glacials and interglacials (especially the last glacial–interglacial cycle of some 170 thousand years) is revealing how our planet's climatic system currently works, and the possible implications arising from increasing the atmospheric level of greenhouse gases.

THE CURRENT ICE AGE

From the end of the Pliocene epoch about 2.0–2.5 million years ago (a paltry one twentieth of one percent of the history of the Earth) our world entered the latest, and current, ice age. Before then – before some 3–5 million years ago in the Pliocene – the Earth was about 4°C warmer than today. The climate in this earlier time had been fairly stable for many millions of years. But 2.0–2.5 million years ago the Earth began to enter the cooler Quaternary period and the Pleistocene epoch (an epoch being a sub-division of a period) in which our world had, as it does today, ice caps covering both poles. The Quaternary period

* Actually, it is quite a complicated story and interested readers are urged to seek out specialist palaeoclimatological texts for details of the Earth's climate over the past half-billion years to 2.5 million years ago. The description in the text is a gross simplification.

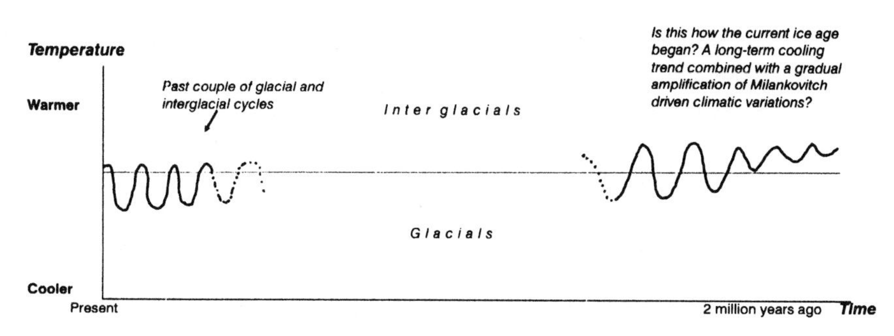

Figure 3.1 Climate changes between interglacials and glacials over the past 2 million years

was, and is, an age of ice – literally an ice age. As mentioned, many do not realise that we are in the middle of an ice age right now, but we are: the current warm interlude we are enjoying, known as the Holocene period, began about 12 000 years ago. What has happened is that the climate in the Quaternary period has been oscillating, constantly getting warmer and cooler in cycles of about 100 000 years in length, and it is still doing this with our present era being in a warm part of the sequence (see Figure 3.1). The maximum temperature difference between the warm interglacials and cool glacial parts of the cycle *globally* is about 4°C (there are greater regional differences); currently the average World temperature in our interglacial is near the top of this range. During the cool glacials the mountain ice caps, as well as those at the poles, expand: while during the interglacials they contract. There have been about a score of Quaternary glacials. The next glacial is, in geological terms, imminent – it is probably less than a few thousand years away, and when we reach it our warm interglacial interlude will judder to an end in a series of cold snaps probably measuring just decades or even less. The previous interglacial took place roughly 120 000 years ago and lasted for about 20 000 years. This previous interglacial (known as the Eem–Sangamon interglacial by US geologists, and the Ipswichian interglacial in the UK) was about 2.0–2.5°C warmer than the current (Holocene) interglacial at the present time. The Earth's climatic patterns and weather during the previous interglacial can be considered to be analagous to that which we expect to experience if atmospheric greenhouse gases continue to build up over the next few decades.

It is thought that this cyclical, glacial–inter-glacial, climatic variation comes about by changes in the way the Earth receives the Sun's rays as it travels in its orbit, an idea championed by the Yugoslavian Milutin Milankovitch[26] (although owing a debt to the work of James Croll in 1864). Subsequently supported by evidence from deep-sea cores, the Milankovitch theory suggests that minute variations in the Earth's orbit act as a kind of pacemaker for glacials[27]. Clearly, if the Earth receives less solar radiation then it will cool; if it catches more it will warm. In fact fluctuations in the Earth's orbit do not unduly alter the overall amount of solar radiation our planet receives, rather the amount hitting the northern hemisphere differs compared to that falling on the southern hemisphere. This difference is compounded in that the Earth's northern hemisphere is dominated by land, while the southern hemisphere is mainly ocean. Land and sea respond differently to heat: the land has a low ability to retain heat compared to the ocean, which is also capable of physically transporting heat (here the North Atlantic Drift, or Gulf Stream is just one example which we shall come back to later). A further complicating factor is that of the geologically recent formation of the polar ice caps. The continents have drifted to their present position, with Antarctica over the southern pole cutting off ocean currents that would have transported warmth, and providing a firm base for that ice cap, while the North Pole is now encircled by land, again impeding the flow of ocean currents. A further complication has been the (continuing) rise of the Himalayan Plateau (due to plate tectonic movement) and the cooling effect that this has had on the planet. These circumstances have contributed to the present (again in the context of geological time) uneasy balance between glacial and interglacial conditions: a balance that is continually being tipped by comparatively minute changes in the way the Earth receives its energy – the so-called Milankovitch orbital forcing component to the Earth's climate.

The Earth's orbit affects the way the Northern and Southern hemispheres receive the Sun's energy in three ways (see Figure 3.2):

(1) Through changes in orbit eccentricity. The Earth's orbit is not perfectly circular, its deviation toward a more eliptical orbit is measured by its eccentricity which changes cyclically on average every 93 000 years.
(2) The angle of the Earth's tilt – its axial inclination. The Earth is tilted by about 23°, but this varies with time with a cycle of some 41 000 years from 22.1° to 24.5°.
(3) The precession of the equinoxes. Just as the axis of a child's toy gyroscope will slowly rotate, so does the Earth's own axis so that the seasons are brought forward by a few seconds each year. It takes about 21 000 years for the Earth's axis to precess once.

Not only does each of these orbital parameters have its effect on the global climate, but each has it to a differing degree. However there are times when more than one of the cycles harmonise, 'forcing' the climate warmer or cooler depending on which parts of the cycles coincide, and so the Earth's climate broadly changes from glacial to interglacial. By combining these three cycles with respect to time and taking into account the effect of each cycle on the amount of sunlight falling on a particular latitude, we arrive at the Milankovitch energy 'wave form'. Evidence from ice cores suggests that the precessional signal is strong, though the eccentricity 93 000 year cycle is

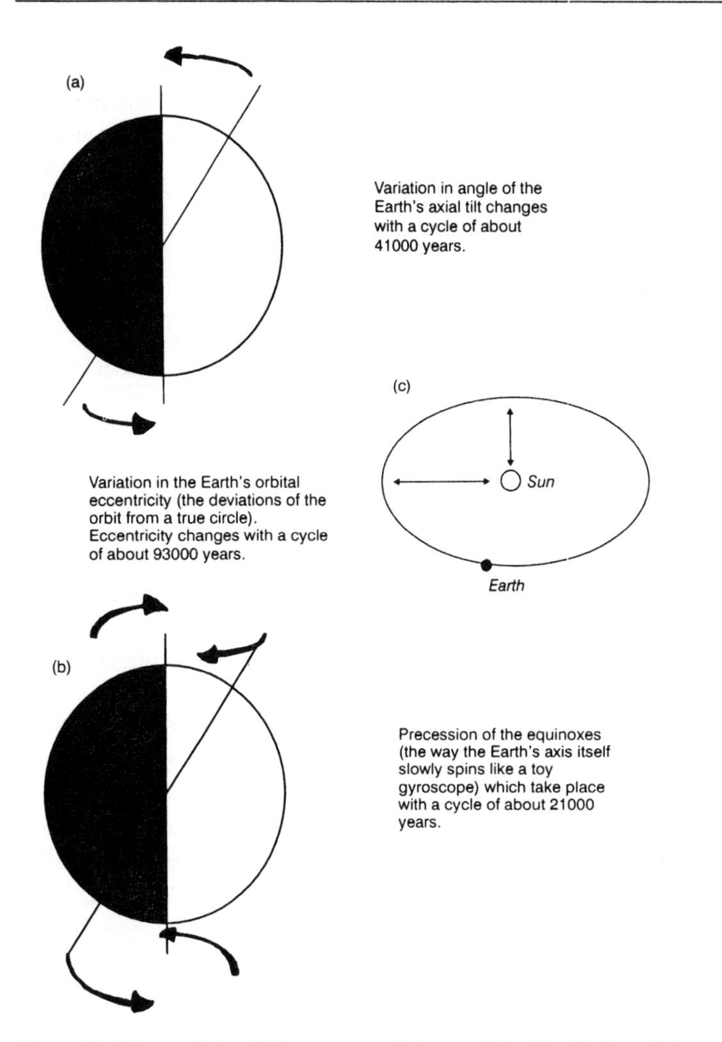

Figure 3.2 Milankovitch orbital parameters of a) eccentricity, b) axial tilt, and c) precession of the equinoxes

dominant. Even so, though the orbital forcing of climate seems to serve to trigger ice ages, the effect is so weak that other factors must be at work. The Milankovitch orbital variations tie in nicely with the timing of glacials and interglacials, but do not provide the necessary difference in heat received from the Sun to completely explain them. Fortunately, the evidence from the ice cores points to other factors[28].

In order for us to find out what caused the Earth's climate to vary so much in the past, events or processes need to be examined that meet specific criteria. The processes to study should be those that:

- continually take place,
- are preserved for us to study,
- and, importantly, are temperature dependent.

Ocean sedimentation is one such process – we can examine the organisms that continually fall to the ocean floor. Polar snow fall is another process. Water, be it liquid or ice, contains different isotopes of oxygen and hydrogen with the H_2O molecule. Furthermore, water made up of heavier isotopes requires more energy, or heat, to evaporate than water made up of lighter isotopes. Similarly, the heavier isotopes are the first to condense out of the atmosphere. The consequence of this physical relationship is that snow containing a higher proportion of heavier isotopes, must therefore, come from condensed water vapour that evaporated from water (the seas) under warmer conditions to that of snow containing a higher proportion of lighter isotopes. By studying the proportion of oxygen 18 to the lighter, far more common oxygen 16, or for that matter the heavier deuterium (hydrogen 2) to hydrogen 1 in the H_2O, it is possible to ascertain past changes in temperatures. Specifically, the proportion of heavier to lighter isotopes is largely related to the difference in temperature between that at which the water vapour was first formed and the final condensation temperature.

In the mid-1980s the Soviet Antarctic Expeditions at Vostok (E. Antarctica) drilled down over two kilometres into the ice and recovered a 2.083 m ice core representing some 160 000 years of snow fall to a time before the beginning of the last glacial. Analysis of the ice[29] revealed that the proportion of hydrogen isotopes had varied with time indicating past patterns of temperature changes very similar to those elucidated from deep-sea sediment cores. What has been found is that past climate changes, as demonstrated by the ice cores, corroborate those calculated from marine sediments. The likelihood is that both are really reflecting the same thing – the global climate. Both these (and other palaeo-geological) records also reflect the past temperature changes predicted by accounting for Milankovitch's three orbital factors, but at the very most the orbital factors could only decrease solar radiation by less than 0.7 Wm^{-2}, or a tenth of one percent of the solar energy the Earth receives: this is not enough to account for the majority of the 4°C cooling during the ice age's glacials. Nonetheless, these orbital variations do seem to provide a pacemaker for the ice age, and ice cores do provide evidence for other factors that reinforce, or amplify, the effect that Milankovitch solar radiation changes have on the Earth's climate.

One of the earliest ice cores to provide a detailed insight into the last glacial's climate, and the first to provide detail back to the previous interglacial, was the 2-km core extracted at Vostok in Antarctica in 1986. Ice cores do not just contain ice, but also minute pores of air that became trapped when the snow originally fell in the past. By crushing the ice under vacuum, and releasing this

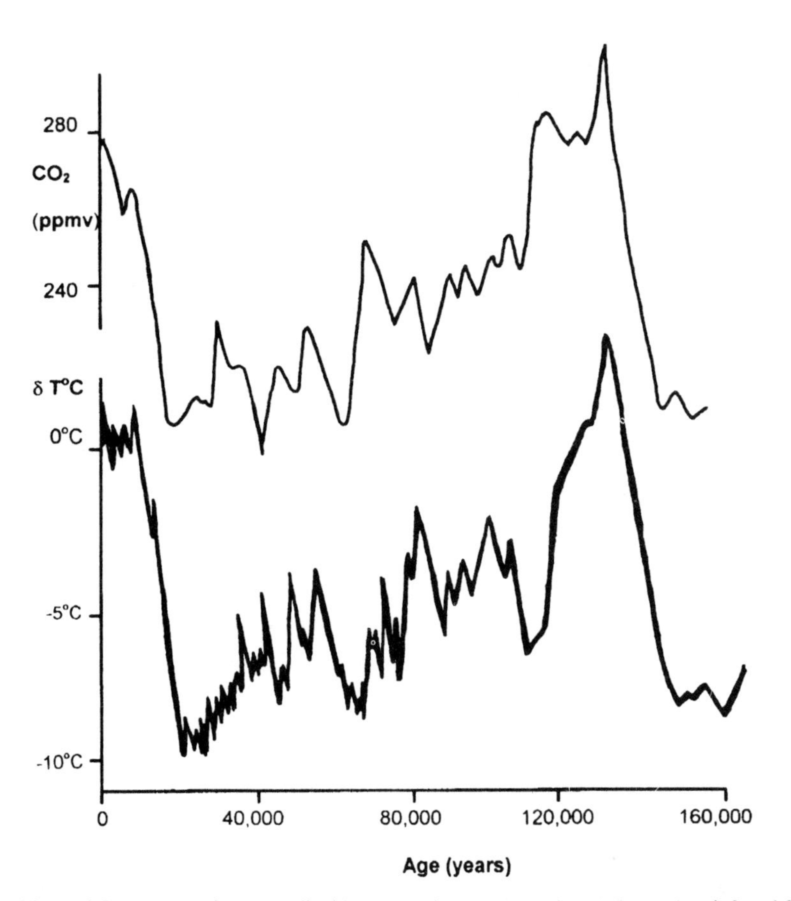

Figure 3.3 Atmospheric carbon dioxide (top) and temperature change (bottom) as inferred from the Vostok (Antartica) ice core over the past 160 000 years to the last interglacial. The temperature change (δ°T) here is the temperature relating to the Antarctic ocean water evaporating and then condensing as snow

trapped air, it is possible to analyse the gas to see how the Earth's atmosphere has changed with time. Measuring the carbon dioxide content of the atmosphere trapped in the Vostock ice core[30] revealed changes that paralleled the cores' hydrogen isotopic changes – as described above. Finally, in addition to the temperature and carbon dioxide plots, a similar pattern of atmospheric methane concentrations trapped in the core was found (see Figure 3.3). Carbon dioxide absorbs shortwave radiation as a so-called 'greenhouse' gas and affects climate to a far greater degree than the Milankovitch effects of historic variations in the Earth's orbit. Methane too, which is a more powerful greenhouse gas on a molecule for molecule basis than carbon dioxide (see Chapter 5 for information on each of the principal greenhouse gases) absorbs and traps shortwave radiation, or heat.

Together the Milankovitch theory and the above climatic records present somewhat of a conundrum. These records (as well as additional less reliable evidence of climate change over a longer period of time) seem to confirm that, whether or not we are in a glacial, or interglacial, their timing *is* governed by Milankovitch's orbital changes. Yet the amount of solar radiation change that this represents is not nearly enough by itself to trigger, or halt, a glacial.

On the other hand, the concentrations of carbon dioxide, and other 'natural' greenhouse gases has changed along with the Earth's climate, but this too is not enough: it could only account for about half of the temperature change experienced between glacial and interglacial times. Furthermore, the *causal* connection between the greenhouse gas and Milankovitch solar insolation changes is far from clear.

What does appear is that the weak Milankovitch effect – the orbital changes affecting the way sunlight falls on the Earth – really is a pacemaker governing the glacial–interglacial cycles, but that somehow it is amplified by other effects including, principally compositional changes in the atmosphere's greenhouse gas concentrations. Unfortunately, the detailed mechanism for this amplification remains unclear (though we will touch upon this again later).

In terms of future climatic change, the importance of obtaining evidence as to past (palaeo) global climate changes, and how they came about, is central to the human-generated, or anthropogenic, greenhouse issue (as opposed to the naturally occurring greenhouse effect – see Chapter 4). It allows us to assess the natural variability of our planet's climatic system. It assists in understanding the causes and mechanisms relating to climate change. Finally, it enables us to validate climate models, and to predict the regional effects of changes in the global system by comparing various scenarios with similar climates from the past. Work in this area is on-going, and is continuing to shed new light and extra detail on past climate change.

The Soviets are drilling a new hole to obtain a core at least a kilometre longer than the one analysed from Vostok above; this one will hopefully reach the bedrock at 3.5 km. By 1993 they had analysed some 2.546 km of core, which represented the accumulation of approximately 250 000 years worth of Antarctic snow fall (though dating the ice towards the bottom of the core becomes increasingly difficult). This has provided palaeoclimatologists with a view through time of conditions during – not just the last glacial but – the last glacial but one. It would appear that the penultimate glacial was more uniformly cooler than the last glacial, and that the intervening interglacial was quite short: about 10 000 years with a peak of just 1000 years warmer than today's climate. In terms of human change, this last may well be of interest. Human civilisation only arose during this current interglacial, one that seems to be a longer and more stable climatic interval[31] (we have already enjoyed 10 000 years of *comparatively* stable climate). This, however, might be purely coincidental.

Other nations are also drilling, or planning to drill, in Antarctica. These include Australia, France, Japan, and the USA. Northern hemisphere ice cores have been, and are also being, bored so we can see how the northern and southern hemispheres interact climatically. One of the key projects here is the Greenland Ice Core Project (GRIP) and the Greenland Ice Sheet Project (GISP) being carried out by European and US scientists, respectively.

Already the Greenland cores have matched those from Antarctica, confirming that glacials and interglacials are global (as opposed to hemispheric) events. Of particular interest are the cores obtained from Greenland's heart, the Summit cores. The Summit is an almost ideal place for ice cores as there is little horizontal movement blurring the ice layers. Additionally, the ice cap is very thick at this point and, after four summers of work, GRIP cores have been extracted that reach the bedrock at a depth of 3.03 km. Again, as with the Vostok core in Antarctica, it sheds light on a period of time spanning more than just the last glacial cycle. While, overall, the Greenland and Vostok cores generally match, there are differences.

Both the Antarctic and Greenland ice cores principally revealed two things. First, that the climate did not gradually move to and from glacial to interglacial conditions, but rather switched rapidly from one climate state to another. The average *global* climatic temperature change between glacial and interglacial conditions is roughly of the order of 4°C or 5°C (regional temperature interglacial–glacial differences can be much larger). Second, both cores exhibited evidence of comparatively brief warming periods from time to time.

There were other differences between the Antarctic and Greenland cores. The Greenland cores show 22 warm events (or interstadials) during the last glaciation (between 20 to 105 thousand years before present), whereas the Vostok core revealed only nine – those that in the Greenland core each lasted 2000 years or more[32]. It seems therefore, that the short warm events recorded in the Greenland cores may represent climatic change that predominantly took place in the northern hemisphere; while those longer ones detectable in Antarctic ice as well, represent global climatic changes. The *regional* interstadial warming inferred by both the Antarctic and Greenland cores is of the order of 7°C, which compares with an average *global* glacial–interglacial temperature difference of 4°C to 5°C. (Note: it is quite possible for a global interstadial to be recorded as 5°C–7°C regional changes in both Antarctica and Greenland but with other parts of the World (say the tropics) being affected less, hence the *global* interstadial temperature changes being greater than global averages are profound in the context of current anthropogenic global warming issues.) Interestingly, a number of the Greenland core interstadials were near times of transition to and from glacial and interglacial periods.

Researchers expected the Greenland cores to be better than the Antarctic cores but, in 1992, the great clarity of both the GRIP and GISP Greenland cores surprised even palaeoclimatologists. Few expected that they would reveal such short-term variation in the climate, not previously seen in the Antarctic Vostok ice cores. That these short interstadials, typically lasting from between 500 to 1000 years, actually took place was confirmed later in another analysis that same year. Intriguingly, the interstadials seem to snap on and off over a period of just a few decades[33]. The favoured hypothesis for the abruptness of these changes is thought to be alterations in ocean circulation transporting heat from the equator possibly related to or assisted by changes in wind patterns (more of which later). However, Greenland ice cores were so perfectly laid down and undisturbed by ice movement that analyses published early in 1993 revealed further surprises.

While the European GRIP team were coming up with their ice core analyses, the US GISP team were working on their second project, GISP 2. GISP 2 served to confirm much of the earlier work, but importantly included a detailed electrical conductivity measurement (ECM) of their core. ECM measures the ability of the core to conduct electrical current which in turn, is related to balance of acids and bases in the core. The ECM increases in the presence of acids, such as sulphuric acids from volcanic sources, while it decreases when the acids are neutralised, by say ammonia (from biomass burning and other sources) or from alkaline dust (continental sources). Consequently an ECM analysis, when combined with other data such as isotopic temperature analyses, can reveal when a major volcanic eruption has temporarily cooled the climate, or how the weather patterns alter (by inference with climate change). In 1993, the GISP 2 results revealed a previously unknown form of climate change. It was already known that there were relatively warm interstadial periods during the last glaciation, and that there were also century-scale returns to glacial conditions during the last glacial interglacial transition. What now emerged through ECM analyses was that there are also even smaller, sharper fluctuations in ice core conductivity of the order of less than 5 to 20 years at times when the climate is generally warm. This represents climatic alterations. (The atmospheric circulation seems to switch between the air carrying high and low quantities of dust, which requires extreme rapid reorganizations in weather patterns[34]). Indeed, isotopic analysis does indicate that temperature flickers coincide with ECM values from cores. It was as if around the time of transition from an interglacial state the regional climate (possibly affecting the entire northern hemisphere) 'flickered' now and then to and from glacial conditions. During interglacial transition, short warm as well as cold snaps between 5 and 20 years seem to take place.

To summarise, the picture that palaeoclimatologists predominantly had in the early 1990s of the last glacial–interglacial cycle was as outlined above (Figure 3.3). The previous two interglacials to our own (Holocene) were comparatively short-lived, less than 10 000 years. The previous interglacial (the Ipswichian or Eem–Sangamon interglacial) to our own took place about 130 000 years ago

Figure 3.4 The extent (stippled) of the northern hemisphere ice caps during the height of the last glaciation. Present-day ice caps are shown diagonal-line shaded

and this, though shorter, was warmer than our current Holocene interglacial. The previous one to that (the Hoxnian or Holstin–Yarmouth interglacial) was again shorter than our own Holocene but cooler and took place about 220 000 years ago. In between these interglacials were cooler glacials during which the polar ice caps expanded, and ice caps formed over much of North America (the Laurentide ice sheet) and northern Europe (see Figure 3.4). The greenhouse gases, methane and carbon dioxide, and (as independently indicated by both hydrogen and oxygen isotopic analysis) the temperature during glacial times dropped sharply from interglacial values. In addition, glacials become cooler in progressive stages over a number of tens of thousands of years, before a comparatively sharp warming to interglacial conditions in under a century. However during periods of glacial–interglacial transitions the climate seems to flicker between two states, providing cool snaps of five to twenty years. Even when in interglacial mode there do sometimes seem to be repeated similar length cool snaps.

In terms of the present-day climate, and the problems associated with alleged human-induced global warming, there are perhaps two lessons to be learned. The first is that the World's climate seems to operate in distinct modes (glacial and interglacial). Second, within these long-term modes there is considerable short-term sharp variation possibly between these climatic modes. The question that then springs to mind is what is the principal factor (or factors) determining which climatic mode the World is in?

As we have seen, the timing of glacials and interglacials is determined by the Milankovitch orbital factors. Yet the Earth has always been subject to variations in eccentricity and obliquity of orbit as well as experiencing precession of the equinoxes. This includes periods before two million years ago and the onset of the Quaternary ice age we are now experiencing. So how come orbital parameters have caused glacials in just the past couple of years? There are three main factors increasing the planet's orbital driven sensitivity to variations in the way solar radiation is received: (i) the rise of the Himalaya, (ii) the blocking and surrounding of the South and North Poles respectively, and (iii) the joining of North and South America. Two are directly related to the way the oceans circulate which, it turns out, are central to governing the climate.

One of the main ways heat is transported away from the tropics is by ocean currents. In the land-dominated northern hemisphere (that loses heat quickly) one of the major currents transfering heat north from the tropics toward the Pole, is the North Atlantic Drift. This water evaporates, releasing its heat and warming the northern hemisphere, while leaving behind cooler and saltier water. The now heavier saltier water sinks and flows back toward the Equator along the bottom of the Atlantic, down around Africa, to rise again in the Indian and Pacific Oceans (see Figure 3.5). One of the principal driving forces behind this 'global conveyor' is the way the water on the North Atlantic's surface becomes saltier as it moves towards the pole. Disrupting the conveyor (for instance by adding freshwater to the saltier water before it sinks) also disrupts heat transportation away from the tropics. Wallace Broecker (of Columbia University), argues that this has happened in the past and that it can happen very quickly (in only decades or years) and so affect the global climate. Subsequently, Gerard Bond (also of Columbia) together with Broecker and some of the Greenland European GRIP, and Antarctica Vostok palaeoclimatologists have put forward an interesting theory that might explain much of how the climatic changes within glacials may work.

With regard to the broader temperature changes in the last glacial, the currently preferred hypothesis that Gerard Bond and his colleagues propose is that during the last glacial the North American (Laurentide) ice sheet periodically collapsed each time, releasing a massive wave of icebergs that melted depositing moraine on the Atlantic floor (these are Heinrich events: named after Hartmut Heinrich). The moraine consisting of rocks ground up by and incorporated into the ice as glaciers flowed off from North America. The collapse of the ice sheet temporarily halted the steady flow of icebergs from the Laurentide sheet, allowing salty water to once more form and so strengthen the 'global conveyer'

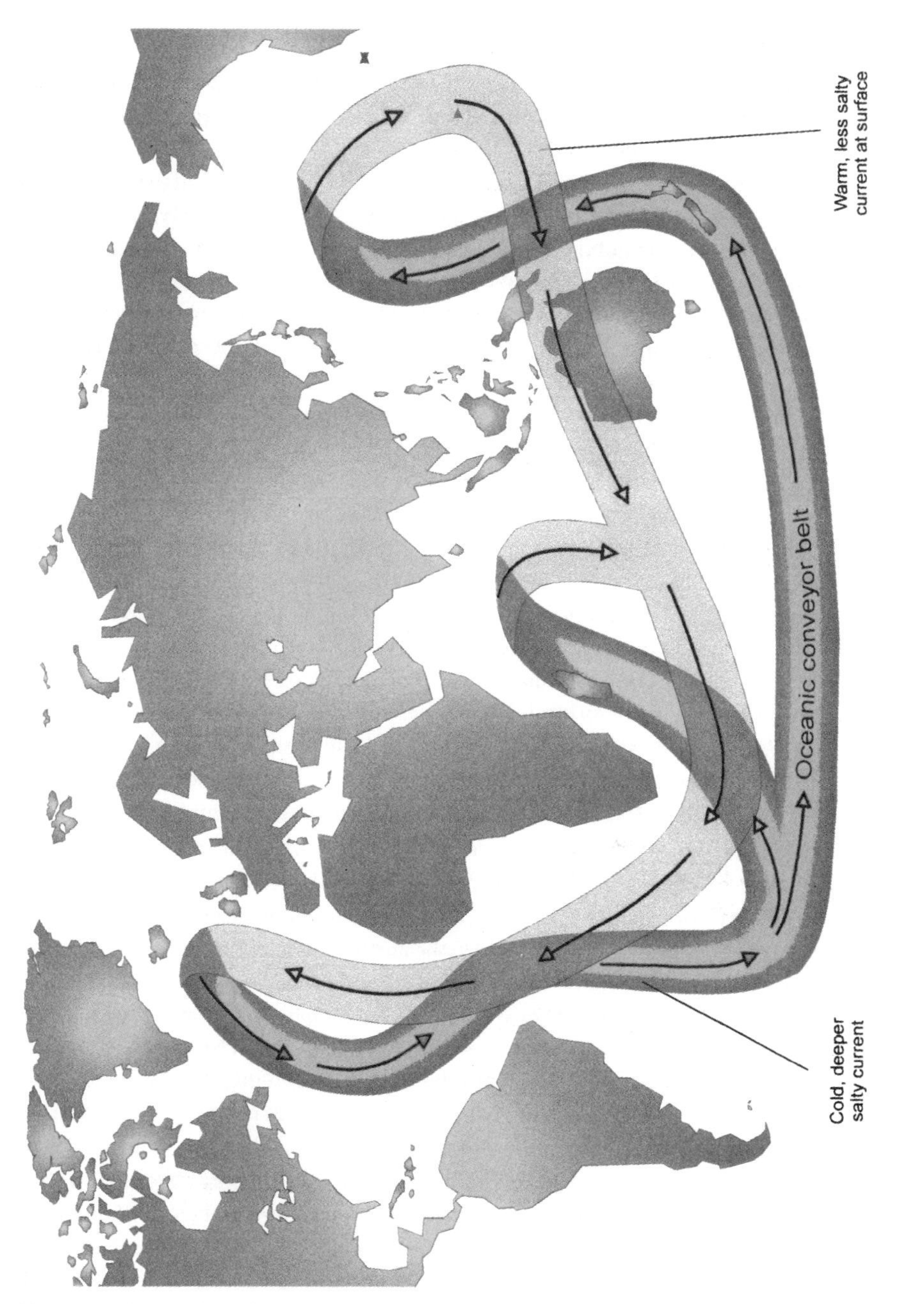

Figure 3.5 The Broecker salt conveyer

transporting heat from the equator, and resulting in a sudden climatic warming. Over the next few thousand years the icebergs began to return, adding fresh-water to the North Atlantic and diluting the saltier water, preventing it from sinking and so weakening the 'global conveyor'. These 'Bond' cycles of climatic change are several thousand years long, within which there are shorter Dansgaard–Oescheger climatic change events, which are possibly caused by the periodic build-up and decay of salt in the North Atlantic surface waters due to changing balances of precipitation and vapour loss between the Pacific and the Atlantic. While the ultimate causes of the Dansgaard–Oescheger events are still unclear, the longer Bond cycles do seem likely to be the result of periodic Laurentide ice sheet collapse[35].

Proof as to exactly why the Laurentide ice sheet periodically collapses has yet to emerge. One favoured theory (the MacAyeal model) suggests that, when small, the Laurentide ice sheet sticks to the rock beneath it. As it grows it reflects more sunlight away, so encouraging further cooling of the Northern Hemisphere and, indeed, perhaps winds that cool the North Atlantic (hence the cooling trend within the Bond cycles). Further, as the ice sheet grows, so it increasingly thermally insulates the rock beneath it, so trapping the Earth's trickle of geothermal heat. Eventually this causes the basal ice to melt, and fairly quickly large areas of the ice sheet will suddenly lie, not directly on rock but, on a lubricating layer of water. It is this lubrication of the grown ice sheet that possibly triggers its own collapse (with icebergs drifting away to melt depositing moraine (the Heinrich events)). With the ice sheet collapse, so its large cool reflective mass departs: wind patterns are affected and may temporarily return to those we see today that encourage North Atlantic evaporation, hence a strengthening of the 'global conveyor', and heat transportation from the tropics[36].

One question that emerges from all of this is whether the changes revealed by the Greenland ice core (that relate to the north American Laurentide ice sheet in glacial times) also apply to the glacial ice sheet over northern Europe on the other side of the Atlantic? In 1995 European geophysicists published their results of analyses of sediments from the Norwegian sea[37]. These showed that the changes in climate, as well as episodes of iceberg calving, largely coincided.

Not only do the 'Bond' and 'MacAyeal' hypotheses respectively explain the warming following Heinrich events (and so the Bond climatic cycle) as well as why the Laurentide ice sheet periodically collapses. but they also support each other and are currently two of the preferred options. Importantly, for human-induced (or anthropogenic) climatic warming, both theories relate to the 'global conveyor'; the existence of which is threatened in a warmer World (we shall return to this in Chapter 4). And yet, despite there being areas of doubt, it is only by understanding how the global climate is formed, both in the present and the past, that we can hope to begin to predict how it might change with circumstances in the future. For these reasons techniques for assessing the climate

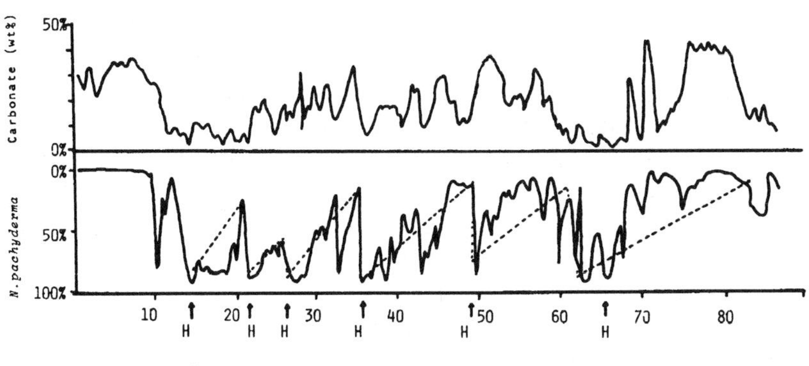

Figure 3.6 Biological reflections of the waxing and waning of the northern hemisphere's polar ice sheet in a series of 'Bond' cycles. The plankton, *Neogloboquadrina pachyderma* thrives in cool water and so can indicate sea temperatures – the abundance scale is inverted for ease in seeing the warming and cooling (bottom graph). The end of a Bond cycle (as denoted by the saw-toothed dotted line) is marked by a Heinrich event. The production of carbonate from deep water plankton (top graph) varies (albeit in a complex way) with periods of lower production coinciding with the ends of Bond cycles

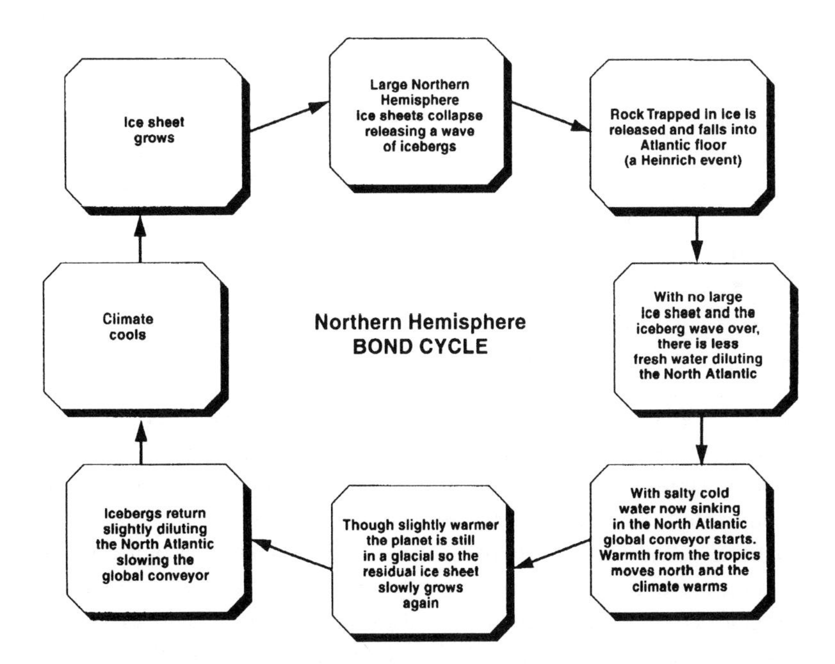

Figure 3.7 Flow diagram of the Bond hypothesis of climate change due to ice sheet expansion and contraction turning the Broecker salt conveyor on and off

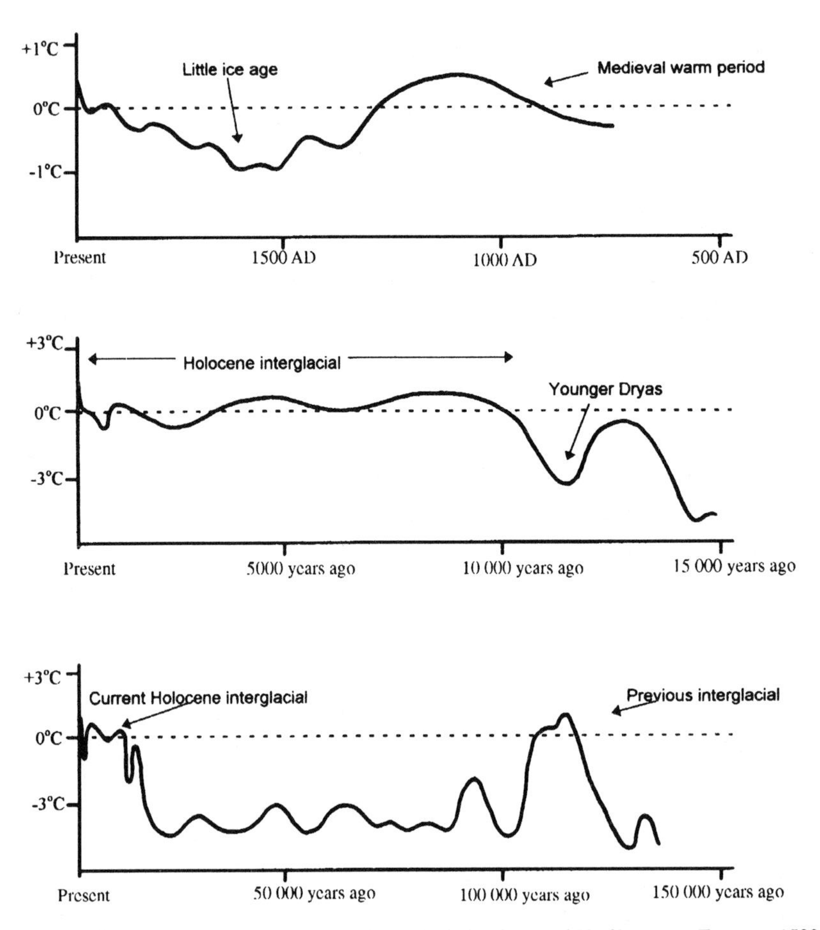

Figure 3.8 The principal features of past climate deviation from a 1950–60 average ***Top:*** past 1500 years. ***Middle:*** past 15000 years. ***Bottom:*** past 150000 years. Note the top graph's temperature axis is expanded

during the last glacial, and its past atmospheric greenhouse gas concentrations, are continually being developed and investigations continue.

Of particular relevance to the current greenhouse issue, is the question of why carbon dioxide concentrations vary between glacial and interglacial times, and considerable effort is being made to ascertain how the carbon cycle functions. One promising avenue of study is that of the relationship between various isotopes of carbon in carbon dioxide that have been metabolised by ancient photosynthetic plankton. Analysing carbonate from such marine sediments containing plankton remains could possibly reveal the past sources of atmospheric carbon and extend climate records far beyond the current time limitations possibly to 2 million years[38], and climate well before the beginning of the current glacial– interglacial cycle, if not the ice age we are in itself.

This is all very well, but from the human perspective the climate of the present interglacial period in which our planet has sported for the best part of 12 000 years is of greatest interest. As mentioned, it is this time that has allowed human civilisation(s) to flourish.

THE RECENT GLOBAL CLIMATE

As mentioned, we are currently in an interglacial in the middle on an ice age. Prior to this, over 2 million years ago, the Earth was about 4°C warmer than today. In that world the sea levels were higher than today, but it was the world in which early species of Man began to evolve. The question as to why sentience subsequently developed cannot be answered with certainty, but it does not seem unreasonable that it was our ice age that subsequently helped spur the evolution of human intelligence. The ice age's series of (in evolutionary terms) rapid hundred-thousand-year glacial–interglacial climatic oscillations provided the need for continual adaptation, as well as the mechanism for the strict weeding out of those species which failed to adapt. Though this cannot be absolutely proved, we do know that the planet only came close to returning to pre-ice age conditions during its interglacial times. The latest of these, our interglacial climate today, bears many similarities to the previous one, the Ipswichian or Eem–Sangamon interglacial some 125 000 years ago. But there are some differences. The Ipswichian/Eem–Sangamon interglacial saw the global climate only some 2.0–2.5°C warmer than today's. Then again, there were other differences with today's Holocene interglacial. The Antarctic ice shield back in the Eem-Sangamon interglacial had partially melted so that the seas were 5–7m higher than at present. Furthermore, the climate of our Holocene has been more uniform.

About 14 500 years ago the main part of last glacial was coming to an end, with the Earth's climate becoming very much like it is today. But the initial warming (known to palaeoclimatologists as the Bolling–Allerod) was short-lived. After 1500 years there followed an abrupt return to a colder climate (the Younger Dryas) that also lasted for about 1500 years, with glaciers reappearing in the UK Lake District so quickly that living trees were caught in their advance[39]. Independent evidence of the climatic lapse has been detected in the USA indicating that the temporary Younger Dryas cooling was planet-wide. Indeed, one might view the Bolling–Allerod warming as a glorified interstadial (or very large climatic warm flicker in the glacial) and not part of the interglacial proper. Overall the Earth was moving out of its glacial phase.

The current Holocene interglacial truly began nearly 12 000 years ago. Our Holocene interglacial's warmest point was some 6000 years ago, when the Earth was up to 1.5°C warmer than today. This period, the so called 'Climatic Optimum', saw a British climate some 2–3°C above that of the present with forests growing up to 300 m above the present tree line. Trees also grew in the Orkneys,

Faeroes and in Iceland, indicating that the climate was also less windy than at present.

As the glacial ended and the ice sheets melted, the sea level rose. About 5000 BC the Baltic and North Sea filled and Britain became an island. (Here, in terms of future global warming, it is perhaps worth noting that it took 4000 years from the onset of the Holocene for the seas to rise enough to flood the continental land bridges.) From about 3000 BC onwards the climate became less stable and between 1000 and 500 BC there was a deterioration of 2°C from which there has never been a full recovery. However there have been partial recoveries.

From the tenth to early fourteenth centuries the Earth went through a comparatively warm period. Although not as warm as the first half of the Holocene, the climate was roughly one degree warmer than currently in the late twentieth century. In the tenth to fourteenth centuries there were Viking farmsteads in Greenland's fjords and vineyards in central England, while wheat and barley were cropped in Iceland. Overall this Mediaeval Warm 'Epoch' also saw a great expansion of European civilisation. Evidence that this regional warming was at least hemispheric, if not global, in nature includes that from California's Sierra Nevada. Preserved tree stumps in marshes and lakes have been radiocarbon dated and the tree ring patterns used to build up a climatic picture. They show that the Sierra experienced severe drought for over two centuries before AD 1112 and for more than 140 years before AD 1350[40]. The implications for today are that, *should* human emissions of greenhouse gases be warming the World, then that area could again suffer from prolonged drought. (It is interesting to note that since the late 1980s that part of western North America has experienced a marked lower rainfall.)

Even closer to the present there have been minor fluctuations in temperature, but of less than one degree centigrade. However, such small changes are difficult to detect and it is not certain whether all the fluctuations seen in the geological record have been truly global or just regional.

These minor fluctuations of 1°C or less are not unimportant as they have in the past affected human activity, indeed the fate of whole settlements[41] and, though the global climate may go through changes of 1°C, there may be even greater variations regionally. Historically, there have been 'disasters' affecting, for instance, whole settlements in the Alps, and it has been suggested that large areas in Scandanavia were simply abandoned within about one generation.

More recently, one of the worst of these little climatic fluctuations – at least for Europe (and, as indicated from treeline data, perhaps the US[42]) – was between AD 1550 and 1850 and was known as the 'Little Ice Age'. Then a tongue of Arctic ice appears to have reached as far south as the Faeroe Islands. So marked was this climatic event, some have argued, that in Britain it played a part in the movement from a rural to urban-based population, so assisting in providing an accessible workforce for the forthcoming Industrial Revolution.

The recovery from the Little Ice Age to today's climate is well documented. It is recorded by over a century of direct observation, including the use of thermometers and rain-gauges, as well as today by such methods as satellite remote sensing. For example, between 1880 to 1930, records at Oxford show an increase of around 10% in what biologists call the thermal growing season (TGS) (the number of days in the year when it is warm enough to grow crops)[41]. It was around this time that humans began to introduce carbon dioxide at a rate above the environment's assimilative capacity, so generating an excess over pre-industrial levels. The question that dogged early climatologists from that time through to the end of the 1980s was whether this increase in atmospheric CO_2 would affect climate? Certainly the climate is becoming warmer, although this in itself is not proof that CO_2 emissions are responsible.

Although evidence for recent climatic change abounds some have argued that such evidence is flawed: in the case of long-term meteorological records it has been suggested that the growth in size of towns (where many recording stations are sited) distort the picture as towns are invariably warmer than nearby countryside. To counter this there is other evidence that confirms that warming is taking place. For instance, research on tree rings of 700-year-old Huon Pine in Tasmania indicates that unprecedented warming has taken place since 1965[43].

The problem facing climatologists is that any signal indicating greenhouse warming tends to be swamped against the natural variability in the weather. So how do climate change signals manifest themselves? There are some indications, though not conclusive. In the USA, between 1975 and 1982 inclusive, over 48 states saw three consecutive winters 1°C cooler than average; yet over the same period of time three winters were also warmer. In 1984 the National Oceanic and Atmospheric Administration (NOAA) concluded that it would take over a thousand years to produce winters as abnormal as those experienced above[44]. The likelihood of the consecutive run of cold winters alone (such as 1976/7–1978/9), they, the NOAA, thought was to have odds of about once every 550 years – so it is slightly surprising, though not impossible, that such an event took place within the context of one to two centuries of reasonably well-recorded meteorology. Cold winters, paradoxically, might well be expected in some regions in an over all warmer world: one possible mechanism being that a warmer world would experience more ocean evaporation, hence regional cloud formation another that normal storm tracks might shift (such as with the later 1996 snowstorms in the Eastern USA). At the time the NOAA's conclusions were non-committal. They felt that either the USA's recent climatic record of winters was a moderately rare event chanced upon in a reasonably stable climate, or that it represented a contribution toward a longer-term climatic change. The NOAA was not alone, most climatologists at that time (pre-1990) were decidedly unwilling to forecast whether global warming was real. Evidence, though, was mounting.

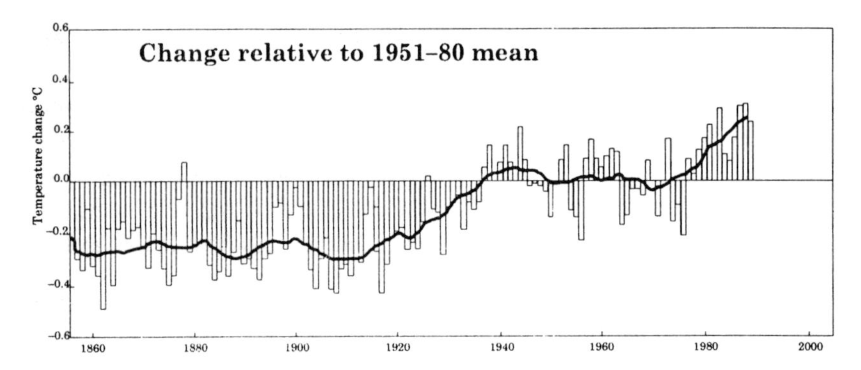

Figure 3.9 Global mean combined land-air and sea surface temperatures both annual (bars) and 5-year smoothed (line) deviation from the 1951–80 mean. From *Bioresources: Some UK Perspectives*. Reproduced with permission from the IoB, London

The mid-1970s through to the mid-1980s were to show a resumption of the warming trend that had been detected over the century's first 40 years. World-wide, the 1980s turned out to be the warmest decade recorded, and even before it ended there was much discussion as to whether this really indicated that global warming was really taking place, as opposed to merely being part of the weather's natural variation?

Some scientists did start making tentative predictions despite the dangers of being discredited. For example, James Hansen of the Goddard Institute for Space Studies said, in the summer of 1988, that barring an improbable event, 1988 would become the warmest year on record[45]. – It was! – It turned out to be some 0.34°C above a 1950–1979 reference period. However, he was only right by 0.1°C, as 1988 just pipped 1987 for the warmest year (until then) of the century. This still was not conclusive evidence of long-term climatic change, and informed opinion was still divided. It was going to take an international panel of over 700 scientists to decide whether or not the scientific community should declare that human activity was affecting the climate and, if so, to what extent? They were going to have to decide whether or not the anthropogenic greenhouse effect was really changing the World's climate? This panel (as we shall see) reported its first conclusions in 1990; a year that itself turned out to be a record-breaker being some 0.05°C warmer than 1988! Other record-breaking warm years would follow.

CHAPTER 4

CLIMATE CHANGE – GLOBAL WARMING

GREENHOUSE THEORY

The term 'greenhouse' effect was coined to describe the way that some gases trap heat within the atmosphere (much the same way as gardeners appreciate that greenhouses trap heat) by allowing it in more readily than out. Greenhouses allow the high energy (high frequency) infra red radiation through the glass, but when this radiation is subsequently reflected or absorbed it is re-emitted at a lower frequency (lower energy) as heat radiation, and is prevented from escaping by the glass. The glass is only transparent to the incoming higher-energy radiation and so the greenhouse warms. Greenhouse gases in the atmosphere have similar properties to glass in greenhouses, hence the term 'greenhouse'. What happens is that the so-called greenhouse gases allow high energy infra red radiation (with wavelengths shorter than 13 μm) through the atmosphere but absorbed, so preventing escape back into space of infra red wavelengths between 13 to 100 μm (see Figure 4.1). However, some argue that the 'greenhouse' expression is misleading, in that it presents Manmade, or anthropogenic, global warming as innocuous, like the common greenhouse whose function is beneficial to horticulturists and gardeners: the 'hothouse effect' might be more appropriate and have a greater perceptual impact.

The idea of greenhouse gases is not new. John Tyndall[46] described the role of water vapour as a greenhouse gas in 1863. A few years later, at the end of the nineteenth century the Swede, Svante Arrhenius[47], argued that human activity generating carbon dioxide could possibly affect the climate by inducing warming of the atmosphere.

Much of the greenhouse effect is natural – it has nothing to do with human activity – and it is possible to estimate the magnitude of this natural effect. The Moon, which is the same distance from the Sun as the Earth, receives a similar intensity of solar radiation as the Earth. Yet, though temperatures on the moon oscillates between 100°C in the day and –150°C at night, the average lunar temperature is only about –25°C (not about 15°C as on Earth). Unlike the Moon, the Earth not only spins faster and has oceans to retain heat but it has an atmosphere trapping heat via the greenhouse effect. The Earth's average temperature is so much warmer than the Moon's that the Earth's natural greenhouse warming, or climate forcing, must approach 40°C.

However, there are greenhouses and greenhouses. Gardeners will often pull

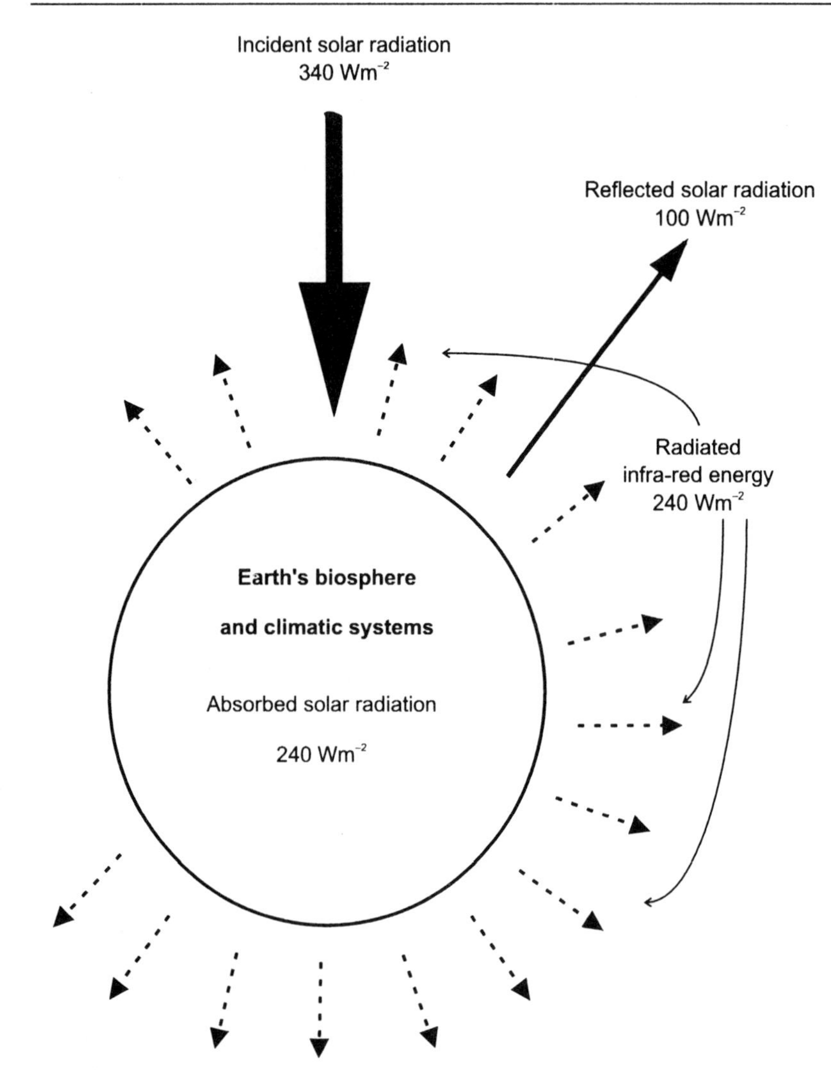

Figure 4.1 A summary of the principal solar energy entering the Earth's atmosphere. Not all the high energy infra-red radiation falling on the Earth is reflected back out into space. The result is warming. Note: the Sun, at the Earth's distance, radiates 1370 Wm^2. However, the Earth is a rotating sphere so that the average energy falling on it is 340 Wm^2

blinds, paint a translucent layer over the glass, or open a window for ventilation in the summer to prevent their greenhouse from overheating. In cooler climes special greenhouse glass may be included that has a greater 'thermal' performance to enhance the greenhouse effect. Likewise, the key question relating to 'greenhouse Earth' is whether its own thermal performance is being changed by altering what acts as its 'glass' – by changing its atmosphere's composition?

The atmosphere is indeed changing, as Arrhenius thought it might, with the proportion of carbon dioxide increasing in recent times because of the burning of fossil fuels. In 1765, prior to the industrial revolution and the subsequent increase in fossil fuel burning, the Earth's atmosphere contained 280 ppm (parts per million) of carbon dioxide, by 1990 it contained 353 ppm. Furthermore, (as discussed in the previous chapter) the atmospheric carbon dioxide concentration, and so by implication the greenhouse warming effect, was less during the last glacial than in the current interglacial[30]. There are also other greenhouse gases, principally methane (CH_4), an assortment of artificially added chlorofluorocarbons (CFCs), and nitrous oxide (N_2O). These gases have all increased in concentration since the industrial revolution (See Table 4.1)

Table 4.1 Summary of principal greenhouse gases (with the exception of ozone, O_3, due to lack of accurate data)

Greenhouse gas	CO_2	CH_4	*CFC-11*	*CFC-12*	N_2O
Late 18th century atmospheric concentration	280 ppm	0.8 ppm	0	0	288 ppb
Current (1994) concentration	358 ppm	1.72 ppm	268 ppt	484 ppt	312 ppb
Current rate of atmospheric accumulation *per annum*	1.8 ppm (0.5%)	0.010 ppm (0.6%)	9.0 ppt (0%)	17 ppt (4%)	0.8 ppb (0.25%)
Atmospheric residence in years or atmospheric content divided by rate of removal	50–200	12	50	130	120

Yet it is one thing to have an anthropogenic greenhouse 'theory', albeit backed by corroborating evidence; quite another to say that human activity is actually causing global warming as a 'fact'. Is there really an anthropogenic contribution to the climate? Though hard evidence can be called upon to support the theory, there are gaps and some of the evidence itself is circumstantial. It is the balance of this evidence that determines the scientific consensus. For example, we can measure the thermal properties of carbon dioxide gas in the laboratory by shining light of various wavelengths through it to see what parts of the infra red spectrum are absorbed, but can we translate this information into an expression of how a change in concentration of such a complex mix and structure as the planet's atmosphere affects its own thermal properties? The climate's own natural variations are so great (as mentioned in the previous chapter) that they might easily smother any comparatively subtle trends that we may discern over a short period. An analogy might be that we suspect that someone may have switched some game-playing dice with slightly loaded dice. Rolling three sixes in fairly quick succession may, or may not, be significant but many more rolls are required for us to be certain if the dice are loaded. Similarly, in discussing the possibility of human-induced climate change we must distinguish between significant and non-significant facts, and between facts and fictions. In short, we must ascertain the areas of uncertainty.

FACTS AND FICTIONS

There are uncertainties as to what degree human activity is affecting climate. Uncertainty, if not clarified, can be used by those with a special interest in maintaining the social *status quo* to counteract the expression of concern, and worse, the implementation of remedial policy.

Several areas of doubt exist, indeed existed especially prior to the initial landmark report in 1990 from the Intergovernmental Panel on Climate Change (IPCC)[20], that report was an informed expression of considered opinion from the scientific community (rather than of pure fact alone). What, for example, is the *exact* effect of anthropogenic atmospheric carbon dioxide on the climate? Here it should be recognised that the human injection of carbon dioxide into the atmosphere began before sufficiently accurate climatic monitoring existed. This makes it impossible to relate accurately, *first hand*, pre-early industrial climate and carbon dioxide data to our present situation. However, while it might not be possible to refer to first-hand data, it is possible to reduce the doubt arising from this absence by examining secondary, climate and carbon dioxide related, data. Much of this data is either inferred (such as by identifying the remains of plants and animals known to prefer certain climate types) or from the bio-geological record, which itself is subject to errors in dating, let alone experimental limitations. While such data presents climatologists with problems, by gathering as much of it as possible and cross-referencing, a fairly accurate picture can be elucidated of the pre-industrial carbon dioxide and climatic conditions.

Another main area of doubt and uncertainty is that associated with predicting possible future climate warming: even if we could establish with 100% accuracy that global warming was taking place to such-and-such a degree, how would this affect the weather regionally? After all in a warmer World there should be more evaporation from the seas, so that some areas might have more cloud cover, more rainfall and even be, on average, cooler. Finally, of special importance to the many inhabited low-lying coastal regions of the World, how will global warming affect sea level.

In the previous chapter we saw that past atmospheric carbon dioxide and methane levels (two of the principal natural greenhouse gases) seem to move in tandem with the climate. Direct atmospheric carbon dioxide measurements have been made over a number of recent decades, and we know for certain that atmospheric carbon dioxide is increasing year by year. Furthermore, by extending the climatic and carbon dioxide records back into pre-industrial times, by using the above non-first hand data, and relating it to recent direct measurements, a coherent picture is formed (see Figure 4.2). From this it is possible to begin to make some fairly confident projections into the future.

From the above we might reasonably conclude that such increased atmospheric concentrations of greenhouse gases *will* affect the Earth's energy budget,

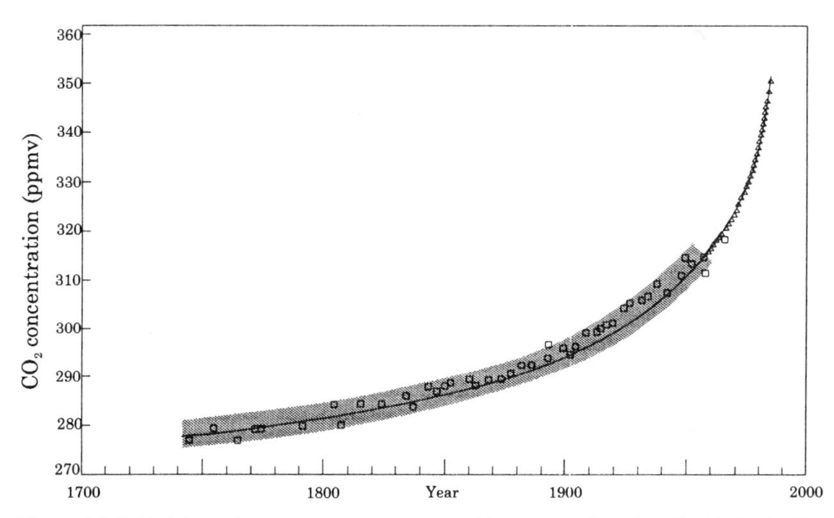

Figure 4.2 Mid-eighteenth century to near-present trend in atmospheric carbon dioxide levels. The squares represent values determined from ice cores whereas the triangle represent direct (Hawaiian) atmospheric measurements. The straight line is an approximation to a smooth curve value and the stippled shading represents likely measurement error. Note how nearly half the increase in the pre-industrial level of atmospheric carbon dioxide took place in the latter half of the twentieth century. Also note how this graph compares for that for energy use and especially the long-term trend in carbon release from energy consumption (see figs 6.5a, b and c). From *Bioresources: Some UK perspectives*. Reproduced with permission from the IoB, London

i.e. how much energy the Earth receives and how much it radiates. But without a more sophisticated analysis the symptoms of this climatic change will be unclear. Even the straight-forward changes-in-atmospheric-greenhouse-gas argument in itself is not absolute proof that the climate is truly warming, only that change is likely.

Direct measurement of the World's climate shows that, if there are any trends to date, they are as likely as not masked by the climate's inherent natural variability (see Figure 3.9b). This has led to some apparently (on the surface at least) contradictory comments within the scientific community. 'Britain shivers in the global greenhouse' was the title of a feature article[48], in the magazine, *New Scientist*, that attempted to chart a way through such (superficial) inconsistencies. It noted that while the World had grown warmer over two decades there was simultaneously a cooling over much of Europe, NE and E Canada and SW Greenland. Over Scandinavia the temperature fell by 0.6°C whereas over the same time period the entire Northern Hemisphere had warmed by 0.31°C.

Discerning such small differences in temperature requires highly accurate measurement. Not surprisingly climatologists themselves have been concerned that they are detecting a real change and not some artifact. For example, it is well known that large urban conurbations generate enough heat to affect weather station readings: in other words, there is enough heat given off by large towns to affect the local weather. But the urban conurbations themselves are not constant, they have been growing in size over many decades so increasing this heat island effect and further alter local weather readings. Weather stations located in such places, and which are contributing data from which the Earth's average temperature is calculated, may mislead climatologists. Climatologists, though, are aware of such complications and, as in this instance, have gone to great pains throughout the 1980s to ascertain the level of error that these may introduce. In this particular case it has been estimated that though urban heat islands may well affect global temperature calculations, it is only a slight effect[49]. Of a 0.5°C global warming trend detected over a century, it is thought that the error due to urban heat islands was an order of magnitude less, at 0.05°C.

Uncertainties in interpreting meteorological data pale against those concerned with our understanding of the mechanisms that generate the climate itself. This impedes the forecasting of the effect on the weather of climate change. As mentioned previously, cloud cover directly affects how much energy is absorbed by the lower atmosphere and the amount reflected. In a warmer World there would be greater evaporation from the oceans. This in turn would tend to reinforce the warming as water vapour is itself a greenhouse gas. On the other hand an atmosphere containing more water is more likely to contain more clouds which reflect sunlight and so result in cooling. The ability of clouds themselves, and winds, to transport energy vertically either to the top or bottom of the cloud layer serves to further complicate matters[50]. Climatologists have, for the time being, accepted that their understanding and quantification of many climate-determining processes is incomplete. It is rare for such uncertainties to be immediately eliminated from calculation once recognised, but they can be partially accounted for. In this instance, prior to the publishing of the initial IPCC report[20], they began to side-step the problem by obtaining satellite data from the Earth Radiation Budget Experiment (ERBE) in order to verify computer climate models: if the models can mimic reality accurately then their forecasts for a hypothetically warmer Earth will have greater validity.

The greatest uncertainty associated with the greenhouse debate arises out of our only having one Earth, and so we have only one set of data. It is simply not possible to set up the standards and controls as required for scientific experiments, so that the human injection of atmospheric carbon dioxide is itself, in effect, an uncontrolled experiment. Furthermore, scientists' appreciation of our one Earth as a 'single whole biosphere' is in its infancy. This largely arises from science's own historical development, but still affects the way climate change issues are examined.

Modern biology grew out of a naturalist approach where individual species were examined. Ecology, the science of communities and how they interact, came later. Consideration of the environment only became a recognised disci-

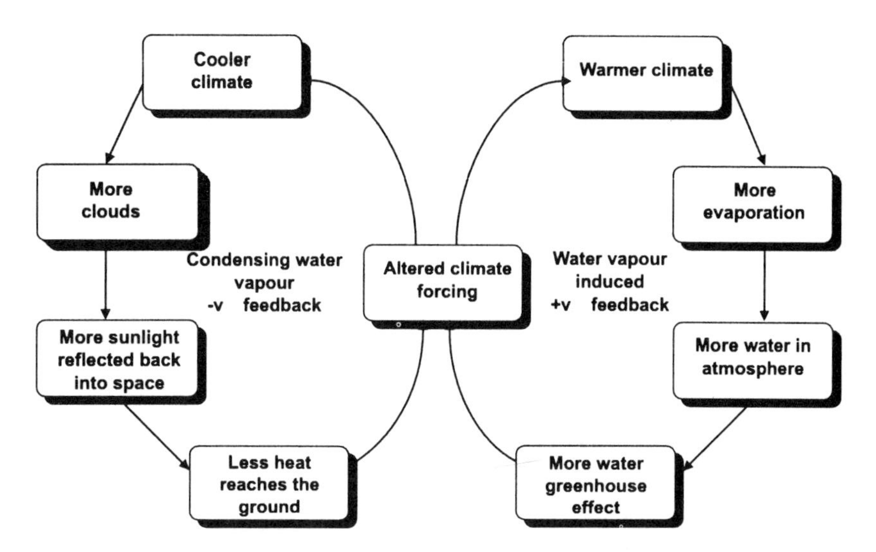

Figure 4.3 Examples of +v (positive) and –v (negative) feedback cycles in determining climate forcing

pline in its own right comparatively recently. In the UK most university undergraduate courses in environmental science were founded in the 1970s and 1980s. Worldwide, environmental science is still very much a young discipline, and the concept of the biosphere hardly explored.

In 1785 James Hutton referred to the global cycles of elements in the soil and their movements between ocean and land. The concept of the 'biosphere' as a term was developed early this century. In 1926 the Russian geologist Vladimir Verdansky concluded in his book, *The Biosphere*, that 'there is no force on the face of the Earth more powerful in its results than the totality of living organisms'[51]. Today we still know comparatively little about this 'totality' let alone the details of how the biosphere and the Earth's climate relate to each other. One recent biosphere hypothesis in particular has generated much discussion, some of it heated.

In 1969 at a meeting in New Jersey, James Lovelock put forward the theory of Gaia as a way of taking a different look at life on Earth. His idea was that the physical and chemical condition of the Earth's surface, atmosphere and oceans has been made fit for life by life itself, via a number of cybernetic processes. Lovelock suggests that the biosphere is still being maintained through homeostatic (self-regulating) processes[52]. With regard to climate, one of the most important biosphere processes is the way carbon is cycled between various forms (be it in plants, atmospheric gases, organic remains (such as peat), or even fossilized (as coal). The way carbon moves about the biosphere is known as the carbon cycle, and it includes the greenhouse gases carbon dioxide and methane

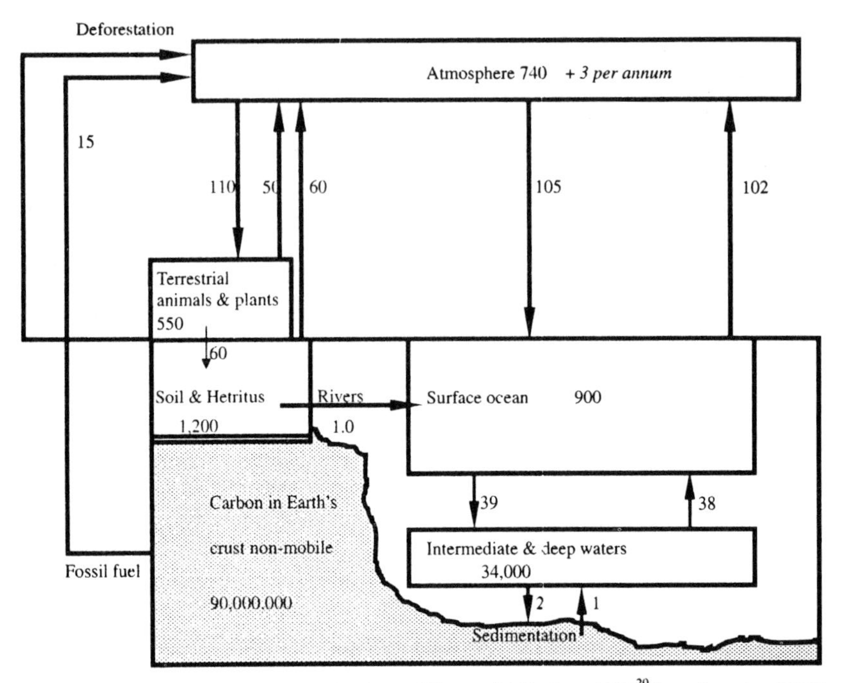

Figure 4.4 Carbon cycle as based on Institute of Terrestrial Ecology 1992[20] but adapted to IPCC 1992 format. Note how it compares with IPCC 1992 (Figure 5.1) and that there are differences. These differences are in part due to the format in which the figure is drawn, as well as due to the *estimates* of the carbon reservoirs and fluxes. The units are gigatons of carbon (GtC = 10^9 metric tonnes = 10^{12}kg) for reservoir sizes and GtC yr^{-1} for fluxes

(see Figure 4.4). Referring to global warming, Lovelock has expressed concern that our loading of the Earth's carbon cycle by boosting atmospheric carbon dioxide might over-stretch the biosphere's present homeostatic systems, so precipitously flipping it into a new steady state – one that may, or may not, be to twentieth-century Mankind's liking[53].

Such speculation by itself cannot be related to reality with any certainty – that comes with the gathering of evidence – but, whatever the merits of Gaian theory, Lovelock is encouraging us to think on a global level. Lovelock also points to the importance of the oceans and the life therein to the running of the carbon cycle and in turn (through the greenhouse contribution of atmospheric carbon dioxide) to the question of human activity and global warming.

The principal forms and routes that carbon takes around the carbon cycle are only known qualitatively. However, quantitatively the exact manner and speed as to how carbon is drawn into and released from many of its various planetary reservoirs is unclear: although it is central to the question of climatic change, it is currently another area of doubt and uncertainty (as illustrated by the two C-cycle versions in Figures 4.4 and 5.1). We know that the variations in atmospheric carbon dioxide concentration between interglacial and glacial times are substantial: the Earth's (pre-industrial) interglacial atmosphere is some 40% richer in carbon dioxide than in glacial times. One possible explanation is that between these climatic states the atmospheric carbon dioxide concentration changes due to changes in biological productivity – in effect, a change in the mass of photosynthetic organisms that absorb carbon dioxide. Here there is yet another area of uncertainty in that it is unknown whether this biological carbon sink is terrestrial- or marine-based.

However, given that atmospheric carbon dioxide concentrations have varied 'naturally' (without burning fossil fuels) during the glacial-interglacial cycle, how then does this relate to the Milankovitch orbital cycles discussed in the previous chapter?

The glacial-interglacial frequency is so closely matched to the Milankovitch radiation curve derived from the three orbital cycles that it cannot simply be dismissed. Yet the size of the changes in the solar radiation falling on the land-dominated Northern Hemisphere are so small that they cannot account for the magnitude of climate change between interglacials and glacials. Over all, the total variation in solar radiation received by the planet over the past million years has varied by less than 0.6% or under 0.7 Wm^2, even though locally solar radiation during the summer at 65° north (a good Milankovich indicator) has changed by about 60 Wm^2. These changes in solar radiation have therefore not made a major contribution to, and so do not solely account for, the climatic difference between interglacials and ice ages. In short, whilst the Milankovitch orbitally induced changes in solar radiation (especially at 65° North) do correlate with glacial timing, they do not explain the majority of the cooling taking place. On the other hand the changes in greenhouse gases (carbon dioxide and methane) do account for between 40% and 65% of the intergalacial/ice age temperature change deduced from the Vostok, Antarctic, ice core[54, 28]. Milankovitch's orbital forcing seems to serve as the 'pacemaker of the ice ages' with other factors, significantly greenhouse gases, amplifying this effect.

Once again uncertainty manifests itself. Whilst we might say that greenhouse gases are thought to contribute to some 40–65% of the interglacial/ice age climate change (we could equally express this as 52.5% with an uncertainty of ± 12.5%) this too is somewhat of a fiction: the understanding is still too imprecise. Researchers in this area are careful when presenting figures. Papers on the subject frequently preceed numerical conclusions with words such as 'estimate', 'approximately', or 'about'. This is the way that numerical conclusions are qualified and an acknowledgement made as to the gaps in our data and understanding as to how the planet-wide climatic system works. But this does not help politicians (who ironically often make qualified statements of their own, albeit for other reasons) or the public to understand the scientific arguments put forward in the climate change debate. There are (as we shall see in

Chapter 12) differences in perception of the issue. In science the presentation of uncertainty and/or experimental error is cherished as a sign of experimental rigour. Whereas in politics qualifying terminology is frequently seen as a means of providing room for future political manoeuvring. Not surprisingly it all too easily adds up to a recipe for confusion.

If there are gaps in the data and understanding as to the exact role of greenhouse gases in the past, the carbon cycle, or the overall global climatic system and its homeostatic systems, then intuitively one might think that direct measurement of today's climate would serve as a more accurate pointer to the relationship between greenhouse gas concentration and global warming. While direct observation of contemporary changes does provide harder evidence that changes in greenhouses gases do alter the climate (as opposed to the 'circumstantial' evidence of changes in the remote past), there are still problems. As discussed in the last chapter, the current global warming signal monitored over a few decades is so small that it is easily swamped by natural climate variations. The reason why the Vostok Antarctic cores tell us so much is that they represent a timescale over a thousand times longer than that of the history of direct climate monitoring, and that over this timescale the interglacial-glacial changes in average global temperature was a factor of five times greater than the 1°C thought to have already taken place since the beginning of the industrial revolution. In other words, the benefits of direct observation have been offset by the search for a small signal which is, in turn, impeded by a small 'signal-to-noise' ratio due to the natural variability of weather. The real question of human-induced, anthropogenic, climate change is not only to find out if this small, hard-to-see signal exists, but also whether the small signal is truly part of an increasing trend in the longer term? This last could only be addressed once we had determined the validity of the small signal in the first place.

The year 1988 was a very special one. Not only was that the year that the Intergovernmental Panel on Climate Change (IPCC) was formed jointly between the United Nations Environment Programme (UNEP) and the World Meteorological Organization (WMO) but, as mentioned in the previous chapter, at the time it was the warmest year on record[45]. But does a series of record-breaking warmest years form a trend? Clearly, by definition, the second year after records began would have been a record breaking year, albeit either the warmest or coldest (providing of course that the World's temperature did not remain constant). Common sense tells us that subsequently there would be other record breaking warmer and cooler years simply due to natural variation. The question is whether or not there is either a stable, positive, or even negative trend over many years embedded in the meteorological data? What was needed was statistical analysis. Fortunately, by the end of the 1980s statistics had developed suitable techniques, and by then there were enough years' worth of meteorological data so that a trend of less than 1/100 of a degree centigrade a year could be deduced. At that time the World's scientific community had yet to come to any firm conclusion as to whether global warming was actually taking place. The IPCC was to provide this in 1990, but just prior to their report's publication the result of a time-series analysis of World climatic data gathered between 1880 and 1988 was announced in the journal *Nature*[55]. This analysis showed that warming, as opposed to cooling, was actually taking place with a statistical confidence of over 99.99%. (This means that if similar climatic data was collected over a similar run of years some 10 000 times, then 9999 of these could be explained by warming and only 1 by natural variation covering an otherwise stable climate or one slightly cooling.)

The researchers noted that, all things being equal (if greenhouse gas atmospheric concentrations were constant over that time), the Milankovitch orbital forcing of climate over this time should have resulted in a cooling trend of some 0.0004°C per annum – in other words the Earth, in the absence of the human addition of greenhouse gases would be beginning to leave our interglacial and moving toward a glacial mode. Instead, their climate record analysis revealed a warming trend of 0.0055 ± 0.00096°C a year, or between about 0.5°C and 0.7°C over the past 108 years. Since this time series analysis, further evidence that global warming is taking place was provided by the UK Meteorological Office and the University of East Anglia's Climatic Research Unit. In 1991 they announced that 1990 had replaced 1988, beating it by 0.05°C, as the warmest year on record[56] – the warming trend was continuing. However, due to overlap between the time series analysis and the IPCC 1990 report, neither were able to refer to or build on the other. Even so the IPCC was able to come to some firm conclusions as well as qualify some areas of doubt.

THE IPCC BRIEF

The first IPCC working group (there were three in all) set out to present an assessment of the scientific information on global warming to date and to provide a consensus of the international scientific community as to whether global warming was taking place and, if so, to what degree[20].

The IPCC truly had a formidable task. One major problem was timing, or rather the lack of time. The UNEP and the WMO wanted as much information as possible quickly to meet the demands from World leaders and the 1988 international Toronto conference on climate and the changing atmosphere. So while the first working group was compiling its report, the IPCC working groups II and III were to examine the likely effects of climate change and to outline possible response strategies, respectively. Unfortunately, as all three groups were formulating their reports simultaneously, in parallel (as opposed to in series), so groups II and III did not have the benefit of the first group's conclusions; only their initial thoughts. It was therefore virtually inevitable that during the course of the first working party's deliberations that they would modify their position so undermining the basis on which groups II and III were working. These last two reports began by taking as their starting point a different set of criteria to that ultimately provided by group I, but nonetheless they were successful in providing a broad picture as to the socio-economic impacts and the response options to global warming.

Despite the hurdles placed before its fellow working groups, the principal IPCC report from group I was successful in arriving at a conclusion that met with the approval of the majority of other scientists concerned with the greenhouse question (170 scientists from 25 countries worked directly on the report, while a further 200 commented on its early drafts). In making its conclusions, the IPCC had to clearly state the degree of certainty. Furthermore it could only sift through the then existing data, and not engage in new research to fill in the gaps. Much uncertainty remained, though the principal areas of doubt were clearly identified. (However, the picture would slowly improve with the subsequent publication of IPCC reports in 1992[57] and 1995/6[58].)

IPCC WORKING GROUP I PRINCIPAL 1990 CONCLUSIONS

The IPCC's principal conclusion was made with surety. It was 'certain' that 'emissions resulting from human activities are substantially increasing the atmospheric concentration of greenhouse gases ... [and that] these increases will enhance the [natural] greenhouse effect'. They concluded with certainty that: there has been a real, but irregular increase of the global surface temperature since the late nineteenth century; there has been a marked, but irregular, recession of the majority of mountain glaciers over the same period; and as the lower atmosphere and Earth surface warms, the stratosphere cools.

The IPCC was less than certain when it came to other matters. Nonetheless, they were able to 'calculate with confidence' that: such is the known variation between the various greenhouse gases in their ability to alter climate that their relative effectiveness can be estimated, and that carbon dioxide has been responsible for over half the enhanced greenhouse effect in the past, and is likely to continue to do so in the future; the atmospheric concentrations of the long-lived gases (carbon dioxide, nitrous oxide and CFCs) adjust only slowly to changes in emissions, so that, were the (1990) rates to continue, we would be committed to increased concentrations for centuries ahead; postponing control of emissions will mean that even greater reductions will be required in the future for concentrations to stabilise at a given level; and the long-lived greenhouse gases would require immediate reductions in emissions of over 60% if we are to stabilise their concentrations at 1990 levels (though methane would only require a 15–20% reduction).

Based on the models available to the IPCC, at the end of the 1980s, the panel was able to predict that under a Business-as-Usual scenario (one with no controls on carbon dioxide emissions) that the resulting rate of climate change would be greater than anything the planet has seen since the ending of the last glacial. Under such a scenario, the 'likely' increase in global mean temperature would be about 1°C above the 1990 value by 2025, and 3°C before the end of the twenty-first century. However the IPCC qualified this in that it was unlikely that the warming would be steady. Furthermore (although predicted with low confidence), southern Europe and central North America would experience a greater-than-the-global-average warming, as well as a lowering of summer rainfall.

In short, and most importantly, the IPCC dispelled the ambiguity as to whether or not global warming was a likelihood – here their view was clear. The principal uncertainty of the scientific community in 1990 lay not in whether warming is taking place but in the extent of the warming and the environmental (hence socio-economic) effects of forthcoming climate change.

Effects under the 1990 Business-as-Usual scenario

As the World entered the 1990s, the issue of global warming received less attention from both the public and politicians. Partly, and not entirely unrelated was the then issue of the Gulf War and the fact that the Middle East contains over 60% of the World's proven reserves of oil[1]. Greenhouse reductions were at that time not being discussed and the World was continuing down the IPCC 'Business-as-Usual' path. Clearly this business-as-usual scenario is a broad one, but it is based on firmly established trends of economic activity and population growth (which will be explored in more detail in Chapter 6). There are many anticipated environmental and economic impacts associated with climatic warming. But there is little detail as to the exact nature of these effects though we can make some broad comments.

Sea-level rise

One of the most frequently discussed effects of global warming is that on sea-level. Past changes in sea-level worldwide have invariably been associated with climate change. During interglacials, when the World climate was periodically, for several thousand years, comparable to now or even 2–2.5°C warmer. During the last interglacial the sea rose by about 5–7 m over its current level. However, it must be remembered that land also rises and falls. This movement can be quite dramatic when the Earth's crust, floating on a plastic mantle, is burdened by, or relieved of, a thick ice cap. Such (isostacy) changes can mean that regionally *apparent* sea level changes can be several tens of metres over many centuries. But what would be the likely sea-level rise due to global warming over the next three decades to the year 2030?

The IPCC's 'best estimate' 'Business-as-Usual' calculation in 1990 was that worldwide the seas will most likely rise by some 18 cm, with an error of about ± 10 cm: this is calculated over a 45 year period between 1985 and 2030[20]. Two years later when the IPCC came to revise their predictions in the light of new data in more complex future scenarios, they found that hardly any modifications to their sea level predictions were required[59]. However, the effect of global climate on sea level is complex.

First, as noted above, a worldwide change in the sea-level is different from a local change; for climate can affect where the shoreline will be formed, albeit with somewhat of a time delay. Locally, on continents that sported ice sheets during glacials, the land today is rising isostacally (in much the same way as a boat rises in the water once its cargo has been removed only, instead of displacing water, continental rock displaces other crustal matter), so effectively lowering the local sea-level. (Scandinavia, and the north of the British Isles is still rising as it recovers from having the weight of an ice sheet on it from the last glacial.)

During glacials when water was locked up in large ice caps, the seas worldwide were lower than today exposing much of the continental shelves and creating land bridges between many of the now-isolated land masses including between Britain and the rest of Europe.

The World's climate imposes a number of effects on the average sea-level worldwide. First, warming causes water to expand and lower its density, this is sometimes known as the 'steric' rise. The second, melting ice from a variety of sources, principally glaciers and ice caps, returns water to the seas. Without allowing for any contribution from melting ice, the IPCC's 1990 'best estimate' 'Business-as-Usual' scenario for the contribution from steric rise in sea level, anticipated between 1990 and 2030, is around some 55% or 9.9 cm ± 5.5 cm[20]. The glaciers and small ice caps will contribute some 38% to the anticipated total rise of 18 cm in the seas between 1985 and 2030. The majority of valley glaciers have been retreating over the past 100 years and many geomorphological studies suggest that this has been a general trend since the end of the Little Ice Age. Between 1900 and 1960 the retreat in glaciers contributed about 2.8 cm to the sea level and global mean temperature rose by approximately 0.35°C. This trend altered between the 1960s and the mid-1980s when the trend temporarily reversed with a short cooling spell.

Finally there are contributions from melting of the large ice caps of Greenland and Antarctica. However, as we have seen before, the greenhouse has its surprises and complications – both ice caps respond in quite different ways. Both caps have the same magnitude and average thickness (some 2 km) but the Greenland cap, the smaller of the two albeit the largest cap in the Northern Hemisphere, contains only 2.95 million km^3 of ice. As such if all of it were to melt it would contribute an additional 7 m to sea-level. The IPCC 'Business-as-

Usual' best estimate calculation for the period between 1985 and 2030 is that melt water from Greenland would add some 1.8 cm to the seas, but here the IPCC have included a wide margin of error by providing a minimum estimated contribution of just 0.5 cm and maximum of 3.7 cm. The IPCC could not provide a tighter estimate having found that all the original work on the Greenland cap contained considerable uncertainty as to the ice sheet's state of heat balance: specifically whether or not the sheet has attained thermal equilibrium since the end of the last ice age. (The rate at which ice melts is not just related to the temperature of its surroundings, but also the temperature profile throughout the ice itself.) This is complicated again by our lack of knowledge as to the exact nature of the processes governing the sheet's iceberg calving rate under normal conditions, let alone relating this to climate warming. Other uncertainties, such as how surface albedo and cloudiness will affect the cap's energy budget further compound these difficulties. Even so, both the IPCC's high and low estimates point to the Greenland cap having a positive effect on the sea level in a warming World. Surprisingly the opposite is true for Antarctica, at least up to the year 2030.

A warmer Earth will see an overall increase in ocean evaporation and, in turn, rainfall. Early in the twenty-first century it is likely that precipitation over Antarctica will increase faster than the increase in ice melting. Antarctica - contains such a massive amount of ice that it is unlikely that it has yet completely adapted to present conditions since the end of the last ice age – it still has to attain thermal equilibrium. Yet if all (both grounded and partially floating, and completely grounded) ice were to melt, then the World's seas would rise by 65 m. The present global warming can only, in the short term at least, serve to enhance the existing post-glacial trend of non-thermal balance and drive the continent's ice further away from thermal equilibrium. Still, as Antarctica is so vast, and *with all other things being equal* (which they probably are not, such is the uncertainty), in the short term a warmer World will experience higher evaporation from the seas, hence precipitation falling as snow in Antarctica will become temporarily locked up as ice. Consequently, the IPCC's best Business-as-Usual estimate from 1985 up to the year 2030 is that Antarctica will actually cause the sea-level to go down by 0.6 cm. At the very least the IPCC think that Antarctica will have no effect, or at most it could cause a sea-level decrease of 0.8 cm. However in the longer term, into the twenty-second century, Antarctica will have a positive effect on the sea-level. Not withstanding this prediction, the ice shelves surrounding the continent could surge forward in fits and starts. Much more study is required if we are to ascertain the likely affect this continent will have on our great-grandchildren's World.

That more research into the stability of Antarctic ice is needed is aptly demonstrated by what we do know: frustratingly this is enough to raise some disturbing questions. As already stated, Antarctica's size, and volume of ice, is so great that parts of its ice shelves are thickening and others thinning, in response to the end of the last glacial some 10 000 years ago. Consequently, *all other things being equal*, the near-term future of its larger Western ice sheet is already determined and is unconnected with the anthropogenic greenhouse effect[60]. Assuming that the World's two other major ice sheets, the Greenland ice sheet and East Antarctic ice sheet are fairly stable and given that the global sea level was possibly some 6 m higher than today during the height of the last interglacial (120 000 years ago), it seems likely that much of this 6 m rise is due to the collapse of the partially grounded part of the West Antarctic ice sheet. (By collapse, this could mean the sheet's thinning so that it completely floats, as opposed to completely melts.) Both geological evidence and computer simulations suggests that this happened quite quickly, though much depends on the geology on which the ice rests[61]. But all other things might not be equal! There is both magnetic and satellite evidence to suggest that there is an active volcano beneath a critical part of the West Antarctic ice sheet, and it *might* possibly be that this could act to precipitously trigger its collapse[62]. The counter-argument is that while it is likely that there is a volcano under the ice sheet, a collapse of the whole ice sheet would have already had to have begun before the volcano could make a significant contribution to further collapse. Finally, there is geological evidence that 3 million years ago the East Antarctic sheet melted. This could happen again if global warming continued for some centuries and a Pliocene-type climate returned[63].

Returning to the short-term prospects. Combining all the positive factors, of thermal expansion and melting ice, while allowing for the negative Antarctic factor, the IPCC best estimate calculation was for a sea rise of some 18.3 cm between 1985 and 2030 under the Business-as-Usual scenario. Given that the IPCC have estimated that sea level has already risen by some 10.5 cm over the previous hundred years, the IPCC are concluding that the seas will be rising roughly twice as fast over the next half a century. Extending their calculations to the end of the next century, the IPCC have a Business-as-Usual best estimate rise of some 66 cm, with a worst- and least-case estimate of 110 cm and 31 cm, respectively. Even if we adopt controls to lower our emissions of greenhouse gases, sea level rises of up to 50 cm can still be expected by 2100 (see Table 4.2).

In short, the IPCC conclude that sea level rise is unavoidable. Yet though we may be committed to rising seas we can take steps to mitigate matters. Further, in terms of human affairs at least, the rise will be slow in occurring over many decades. This last, as we shall see, presents policy-makers with a dilemma in that this effect of global warming may not appear to require urgent consideration.

Table 4.2 IPCC Business-as-Usual scenario for sea level rise (cm) from 1985

Year	*2025*	*2070*	*2100*
		Rise (cm)	
High	29	71	110
Best estimate	18	44	66
Low	8	21	31

Agriculture

Perhaps the greatest effect of global warming will be on the pattern of agricultural production worldwide. Though the theoretical ability for our planet to furnish us with food may not necessarily be reduced, it is likely that there will be some considerable changes in where we get our food and that the resulting dislocation could well impair food supply, or at least alter costs.

In a warmer World the most important single effect on plant growth will be that of the increased evapotranspiration. This, though, belies a far more complex picture of positive and negative effects. Evapotranspiration, the combined evaporation of water from the soil surface and plant transpiration, can be considered to be the opposite of precipitation. Yet, as noted above, in a warmer World evaporation from the oceans will also increase, resulting in greater rainfall in some regions. On the other hand even if a region had the extra rainfall to exactly counteract increased evapotranspiration, there is one other key factor that will affect plant growth. The increased concentration of the very CO_2 responsible for over half the greenhouse warming, will also have its effect on crop growth and production.

CO_2 is the gaseous half of the two physical inputs required for the photosynthetic reaction forming plant carbohydrates. Here, in line with the law of mass action (which says that the greater the concentration of reactants then the faster will products a chemical reactions form) so an increase in CO_2 will encourage photosynthesis. Because, unlike with a homogeneous chemical reaction, photosynthetic intermediates vary with species and the reaction has many enzyme-catalysed steps, the increase in photosynthesis is not directly proportional to CO_2 concentration. However, in many instances CO_2 is a limiting factor. A doubling of CO_2 may sometimes increase photosynthesis by between 30–100%.

Plants can be divided into two principal groups on the basis of the way they metabolise carbon. C_3 species (where the first product in metabolic reactions has three carbon atoms) can respond to increased levels of CO_2. Examples of C_3 species of importance to Man include wheat, rice and soya. The other principal plant group, the C_4 plants, including maize, sorghum, sugar cane and millet, are found in hot countries and are not so responsive to increases in CO_2. C_4 crops account for roughly a fifth of the world's food supply. Maize alone accounts for about 14% of the total World food production, and roughly three quarters of all grain traded[64].

Table 4.3 Mean predicted growth and yield increases for various groupings of C_3 species for a doubling of CO_2 concentration from 330 ppm. Table summarised from Parry[64]

Fibre crops	104%
Fruit crops	21%
Grain crops	36%
Pulses	17%

Increasing CO_2 levels (without changing other factors such as rainfall) results in a reduction in stomatal opening (the leaf pores through which gases – water vapour, oxygen and carbon dioxide – move). A doubling of CO_2 concentration can reduce stomatal aperture by up to 40% for both C_3 and C_4 plants so potentially restricting gaseous exchange. However, C_3 plants would still have the ability to photosynthetically fix extra CO_2. In both C_3 and C_4 plants the reduced stomatal apertures mean that water loss per unit leaf area is reduced. This might seem like a boon in areas that are dry, or will become so in a warmer World, if it were not for the tendency for increased CO_2 to increase plant size. In some instances it might be that the reduced water loss per unit area of leaf is offset for the individual plant as it may have more leaves (hence a greater total number of stomata). Further, as mentioned above, the raised temperature associated with the increased CO_2 in the greenhouse World may in turn increase evapotranspiration, and so offset any improved efficiency in water use. How these factors will all balance out varies from species to species and specific local/regional conditions. Much work is currently being conducted in this area, though we can still make a few predictions as to how this is likely to affect agriculture in various parts of the World.

According to the IPCC, the greatest temperature changes to take place under an increasing greenhoused World, will be in the temperate C_3 crop zones. The least temperature changes will take place in the tropics where C_4 plants grow. So providing they continue to receive water, the greatest increases in photosynthetic productivity in the greenhouse World could well take place in temperate C_3 zones. However, it must be remembered that plant respiration (output of CO_2) also increases with temperature. Where there is an overlap between C_3 and C_4 species in lower latitude temperate and sub-tropical regions, the C_3 species could again benefit by out-competing C_4 species. This may well be a crucial factor for agriculture as 14 of the World's 17 most nuisance weeds are C_4 species.

The situation is reversed in the tropics where C_3 weeds could prove harder to control in a C_4 crop though much remains unclear and depends on exactly how individual species respond to a CO_2-rich and climate altered World. Here, despite our lack of detailed knowledge one thing is likely, that it may be necessary to change the pattern of agriculture as climate alters. Once again, through an 'accident' of human geography, nations relying on tropical agriculture will tend to fare worse than their temperate counterparts if only because the world's rich nations tend to be centred in the northern temperate climes. Conversely, most tropical countries are poor and will find it harder, lacking financial - resources, to adapt to changes – whatever these may be. For example, the current trend in India towards growing wheat, rice and barley, rather than C_4 maize and millet may continue and increase. And should this greenhouse discrimination against C_4 plants hold true, then in future the USA might see a drift away from cultivating C_4 maize with more C_3 species cultivated instead. More C_3 soya, which some USA farmers already cultivate, may be grown. As mentioned, with C_4 crops accounting for a fifth of World food product, and maize itself some 14%, this issue is not one to be taken lightly[64].

In short, though it is known that plants respond to increased levels of CO_2, and that this response varies from species to species, it is most likely that either the cultivation of C_4 species will decline, and/or that more effort will have to be put into growing such crops. Whatever the outcome, future agricultural practice will not just be determined by increased CO_2 but by the interaction of the variety of new factors found in the greenhouse World. Of those affecting agriculture, aside from CO_2, the regional changes in temperature and rainfall will play a major part.

For plants, and again varying from species to species, a minimum temperature is required for growth. The transition to the greenhouse World will see temperature changes at their greatest outside of the tropics above a latitude of about 45°. Outside of this (somewhat arbitary) 45° zone, where winter is a marked season, plant growth is severely restricted for at least part of the year.

The period during which plants can grow is known as the Thermal Growing Season (TGS). Typically most plant species require a minimum TGS of 90 days. In areas (outside of the tropics) where the TGS does not last all year, the growing season may be extended in a warmer World. In places this may allow some marginal crops to become a regularly available crop option. Alternatively the longer TGS may allow new species to be cultivated where they were not possible before. The effect of temperature rise therefore varies both with region and species of crop. For example, per °C rise in mean annual temperature, the TGS for wheat will be increased by about 10 days in Europe, but only 8 days in Japan.

A longer TGS might seem like good news for many farmers, unfortunately the picture is more complex. Some species are more productive with the spread of seasons associated with a lower mean annual temperature. For example, spring wheat will produce dry matter, from heading to ripening, faster under warmer conditions but the actual grain yield is reduced. Some farmers in the greenhouse World may find their crops growing more quickly and harvest them sooner but obtain lower yields. Though there may be extra growing days available, there may not be enough days remaining in the TGS to plant another crop (albeit a different one). Generally speaking, if warming is limited to about 3°C (*cf* IPCC Business-as-Usual scenario to the end of the twenty-first century) then the TGS in the mid-to-high latitudes will increase. However if the temperature rise is greater than this (as it may well be in some regions prior to, and in others after, the year 2100) then the increased evapotranspiration combined with water availability may curtail crop productivity.

In the main, the overall rider as to whether greenhouse-induced temperature changes will increase local crop yield depends on whether the crop is currently temperature limited. As mentioned, such limitation currently takes place in high latitudes, not so nearer the tropics. As the IPCC predict that the warming will be greater in these higher latitudes then here is where yield increases are likely to be greatest. To continue with our example of spring wheat, yields in the European region of the USSR are likely to increase by 3% per°C rise in mean annual temperature, that is with all things being equal – including the assumption that there will be no changes in rainfall. Taking rainfall into account, a more complex picture emerges as typified by work on spring wheat grown in Cherdyn forest zone west of the Urals[64].

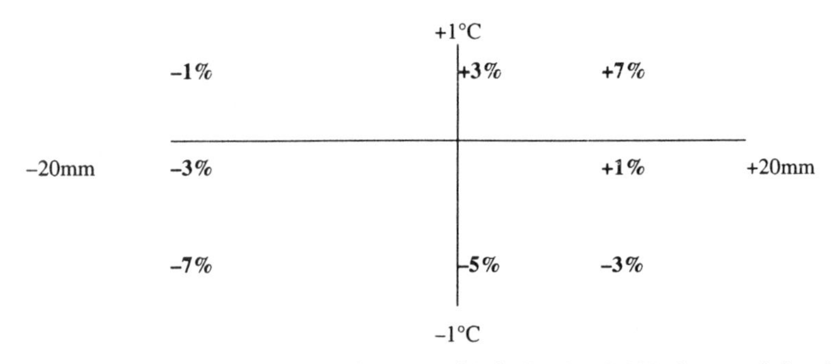

Figure 4.5 Percentage changes in spring wheat yield in Cherdyn, USSR, due to variations in temperature (y axis ± 1°C) and rainfall (X axis ± 20 mm) (Redrawn from reference[64])

The effect of temperature change on animals is two-fold. First, animals rely on plants for food so that the changes affecting crops will have an indirect effect on the farming of animals. Second, there are the direct effects themselves. Here, young animals are especially susceptible to the temperature of their surroundings. Where the temperature change in the greenhouse World approaches their threshold then their productivity decreases. Such an effect might be seen in existing livestock species such as piglets, chickens and lambs, in places like the south-central USA and southern USSR. As with plants, at high latitudes it may be possible to push back the boundary where the farming of animals is possible.

However there is one response available to farmers whose plants and animals are affected by greenhouse changes, that is increasing the energy inputs into their farming. Farmers whose crops are suffering through lack of water may be able to provide pumps and channels for irrigation. Similarly air-conditioned housing for animals (especially when young) may be an option in areas where the temperature has risen too high. Such options, though, have their costs (in fossil fuel burning terms) and so are likely to be market/subsidy constrained.

Again, just as the poorer southern and tropical countries will find it difficult to cover the cost of the dislocation incurred by changing their agriculture (even if their new practices are no more expensive than the old), they will find it even harder to move to more expensive systems. The high energy input option can only be realistically considered by the wealthy northern (largely temperate zone) nations.

Given that the greenhouse effect changes in temperature and precipitation will both have their individual agricultural impacts, there will also be a combined, synergistic, impact. The increase in plant growth and the resulting extra demand for water (where available) has already been discussed and that alone will increase evapotranspiration, but the increase in temperature alone (assuming for the moment no plant response) will in itself increase evaporation from soil, so further enhancing evapotranspiration and reducing water availability. In mid-latitudes soil evaporation alone increases by about 4–5% for every°C increase in mean annual temperature. If, the current temperate temperature zone increases by 2°C by the year 2100, then evaporation could increase by 9%. This last assumes no changes in precipitation, however in the 1970s and 1980s the southern half of England, for instance, has experienced a number of years of summer drought and lower annual rainfall. The water hose-pipe restrictions have been in force each year since 1991 in virtually all of the southern UK water authorities. If this weather is part of a real 'greenhouse' climate trend then mid-summer evaporation could be so high as to result in a major soil moisture deficit and a considerable increase in demand for agricultural irrigation.

Water availability is a major factor affecting crop production worldwide. Generally a crop's dry weight yield is proportional to the amount of water it transpires while growing and maturing. A fairly firm relationship between rainfall and productivity, and in turn yield, has been established in mid-latitude cereal regions including those of the USA and the Ukraine. It has been suggested[64] that, with an IPCC 2°C rise, wheat and maize yields from such bread-basket regions could be lowered by an average of a fifth. Even if yields were reduced by just half this, there would still be a major (at the very least dislocating) effect on the World economy, let alone how the World population is nourished. The IPCC's second working group (on the impacts of climate change) notes that if European Community and North American (including Canada) yields decreased by 10% or 20% then World market prices would rise by 5%

and 7%, respectively. This, though, assumes that the World market is buffered through other countries being unaffected by greenhouse changes. If yields also decrease in Africa and SE Asia by 25% then World market prices could rise between 20% and 25%. Such figures are at best 'what if' guestimates that can only serve to outline the range of the possible effects on agriculture. A general picture of the likely changes to be expected in the greenhouse World is required, one taking in all the varying factors (temperature, CO_2, precipitation, crop response etc,) into account together with their interactions.

Many experts agree that while increased CO_2 by itself will generally benefit crop production, this will be offset by other greenhouse effects. Parry, who co-chaired the agriculture and forestry section of the IPCC second working group report, and his colleagues have provided in that report a glimpse of how World agriculture will fare in the greenhouse World. This is, though, only a glimpse due to the uncertainty arising from the non-linear response to warming as well as the regional variations in climatic change. It should also be noted that any impacts on major producers, such as the USA, will not solely arise from their internal greenhouse agricultural impact: their role will also be affected by changes in other countries[65]. Parry himself broadly concludes that, high mid-latitude production of staple crops is likely to increase. However the major mid-latitude grain belts – the US plains, Canadian Prairies, Soviet Ukraine and the northern European lowlands – together with their southern hemisphere counterparts – the Australian wheat belt and Argentine Pampas – will probably find that water availability becomes the determining factor. Already in the 1980s the mid-continental regions of the USA, Canada and Ukraine have had drought problems. In 1989, for example, the USA suffered two consecutive years of drought (and California was in its third dry year, a phenomenon not usually anticipated more than once in 400 years). May that year saw US grain reserves down to 550 bushels, or half the stockpile at the end of the previous year representing 20% of the US internal annual demand – it was the second lowest figure since the end of WWII[66]. If such droughts are part of a greenhouse trend, then USA grain productivity will decline. This last might be partially off-set by increased irrigation though here there will be cost implications and both these (the reduced supply and irrigation) will have their effects on world market prices. Meanwhile the world market itself will be beleaguered by the effect of change and the problems of overcoming industrial inertia. (Just one example of the many changes that the food industries will have to tackle or genetic modifications necessary, arises from the change in dough strength from wheat varieties grown at different temperatures[67].)

In the USA (whose contribution makes up about 40% and dominates the World grain market) grain production could decline in the next century by 10% or more. This is unlikely to be completely offset by the increase in production expected for the UK and lowland Europe. In short, and in all probability, the effect of climate change on agriculture worldwide could well have grave consequences for global food supply and food security.

Ecological and physical effects and wildlife

Greenhouse changes affect terrestrial ecosystems literally from the ground up as soils adapt to the different climatic regimen. Changes in the soil add to those from climatic change (temperature and precipitation, etc, discussed above) in affecting the distribution of plant species. Again, as the primary producer plant species within an ecosystem change, then so do the secondary producer animal species that rely on plants for food and habitat. Marked changes can be expected in entire ecosystems in many parts of the world. Yet because these changes are unprecedented (in human times) we cannot make accurate predictions; although from our existing understanding of ecological processes it is possible to make a few broad generalisations.

The current relationship between climate and the topological relief (the physical shape) of the land to the type of soil supported is known and fairly well understood. However these relationships have been established for thousands of years during which time the global climate and CO_2 levels have been relatively stable. The recent human-induced changes in climate are incomparable

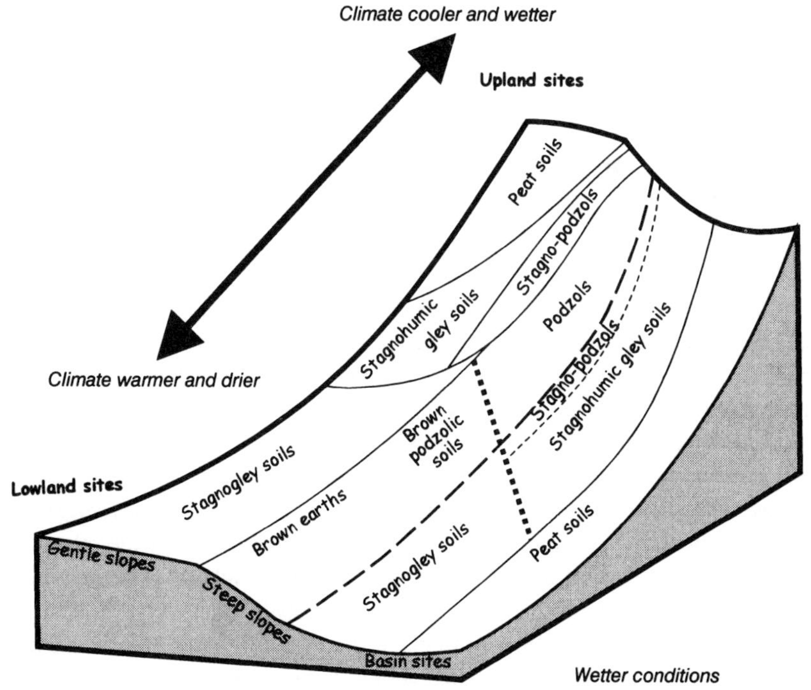

Figure 4.6 The relationship between climate, topography (shape of the land) and soil type. Adapted and redrawn from reference 68

in nature since the ice ages, and so their effect on soils and how they relate to climate and topography are less certain. Based on current relationships, we might expect in the UK that upland peat soils will revert to stagnohumic gleys – a sticky organic-rich soil – as the climate becomes warmer and drier. Similarly, soils that have in the past been subject to high rainfall (hence have been depleted in chemical bases through leaching) will not necessarily become less acidic in the short term. In summary, previously observed well known soil interactions may not apply as we enter the greenhouse World[68].

Given the above, and assuming a mean annual temperature rise for the UK sometime over the next century with no great change in rainfall, we would expect to see soil processes altering, including those of decomposition, hydrological pathways (water movement) and mineral weathering. The responses to these altered soil processes will vary from soil to soil, and region to region depending on the change in climatic regimen. Soils in the south-east UK will probably change to those currently associated with warm temperate areas, especially those of the Mediterranean. Soil moisture deficits (the amount of water absorbed by the soil before it can percolate away), and specifically the mean maximum soil moisture deficit (PSMD) could increase. PSMDs currently exceed 200 mm in parts of east Essex[69] and these could increase if the temperature increased without a corresponding rise in rainfall. As discussed in the previous section, this could result in the need for extra irrigation to sustain the continuous growth of sensitive crops[64], and it could also have an adverse impact on human water supplies. Yet again, the long periods of summer drought that the southern UK might anticipate in the greenhouse World could irreversibly crack lowland clay soils as well as upland peats which would result in greater aeration, increased biological activity, and nutrient release. All these changes are likely to affect land use, regional rural economics, and be of some consequence to town and country planners[68].

One dimension to soils of particular importance to the greenhouse issue is that they themselves are a source of several important greenhouse gases: namely carbon dioxide, methane and oxides of nitrogen. As their release is in part determined by the climate, there is yet another climatic feedback mechanism. Soil micro-organisms release carbon dioxide as a principal product of aerobic decomposition and there is a surprising degree of consistency of carbon dioxide release across a number of ecosystems even though the soil carbon content can vary significantly between various soil types. Typically, worldwide release is about 700 g of carbon m^{-2} yr^{-1}, though this can be substantially increased through changes in land use and disturbance. Globally, soils serve as a major reservoir for carbon, forming a store the order of magnitude of which is some 2000 billion tonnes[68]. This compares with a current (1991) annual increase in atmospheric carbon of 3.5 billion tonnes, or about half a percent of the atmospheric pool. So if by some chance the worldwide soil pool were to decrease by

0.1% through atmospheric release then there would be approximately a 1 ppm increase in atmospheric carbon dioxide. Indeed, it has been suggested that natural soil evolution during comparatively stable climatic periods, be they glacials or interglacials, have altered the global soil carbon budget to create a critical state providing yet another factor (in addition to ocean circulation current changes) enabling the World to switch quickly between these two climatic modes. However, as soil carbon release is affected by temperature – with reference to our above climatic feedback mechanism – and as there are similar temperature relationships associated with the production of methane and the oxides of nitrogen, it could be that soils have a significant effect on the global climate. Here, as we enter the greenhouse World, some scientists have raised a disturbing thought. Arctic tundra may not survive in the greenhouse world, yet some 14% of the Earth's soil carbon is stored there and a significant proportion of this carbon could well be released.

If soils change then, not withstanding other climatic impacts, so too will the plant species dependent on soil type, and so in turn the fauna relying on the plants for food at higher trophic levels. Inter-and intra-species competition will alter and marked changes in ecosystems will be seen. One response to this will be that some species will migrate as their habitat shifts with the climatic belts. In a warmer World many species might migrate either toward the poles, away from the equator, or vertically. In the UK a difference of 2°C in the mean summer temperature equates to a 400 km latitude or a 300 m change in altitude. During the warmest part of this current interglacial to date, some 4000–8000 years ago, Europe may have been about 1°C warmer than present. Then Scots pine (*Pinus sylvestris*) is known to have grown in the Cairngorm Mountains 170 m above the present treeline of 620 m above sea level. More recently, in Canada the range of white spruce (*Picea glauca*) has expanded northwards into the Hudson Bay area due to the warming that took place between 1920 and 1965. However, such migration may not always be a viable option.

Species migration can be prevented in a number of ways. First, there may be an absence of a suitable corridor down which migration can take place. Here the extreme example is found with island species where frequently the distance is simply too great (either between islands or the islands and mainland) for species to jump. Ironically it is the very isolation that gives many islands their unique ecology – by preventing potential competitors invading – that may doom them. Equally, not all ecological islands are surrounded by water. Species associated with an upland area may be surrounded by a 'sea' of lowland. Secondly, migration corridors may end abruptly. Typically the high latitude boundaries to continents terminate possible migration away from the equator. Similarly, mountain species are limited by the height of the land. In both of these instances plants have 'nowhere to go'.

Given that the ends of continents and mountain tops will become barriers to species migrating in a warming World, by studying the biology of these regions we might be able to detect ecological responses indicative that climate change is taking place. One study did just this by looking at the biodiversity on Swiss and Austrian Alpine mountain summits[70]. In 1992 they studied the biodiversity (species richness) of 26 summits above 3000 m to compare present-day biodiversity with that historically recorded by naturalists over the past century. The researchers were so concerned that they would falsely detect increased species richness due to the inexperience and lack of rigour of early biologists that they slightly weighted their present-day study against rare species on the basis that earlier biologists might have missed them. This meant that if there really was ecological change taking place, then it would have to be so great that it overcame the margin of error safety built in to the present-day analysis. Yet their results did show a change, an increase in biodiversity as species from lower elevations more rich in biodiversity migrated upward. Over the past century since the industrial revolution, it appears that alpine species have migrated between 80 m and 100 m upwards.

The greatest array of barriers preventing migration, indeed threatening the very survival *in situ* of the largest number of wildlife species and their habitats, have been created by one species alone – *Homo sapiens*[71]. The individual examples of the way 6 billion humans worldwide (possibly 10 billion by the end of the twenty-first century) jeopardize wildlife, even to extinction, are many. These range from threats to tropical rainforests of the east[72] and west[73], to species and ecosystems at the Earth's poles[74]. Not since the extinction at the time of the dinosaurs (65 million years ago) has this pressure on wildlife been as great as it is today. Compounding it with the threat from climatic change (itself a result of human activity) can only serve to increase the rate of the current 'mass extinction' now silently taking place worldwide. So great is the fall in population of many species, and the dwindling area of habitats, that environmental scientists and conservation biologists face critical decisions as to the best way to save individual species and communities. Of these, the conservation debate that is perhaps raising the greatest moral and ethical dilemmas is whether one should remove species from the wild where they are dying out to breed up a population in captivity, or alternatively continue to try to protect species in the natural environment and hope that population and habitat declines will reverse? The problem is that some species, especially animal species, do not breed well in captivity and so employing artificial reproduction techniques is sometimes the only option if a captive population is to be maintained without continuing to rely on an influx from the wild. For example, sperm banking could possibly be used to maintain genetic diversity[75]. This one example alone illustrates the emergency nature of the measures now being taken, hence the seriousness with which biologists view the current situation. However, in this instance while reproductive biology is steadily developing it has yet to be successfully applied to many endangered species. But even moving individual animals or plants from one location in the wild to another can in itself pose problems arising from the disturbance in ecological balance (*cf* Rhododendron which was introduced to Wales, where it has no natural competitors and so in places it is displacing indigenous species and choking woodlands).

In a climatically changing World, the zones species can occupy will move and, if they are to survive, the species must migrate with them. Yet individual species will react to climate change differently, so we can see new mixes of species arise; new ecosystems with opportunities for some but not others. First though, the species has to migrate.

Not all species lack a migration corridor. Even so there are doubts as to whether all of these will be able to migrate as the climatic belts shift. Latitudinal migration may prove difficult for those plant species that are genetically programmed to flower only when there are the appropriate number of daylight hours (photoperiod). Fortunately it has been shown that the photoperiodism of plant species does not totally restrict their possible migration. There have been studies conducted on plant species found in both the arctic and the alpine tundra (between which the annual pattern of day length differs considerably). These have shown that (if they are photoperiodic) such species have local genetic races with each race adapted to the latitude where it grows. Here the question is whether the individual strains occupying the latitudinal boundaries to the species zone of occupation produce enough variation (in photoperiodism) to evolve with the climate. For this and other reasons the question that arises is whether the change in climate as we enter the greenhouse world will proceed so fast as to outstrip the ability of some species to adapt.

Species have migrated before due to climate change. The last period of mass migration worldwide took place at the end of the last glacial. By examining the type of pollen preserved in the absence of oxygen in clay, peat, or silt sediments and applying radio-isotope dating techniques, it is possible to date when and where many plant species lived. It turns out that each species has its own unique migration history. For example, in the USA it is believed that the eastern white pine and the eastern hemlock survived the last glaciation in the eastern foothills of the Appalachians and on the adjacent coastal plain. In contrast, the chestnut and maple rode out that glacial near the mouth of the Mississippi. At the end of the last glacial these two species moved forward from their different starting points at different rates as the climate warmed. Furthermore, species do not all respond to the warming at the same rate. It is thought that the eastern pine began to migrate 1000 years before the eastern hemlock. The eastern white pine migrated probably at a long-term average rate of between 300–350 m a year; the eastern hemlock at 200–250 m a year; the maple at about 200 m a year; and the chestnut at 100 m per year[76]. However if, in an anthropogenically climatically warming World, a 2°C temperature rise is experienced in North America over the next 50 years (which is broadly in line with the IPCC Business-as-Usual estimate for N. America) then there will be about a 200 mile shift in the climatic zone which will require a migration rate of around 0.5 m an hour. Even though this estimate may be somewhat pessimistic, it is still a few orders of magnitude greater than past known migration rates at the end of the last glacial.

There is other corroborative evidence for such concern. For instance, researchers from the Durham University have determined that 4000 years ago the climate fluctuated causing pine trees, *Pinus sylvestris*, in northern Scotland to advance and retreat about 75 km at a rate of between 375 to 800 m year^{-1} (a similar order of magnitude to trees migrating as the last glacial ended in the USA cited above) but that this is still more than an order of magnitude less than they will need to migrate in a Business-as-Usual climatically warming World[77].

Meanwhile, in the USA researchers from the NOAA, University of Oregon and Brown University have, through pollen analysis, linked past climates and species and then tied these to a computer forecast of North America's climate in a World with twice as much carbon dioxide as in pre-industrial times. Under the IPCC 1990 Business-as-Usual scenario this would take place fifty years or so from now in the middle of the next century. Of course computer models are inexact, so the researchers looked at four models. Their results also indicate that under the IPCC Business-as-Usual scenario the climatic pressure (ignoring other human pressures) for vegetation to change was great and could indeed be greater than at the most rapid phase of warming at the end of the last glacial[78]. The only more rapid changes took place earlier in the glacial itself, these being the end of glacial climatic flickers, and these were very short-lived. What we are faced with in a Business-as-Usual world is a more sustained climatic change lasting many decades, indeed centuries, with the transition having very gradually begun over the past century, and only now (since the 1960s) possibly starting to accelerate!

The IPCC itself notes that a 1°C warming requires most tree species to expand about 100 km northward. It may well be that the most critical factor affecting the plant communities, upon which wildlife in turn depends, is not that globally climatic change is taking place, rather that it is happening faster than in the past (with the exception of the pre-historic glacial flickers) and far more rapidly than the known ability of some species to migrate.

Greenhouse impacts on the coast

The fate of many coastal ecosystems is also of concern. Worldwide coastal marine ecosystems occupy some 8% of the World's surface but within this, between the tides, there are the inter-tidal ecosystems and these in turn are sometimes adjacent to low-lying land: both these ecotones are vulnerable to impacts from greenhouse-induced changes. Unlike purely land-based ecosystems, the greenhouse impact on coastal ecosystems is dominated not so much

Figure 4.7 Parts of UK under threat from seal level rise

by changes in the climate but rather in the rise in sea-level. The IPCC best estimate sea-level rise prediction[20] of two thirds of a metre by the end of the next century means that some coastal habitats of ecological importance will be completely drowned or translocated inland. In Britain, the Institute for Terrestrial Ecology have published a preliminary report as to the likely effects of sea-level rise on UK coastal ecosystems[79] and the IPCC have taken a more general appraisal worldwide[20, 57]

The UK with its varied coastline provides many examples, all be they modest in scale of the impact of sea level rise worldwide. Regarding wildlife, just as inland terrestrial species require time to adapt and/or migrate, so does the wildlife living on and next to the thin ribbon of intertidal (or littoral) habitat separating the land from the sea. Life scientists excepted, most people who have visited the seaside may be excused for thinking that the zone between high and low tide mark seems a poor habitat for plants and animals. Yet this zone often houses a rich variety of species with a large productive turnover. This zone has two qualities that help life. First, being shallow, sunlight has no difficulty in penetrating the entirety of the water column. Multicellular plants are still able to photosynthesise while being anchored to stones or rocks. Second, unlike some parts of the open seas, there is no shortage of nutrients be they brought by run-off from the land or stirred up by the action of waves. If plants and animals can overcome the difficulty of being alternately exposed and submerged by water they have everything else they require to thrive. Salt marshes and, in particular, mud flats support so many plants and animals that significant numbers of other non-littoral (non-intertidal) species can often live off these. Here estuaries are particularly important. In the UK estuaries such as the Wash, Greater Thames, and Morecambe Bay provide feeding grounds for 200 000, 183 000 and 157 000 waders, respectively, and 63 000, 66 000 and 24 000 wildfowl, respectively[80]. Whether such numbers can be supported in the future will depend on whether changes in sedimentation and plant growth as the sea rises prevents the area of such habitats from disappearing: the answers are as yet unclear. One threat to these habitats could stem indirectly as a result of climate change-induced sea-level rise from human-built storm surge barrages designed to protect low lying land around estuaries from flooding, such as in the UK the extensive fenland surrounding the Wash and Kings Lynn. These barrages would restrict the local tidal range and affect the salinity of the estuary's waters; both would significantly alter the invertebrate composition of these ecosystems.

One solution to preserve the present extent of coastal ecosystems would be to allow them to migrate, where possible, inland. Such abandonment of coast would obviously involve the loss of terrestrial land and inevitably some of this would be of agricultural and some of ecological value. Here, there is often the possibility of compromise by minimising the land lost through constructing brand new sea defences further inland. The Institute for Terrestrial Ecology note[79] that the capital cost of just adapting existing UK sea defences to counter a rise of between 0.8 m and 1.65 m (anticipated least and worst case Business-as-Usual scenario by the IPCC in 100 years) would be £5 billion (assume 1989 prices), this gives us an idea of the order of magnitude as to the cost of building new defences elsewhere. In 1987 the USA Congress was told that the cost of building its coastal defences was of the order of $111 billion[81].

It has to be remembered that the level of the sea has not always been constant. Worldwide the sea-level has varied by several metres over the past hundreds of thousands of years and regionally land rises or falls due to either plate movements or the change in continental ice cap volume that have at times locally compounded these worldwide changes. In the UK the Norfolk Broads are the flooded remains of peat diggings by humans a few centuries ago: sea-level rise is not new. The main difference now as we enter the greenhouse World of the twenty-first century, is that in the past natural ecosystems had plenty of elbow room to adapt/migrate. Today the World's wildlife is hemmed in by human settlements and land tailored to human needs. A perhaps extreme example is provided by the Maldive Islands which have often been cited as the nation perhaps most vulnerable to rises in sea-level. Here the IPCC notes that with a 2 m rise (which we might expect to take place toward the end of the twenty-second century) half of its total land area would be vulnerable to flooding. Clearly humans, human-managed species, and the indigenous wildlife are going to have to move into and share the remaining half. Of these three groups it would be naive to suggest anything other than the greatest losses will be experienced by wildlife. Though elsewhere the fraction of land lost to a country from rising seas will be less, the problem is the same in that both humans and wildlife will frequently compete for the same remaining land resources. Clearly, unless humans on occasion deliberately give way, it will always be the wildlife that is squeezed out.

RISK ANALYSIS

Human activity is relentlessly pushing us through the door into the greenhouse World of the twenty-first century. Specifically, it is the developed nations' carbon-generating activities pioneered on an extensive scale as if high energy, carbon-generating consumption per capita were going out of fashion (which, of course, it is: fossil resources are finite). The industrialised urban spaceman has and is enjoying the fruits of big business (which includes many small businesses acting together in a big way). It therefore seems appropriate to look at the question of global warming employing a technique favoured by business itself – that of risk analysis.

Risk analysis in essence consists of weighing up the risks of various options in an attempt to identify a preferred course of action -- ideally the one with the least cost. This is not to be confused with cost-benefit analysis which is usually restricted to examining the costs and benefits of a single option or project, be it the construction of a new railway line, supersonic passenger plane, or power station.

Unfortunately conducting a risk analysis not only involves us in taking stock and seeing what is at risk, but then assigning probabilities as to the loss of all or part of that stock. Such questions are well beyond the scope of this work but we can begin to appreciate the scale of the problem by identifying part of what is at risk from global warming.

A cursory examination quickly reveals that the stakes are high. On one hand we have the material 'well-being' of over six billion people (and counting), and on the other the various costs to maintain this well-being. Such costs include those of providing food, shelter, transport, and culture – the demand to provide for wants (providing psychological 'well-being') can often be as urgent (from a wealthy population) as the demand to provide for needs. (In, for example, the USA the demand for a new opera house may be met though housing for New York's destitute may not.) Important in the global warming stakes at risk are the costs associated with the non-direct 'externalities' such as the damage to human health from urban smogs – NO_x compounds do not just contribute to global warming but also the degradation of health locally. Then of course there are the costs affecting the generations to come – even if we, by some miracle, stopped pumping greenhouse gases into the atmosphere today, not only would the climate continue to be greenhouse forced for decades to come but we would have still deprived the future generation of the fossil fuel resources we have already consumed. (This question revolves around whether we have spent these resources frivolously or wisely in providing a sound infrastructure for those who follow us.)

Naturally, human well-being does not just depend on Terawatts of energy. Humans rely entirely on other living organisms, not only fundamentally for the maintenance of the environment including its abiotic components such as air but also for clothing and day-to-day basic necessities such as (amongst many other essentials) food. But this is only one step in a chain of reliance. To take our food example further, while much of our food today comes from managed domesticated species these have all, without exception, arisen from wild species. Today the majority of the many new crop cultivars are still created through cross-breeding with wild strains. Even the new genetic biotechnologies primarily rely on transferring genes from wild species. The commercial potential of just a single species is exemplified in the 1980s by one wild wheat plant from Turkey that was crossed with a commercial strain. It so improved the latter's resistance to disease that early 1990s sales in the USA alone stood at $50 million a year. So what of the rest of wildlife? There are between an estimated 5 and 30 million species on Earth. Of these only 1.5 to 1.7 million are known to science. Man has used about 7000 types of plant for food, though in the main relies on less than two dozen species, yet there are at least 75 000 edible plants available. The pharmaceutical industry too relies heavily on plants to supply them with half of the raw materials they require and all have their origins in the wild. Estimates of the annual sales for animal and plant derived medicines stand at between $8–$20 billion (at 1996 prices).

Wildlife is therefore not simply an amenity or some luxury, it is a necessity for human life. Consequently, anything that affects many species worldwide (as discussed in the previous section) cannot be simply ignored. Here the potential value of the stock at risk has a multi-billion dollar tag, but if the stakes are so high for future foods and medicines arising from wildlife what of the agricultural system itself and the food it supplies?

The direct effects of global warming on the World's food supply, as we have seen, are fundamental. Just as the climatic belts shift, so too agricultural zones will move. This last is a virtual certainty, what is uncertain is whether we will be able to bring the new regions favourable to agriculture into production without any disruption in food supply, or whether or not the total theoretical capability of these regions to supply food will be as great as those currently in use? Finally, these questions have to be placed in context of an increasing World population. Though we will return to this last in terms of energy demands in Chapter 6, there are doubts as to the security of human food supply in a warming World with a growing population. Such concerns are exemplified by a Stanford University study[82] the conclusions of which not only highlight the risk to food supply from global warming but also confirm the existing trends as to the increasing gulf between the developed and less-developed nations. This study uses grain production as the key staple food indicator to food supply and concludes that even if agricultural production manages to adapt (meeting unfavourable conditions in currently productive areas and/or changing the pattern of where various types of agriculture are practiced) and keep pace with population growth, then the number of hunger-related deaths could double over two decades to 400 million deaths, compared to the estimated 200 million hunger-related deaths estimated to have taken place in the 1970s and 1980s. Even if the climatic changes are favourable to World grain production, then deaths could increase four-fold. At first this seems paradoxical but the study suggests that, all else being equal, a fairly constant proportion of the population starves so that as population grows so does the absolute number of those starving. In the study's increased agricultural production scenario the population also increases in response which results in more mouths to feed and inevitably an increase in the absolute number that will not be fed. In an unfavourable scenario things just get worse. In short the study concludes that global warming enhances the current development problems facing, what Brandt[3] calls the poor nations of the south. The solutions, the study's authors suggest, lie in managing human population growth and this in turn will help reduce the rate of climate change. It may seem somewhat whimsical, if not callous, to pluck million-death figures from a theoretical study but this is just one of many perspectives that point to global warming having an adverse effect on millions of people worldwide. For example, both the low-lying fertile areas of the Nile and Ganges deltas are threatened due to human interference to the water flow of the respective rivers up-stream and this is compounded (if not by flooding alone but salination effects) by the rise in sea-level from global warming anticipated by the IPCC[20]. Here alone the livelihoods of 46 million people are threatened.[83].

The impact on food supply is not the only health effect likely to arise from global warming, even those well fed populations with secure food supplies will feel the effects of a changing climate. Following the publishing of the 1990 IPCC report[20], a summary of likely health impacts from climate change appeared in the *British Medical Journal*[84]. Greenhouse health impacts are not restricted to exotic health disorders but in the developed nations are likely to affect the incidence of the most common causes of death such as heart attacks. The incidence of heart attacks in USA cities shows a minimum when temperature is between –5°C and 25°C with increases in mortality above and below this range. Similarly, there are associations between increased mortality and short spells of relatively hot weather in London, though these may in part be offset by reduced deaths in winter. Either way such effects are felt by the elderly whose absolute numbers, as well as proportion of the population, are increasing in many of the developed countries including the UK and the USA. Climate may also affect the incidence of other disorders positively as well as negatively. The frequency of respiratory disease could well be affected in two ways. Pneumonia, acute bronchitis and bronchiolitis are more common in winter whereas asthma and hay fever peak in the summer. Changes in the climate may also influence the spread of communicable diseases. Many of the now tropical diseases such as malaria, plague, typhus and yellow fever occurred in the past in the temperate zone but a warmer World could make it easier for some of them to return, or at least migrate from the equator. The insect vectors that spread many of these diseases are sensitive to temperature and the effects will not be restricted to the Third World.

Today over 2 billion, 40% of the World's population, are threatened by malaria[85]. Of the four species of malaria *Plasmodium* species, *Plasmodium vivax*, is the most widespread and until recent decades could be found in parts of the USA and Europe; the last outbreak in Britain (of the then-called 'marsh fever') took place in the early 1950s. Today antimalarial drugs are less efficient as the species acquires resistance. Add a favourable climate and parts of Europe could return to being a zone of infection. Meanwhile, in the tropics warming could result in the most dangerous malaria species, *Plasmodium falciparum*, migrating to formerly cooler highland areas; places where the local human population have not had previous exposure, and hence not acquired some resistance. *Plasmodium falciparum* could even possibly return to the warmest parts of Europe around the end of the next century under the climate change forecast by the IPCC 1990 Business-as-Usual scenario. In Australia an increase in mosquito-borne diseases such as Australian encephalitis could be expected. Again, in the USA five mosquito-borne diseases have been considered to be potential risks following a warming of the climate, including malaria, dengue fever and yellow fever. Indeed, the incidence and range of disease already found in developed countries could increase: for example in the USA Rocky Mountain fever and Lyme disease could spread northward.

These are just a few examples of the threats to health from *global* warming. However the threats are aggregate or cumulative; the net result of small changes as opposed to the large individual effects from *local* pollution incidents (Figure 4.8)[86]. Importantly, all these changes have their costs, be they in terms of reduced health or in terms of those incurred through prevention and treatment.

The above examples illustrate that climate change has its costs, or rather has its stakes. These are the costs, to be set against the benefits, should we go down the Business-as-Usual route. Quantifying these global warming costs accurately is quite an endeavour but it is a task that policy makers in the 1990s will have to tackle if greenhouse policy is to be addressed. To begin with, while we cannot predict with precision effects of global warming we can hardly come up with a bill, hence pay for preventative and ameliorative measures. Even so, just as the broad impacts can be identified, so too can the approximate magnitude of climate change costs. Remembering the UK example for the estimated costs for sea defences of £5 billion[79] (a non-depreciated estimate at 1989 prices), and that the cost quoted to Congress in 1988 for USA sea defences was $111 billion[81], it

Figure 4.8 The relationship between local and global impacts on health. Local pollution incidents tend to affect a few people acutely, whereas global impacts tend to affect many people slightly. Based on McMichael *et al.*[68]

would seem reasonable to assume that the approximate order of sea defence costs worldwide would be in the order of thousands of billions of dollars (or $\$10^{12}$) over a century. Of course there is nothing to compel the absolute payment of this cost, we need not bother to improve sea defences, save that money and lose the land: though that too has a cost be it in terms of lost agricultural productivity, human dislocation, and such.

Then again there is the cost of the loss of wildlife species. Bearing in mind that the pharmaceutical industry annually turns over some $8–$20 billion (1996 prices) from plant- and animal-derived species, then in real present-day terms turnover over the next 100 years will be 100-times this figure. Of course, not all of this one-century turnover will come from new drugs extracted from new plants taken from the wild, but we might reasonably expect a few percent, say $100 billion worth to be so generated. These costs have to be added to the cost of adapting agriculture to the greenhouse climate, not to mention the cost of increased human deaths of the order of hundreds of millions over the next century as a dislocated world agriculture increasingly fails to meet the needs of a growing Third World population. (Though this last is difficult to cost. Is a Third World life more or less valuable than a life in a developed country? These questions shamefully still remain unaddressed and so this will probably continue to remain unanswered for the foreseeable future.)

The above debate hardly brings us close to an exact price tag associated with, say, a 1°C increase in temperature worldwide, but it does give an indication as to the 'seriousness' of greenhouse costs. At best it gives us an indication as to the order of magnitude of annual greenhouse costs to human societies worldwide. From the various estimates that have been given for specific impacts a cost magnitude of a thousand billion dollars a year ($\$10^{12}$) is not beyond the bounds of possibility.

It may seem difficult to grasp the significance of a $\$10^{12}$ annual greenhouse cost in fossil fuel terms, albeit that this is a rough estimate. Yet it is possible to calculate from World fossil fuel production figures[1] the magnitude of the annual turnover of those producing the raw materials for the principal greenhouse gas, carbon dioxide. It turns out that the unrefined fossil fuel business (getting the fuel out of the ground) also has an annual turnover of some $\$10^{12}$ worldwide. The carbon business truly is big business. It is so big that other large price tags in the World economy are small by comparison. For instance, the magnitude of the developing nations debt to developed countries grew from under $\$10^{11}$ dollars in 1970 to over $\$8 \times 10^{11}$ by 1987[4] – as an aside it is intriguing to note that this debt, having not been met, is now being written off by many of the industrialised nations'.

The stakes at risk *both* to the energy industries and to our global society are therefore not only high but comparable and it is most likely that one will be a significant fraction of the other. Indeed, because of this, and the fact that the fossil fuel business is so large, these industries will be wary of new greenhouse-combating energy policies; especially if they are likely to erode profitability. The energy industries' contribution to the greenhouse discussion will most likely be presented powerfully throughout the 1990s and probably, it would be naive to think otherwise, with a certain amount of bias (we shall address this last in Chapters 10 and 11). It might appear that the human global village is in a $\$10^{12}$ per annum lose–lose situation. Either the energy industries cut back (for that read 'you, I and society buys less energy to power our lifestyle') or we all suffer economic damage from global warming costs. Fortunately this is not the case.

The oil crisis of 1973, during which crude oil prices jumped from around $8 a barrel to $27[1] (1989 prices), served to bring home the World economy's reliance on fossil fuels. Equally, it is well known that fossil fuels are a finite resource so that reliance on them cannot continue indefinitely: ultimately the move from fossil fuels will be forced upon us. The inevitable logical conclusion is just the reason why environmental scientists (or indeed the lay green movement) have long argued for using these resources in a less profligate and more efficient manner. As we shall see, there is such slack in the system that this is not an impossible task. Furthermore, the efficiency strategy has high economic leverage – it will help to solve a multitude of problems with a single investment. Not only will it help reduce the rate of increase in the build up of atmospheric CO_2 (so at least providing time for wildlife, agriculture etc, to adapt), but also reduce acid rain, cut urban air pollution and improve energy importing nations' balance of payments[87]. This argument is powerful indeed in that the departure from the current energy *status quo* is not solely greenhouse dependent.

The IPCC provided a margin for error in their forecasts which leads us to the other argument that paying for greenhouse ameliorating measures for a worst, or even best-estimate warming scenario that may never happen would be money down the drain. On the other hand if these ameliorating measures have other benefits then we have less to lose and more to gain by their implementation. (We shall return to this in Chapter 13).

Finally, in determining whether or not the global warming risk is great enough to adopt a greenhouse policy, there is the risk that greenhouse costs cannot be left indefinitely to future generations: that those yet to be born will have to pay for greenhouse impacts yet to be exactly defined but impacts that are nonetheless being generated now. Paying for such unforeseen intergenerational costs may at first appear impractical if not unprecedented, but similar financial practices already take place. After all it is not unusual to make a will to leave one's property to one's offspring even though those who write the will have no knowledge as to how such transferred wealth will be spent. Two economists, Spash and d'Arge from Wyoming University, have applied this idea of intergenerational transfers to greenhouse warming[88]. They argue that it is certain that much of the greenhouse costs will have to be met by future generations, but that this could still be paid for now by setting aside some of the profit arising from the present-day, greenhouse gas-generating economic activity. In economic terms they want to maintain intergenerational equity. They advocate what amounts to a pension scheme for the future. Just as one might put aside a proportion of one's salary during a working life to meet the costs of living in retirement, so a similar scheme can be created to meet future greenhouse costs. Furthermore, just as the size of one's pension relates to one's income and the contributions paid, so the size of the intergenerational compensation might relate to current wealth-generating activities.

Such discussion is, of course, purely theoretical and so far so good. Nonetheless any businessman invited to make an investment in the future is going to examine the risk to his capital in a number of scenarios, and all of them require his investment to meet a need and so see a return. In the context of investing in offsetting greenhouse impacts, three principal scenarios apply. The first, which we have looked at, is that IPCC and the scientific community have somehow got it wrong and that global warming will not take place. Here, their investment, providing it is spent wisely, could be protected in that energy efficiency and reducing our dependence on fossil fuels both have economic benefits that have nothing whatsoever to do with global warming. Second, global warming might take place exactly as the IPCC predicts so that again the investment is secure in that there will be an economic need to meet the costs arising out of greenhouse impacts. Last, the IPCC could have got it wrong in that though global warming does take place it is far more pronounced than predicted. Here, those nations investing in meeting/offsetting greenhouse impacts once again win as they would have an economic advantage over those countries who have ignored the problem. Naturally, this presumes that the global warming will not be so rapid that there is no time in which to set aside the necessary capital to meet the cost of greenhouse impacts.

As to which of the above scenarios will ultimately take place cannot be divined no matter how much we crystal ball gaze. Human-induced climate change is a human gamble, one that Wallace Broecker (of Columbia University's Geological Observatory) has likened to Russian roulette[89]. However, he points out that unlike Russian roulette, where we know that at least one chamber of the gun contains a bullet, nobody knows exactly what will take place in the greenhouse of the future (i.e. how many chambers are loaded). It is also possible that global warming may be far worse than is anticipated (the nature of the bullet). He feels that climatologists have been lulled into complacency by computer simulations that suggest that global warming will be gradual over a period of about 100 years. In turn, he points to the evidence from ice cores[31] that indicate that the response to climate forcing is far from smooth but takes place in large jumps – larger than the 3°C we have been discussing so far (typically

6°C regionally but not globally). This could be due to changes in the deep ocean currents that drive the surface movement of water and so the way heat is transported about the globe – an idea that rapidly formed the basis of one of the preferred palaeoclimate change theories discussed in the early and mid-1990s. Such sharp switching of states is reminiscent of Lovelock's idea that the Earth can switch sharply between a number of geophysical modes[52, 53] and that such switching can be human-induced. If, as Broeker contemplates, the global climate does suddenly and rapidly grow warmer then the question of setting aside finance could become redundant – we might have to channel what resources we can into meeting immediate costs (as opposed to the intergenerational cost) of greenhouse impacts. However, until that time it would seem to be only prudent to invest in greenhouse policy and fortunately the signs are that this is beginning to take place.

At the moment governments are not investing much in either meeting or ameliorating future greenhouse costs other than on the smallest of scales, primarily focused on scientific research. On the other hand, business concerns are beginning to recognise the need for such investment. One example is provided by the water industry in the UK and the USA. The UK view was summarised in 1991 by Dr Gadbury, a scientist working for Southern Water[90]. When discussing whether or not the UK droughts at the end of the 1980s were greenhouse induced, he said that:

> 'The fears of global warming are relatively recent and the water industry is a very long-term industry. It would be foolish to over-react to what may be no more than a short-term problem, but we are starting to think seriously about global warming. That may mean more [water] storage – it may mean more reservoirs. It may mean some kind of pricing mechanism whereby we discourage people from using vast quantities of water. All these are aspects which we are now taking into account.'

Similarly, hard decisions are now being made in the USA. The 1991 spring saw heavy rains and snows provide a respite to a California that had been suffering a five-year drought. Welcome as the precipitation was, it did not go far to replenish reservoirs and restore fallow farmland. Low on supply, some Californian cities have been looking at introducing desalination plants to obtain fresh water from the sea even though water costs could increase from about $200 per acre foot (or 326 000 gallons) to $1900[91]. Ironically, while the Californian water authorities have accepted the risk that the climate change they experienced may be part of a long-lasting trend (and so have costed policy accordingly), their proposed desalination plants use tremendous amounts of energy which if supplied from fossil fuel combustion would only compound climate change (albeit marginally). Yet from the palaeoclimatological evidence we have already mentioned[40] the Californians may well have good cause for concern. The last times the Earth was as warm as it is becoming today was between AD 900 and 1110, and AD 1200 and 1350 when Europe was experiencing a climatic optimum. On those occasions the Sierra Nevada mountains of California experienced severe drought. Today that region supplies California's cities and agriculture with about two thirds of their water.

Meteorological records over the past few decades, increasing biodiversity on European mountain summits, and ecological changes in northern Canada and southern Chile, do provide hard evidence that the Earth is getting warmer, this in itself does not substantiate the IPCC's considered opinion that they are confident that the Earth will continue to become warmer for many decades to come due to the human emissions of greenhouse gases. Unfortunately droughts in California, dry summers in the UK, heavy January west European rainfall (such as in 1995 that resulted in flooding and the evacuation of a quarter of a million Europeans from their homes) or increased violent storms, do not provide as good evidence of climatic change as meteorological records. Droughts, floods and storms that happen to be what climatologists would expect in a warmer World could be purely coincidence. At best they are circumstantial evidence. Yet all have associated cash costs.

So where does this all take us if the IPCC's 'confident' considered opinion in 1990 (reiterated in 1995) is in fact erroneous? After all the IPCC does not claim absolute certainty! They accept the fact that they may be wrong, but think it unlikely. But what if the IPCC is wrong and human-caused emissions of greenhouse gases are not responsible for warming the World, and that the warming over the past decades is a 'natural' fluctuation? What then?

It might seem that paying for reducing greenhouse emissions – rather than investing in smaller economic growth – would not be worth it. We shall return to that question in Chapter 13. It might also seem that we have nothing to worry about from the climate. However this view would be very short sighted indeed. We already know that the climate is changing, even if the remotest chance that the IPCC is wrong about long-term warming, climate fluctuation would seem to be the only alternative to explain what is happening. If this is the case then we are almost in as bad a situation. As noted in Chapter 1, the World population is growing and at the moment this growth does not show signs of immediately slowing, let alone stopping. Yet the increase in productivity per hectare of rice, the most subsidised rice production in the World in Japan, has stopped. It has also stopped in China and India[92]. Similarly, the improvement in wheat yields per hectare has stopped in the USA and UK, as well as almost halted in China[92]. Water scarcity and soil erosion compound the problem. If we are to feed the more populous World of the twenty-first century then we are going to need something not far off a miracle! Fortunately there is a potential miracle of sorts emerging, genetically engineered crops. Through genetic engineering it will be possible to develop crops more suited to local conditions. Yet if the weather is going to be highly variable on a two or three year cycle, while at the same time following some sort of multidecadal cycle, then predicting where to plant the appropriately engineered crops will be problematic, let alone managing regional agricultural strategies. The cost from a changing climate would still be with us.

Considering the above against the IPCC's confident view, where are we? If there really is no conclusive evidence for human-induced climatic change then we can only look at what little hard evidence we have together with the circumstantial evidence. If, as the IPCC thinks, there is cause for concern then it would be foolish at the very least not to consider the financial risks of climate change.

Of course the IPCC, as represented in its reports of 1990[20] and later, could be wrong in quite another way – the nature of climate change could be worse. Here it must be remembered that the work revealing the climatic flickers at the end of the last glacial, as well as the possible intermittent breakdown of the ocean salt conveyor, was only widely accepted after some time. It could be that the climate could flicker again and not gradually warm. But accepting that there really is a financial risk is not enough (before climate mitigating cash is placed on the table) without understanding the nature of the anthropogenic greenhouse induced- climate change and so costing within a greenhouse policy framework. If we can understand this – the greenhouse gases and the human relationship with them – then we can begin to appreciate the IPCC view that there is genuine cause for concern as to current climatic change and future climatic prospects.

CHAPTER 5

GREENHOUSE GAS FUNDAMENTALS

Whatever the risks and the costs to human activities and environmental impacts, be they large or small, or which ever way the figures are cut, over the last century the Earth has become warmer. It still is. This change in climate has followed an increase in the atmospheric greenhouse gases carbon dioxide, methane, nitrous oxides and chlorofluorocarbons (CFCs). Each of these gases has its own sources and is removed from the atmosphere in various ways–they each have their own sinks. Each has its own heat absorbing properties and each is produced in different quantities. By identifying these greenhouse characteristics and the way these gases behave in the environment we can begin to see how much these contribute to global warming: in other words how these gases 'force' climate.

CARBON DIOXIDE

Carbon dioxide is the one gas many people popularly see as being synonymous with the greenhouse effect itself. One reason for this is because of the large amount released by human activity – some 26 000 Mt (26 000 × 10^6 tonnes) of carbon dioxide a year.[20] Since early industrial times the effect of human activity increasing atmospheric carbon dioxide has been staggering. Given that 1 tonne of carbon is equivalent to 3.7 tonnes carbon dioxide, measured in gigatonnes (tonnes × 10^9) of carbon between 1850 and 1986 some 195 ± 20 Gt of carbon have been released into the atmosphere through the burning of fossil fuels and 117 ± 35 Gt of carbon through deforestation and changes in land use. The IPCC estimated cumulative total released since early industrial times (1850) of 312 ± 40 Gt of carbon.

The amount of carbon that human activity has returned to the atmosphere over the past one and a half centuries has been large, but to put this in perspective this figure is a fraction of, but close to, one percent of that moving naturally between various sources and sinks in the carbon cycle over the whole of that same time. Annually, the carbon cycle represents the movement of some 500 Gt of carbon between major carbon reservoirs (see Figure 5.1). This 500 Gt of carbon moved is, in turn, about one percent of the size of the amount of carbon in the natural reservoirs (excluding that trapped in the fossil fuel reserves or the carbon geologically locked up as chalk or limestone). These carbon reservoirs not only include the carbon trapped in living things and the reservoir

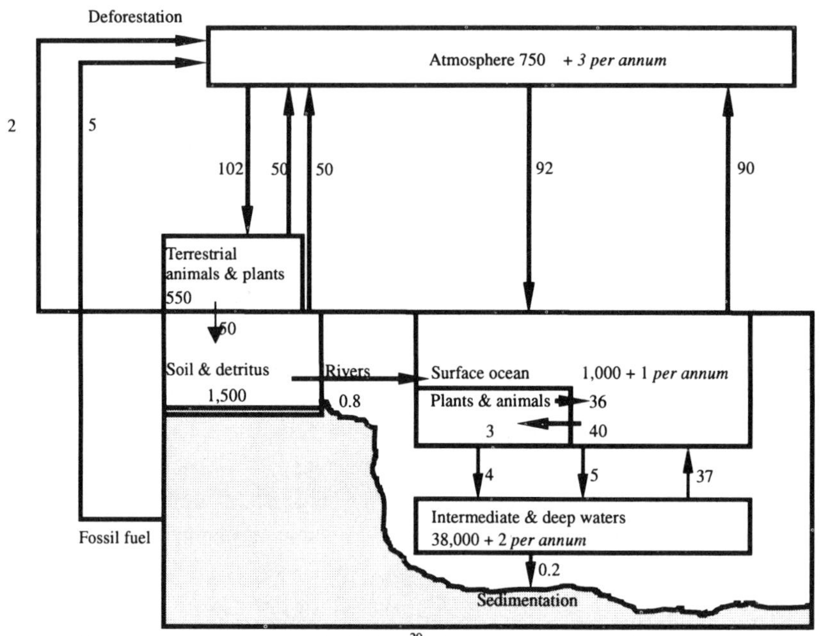

Figure 5.1 Carbon cycle as based on IPCC 1992[20]. Note how it compares with Figure 4.4 and that there are differences. These differences are in part due to the format in which the figure is drawn, as well as due to the *estimates* of the carbon reservoirs and fluxes: work on confirming these estimates continues but both this figure and Figure 5.1 provides an idea as to the general nature of the carbon cycle and comparison of the two (as well as of other portrayals of the cabon cycle in other works) gives an indication as to the areas where there is uncertainty and its degree. The units are gigatonnes of carbon (GtC=10^9 metric tonnes=10^{12}kg) for reservoir sizes and GtC year^{-1} for fluxes. Where '+*x* *per* year' is indicated, the reservoir is increasing by about *x* GtC a year

of atmospheric carbon dioxide itself, but also as carbonic acids in the oceans formed when carbon dioxide reacts with water.

That the carbon cycle is dominated by the opposing actions of plant photosynthesis and respiration can be seen from the direct measurement of atmospheric carbon dioxide. This rises and falls annually in each hemisphere alternatively by 5 ppm as photosynthetic activity increases during the summer months (with plants drawing carbon dioxide out of the air) and decreases during the winter. Drawing a graph of the atmospheric concentration of carbon dioxide illustrates this waxing and waning (see Figure 5.2). The photorespiratory pumping of carbon dioxide into and out of the atmosphere is so powerful that, on average, a molecule of carbon dioxide stays in the atmosphere for only a few years – less than half a decade. This short residence time should not be confused with the many years it takes for the atmosphere to adjust to a new equilibrium if carbon sources or sinks change. The adjustment time is thought to be between 50–200 years due to the slow exchange of carbon between the surface and deep layers of the ocean. Relating this to the greenhouse debate, greenhouse policies

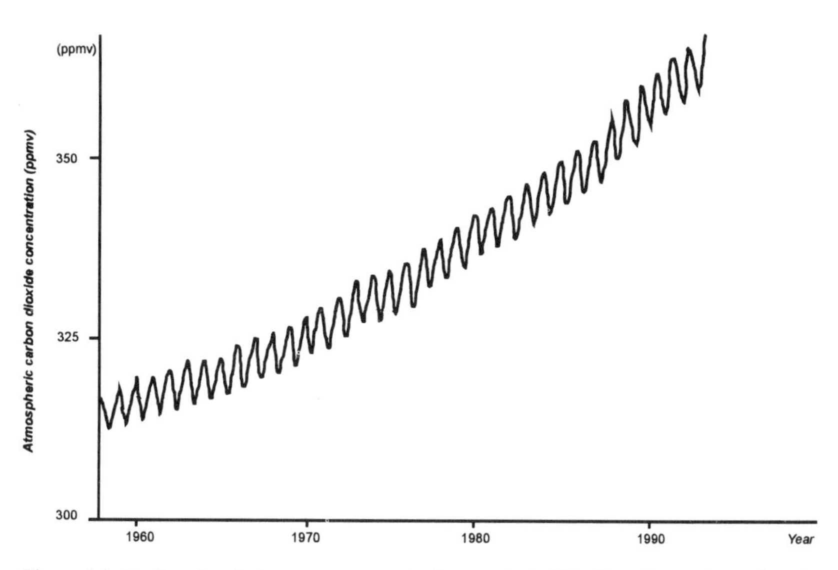

Figure 5.2 Northern hemisphere measurements of atmospheric CO_2. Note the waxing and waning due to increased summer photosythesis (the opposite takes place in the southern hemisphere). The overall upward trend is quite clear

must take into account the time involved for the atmosphere to adjust to changes in carbon fluxes: policies should operate on similar timescales.

Plotting the atmospheric concentration of carbon dioxide against time reveals a second trend – the key trend pertaining to the greenhouse issue – the gradual build up of CO_2 concentration. As discussed in Chapter 4, we know from measuring the carbon dioxide in air trapped in Antarctic ice bubbles for thousands of years that atmospheric CO_2 concentrations were lower during glacials, and that there appears to be a relationship between CO_2 concentration and climate. One question that the Antarctic ice core researchers had to ask was whether or not their measurements of trapped air bubbles could be properly related to the atmospheric concentrations of gases at the time of entrapment. One possibility was that the various gases in the bubbles could have migrated out of the bubbles at various rates so giving an erroneous picture. On the other hand, if these measurements were accurate then a graph of carbon dioxide concentration trapped in bubbles should neatly join up with a graph of CO_2 concentration made from the recent direct measurements. Researchers at the Siple Station in the Antarctic examined bubbles in an ice core to monitor the change in concentration over the past two centuries. They were particularly interested to see whether their data would tie in with contemporary direct measurements. Fortunately, as can be seen from Figure 4.2, this is the case and a smooth transition between the two sets of data can be seen. What is also apparent is that the gradual increase in atmospheric carbon dioxide only began with the onset of industrial activity.

Together, these carbon dioxide measurements from ice cores and direct atmospheric measurements provide a compelling case for the current increase in atmospheric carbon dioxide being of anthropogenic (human) origin.

One problem with the way atmospheric carbon dioxide concentration is increasing is that the sums do not add up! Not all the carbon dioxide from the burning of fossil fuels and wood from forest clearance ends up in the atmosphere. On average each year between 1980 and 1989 some 7.0 ± 1.5 Gt are thought to be released through human activity, yet only 3.4 ± 0.2 Gt of carbon dioxide end up being added to the atmospheric reservoir[20]. Where have the missing 3.6 Gt gone? This is quite a mystery. The current view is that much of this missing carbon dioxide is absorbed either into the oceans in various ways, and/or by vegetation on land. Forms in which carbon can be found in the oceans include: suspended particulate and dissolved organic compounds; particulate and dissolved inorganic compounds; as well as that trapped in ocean sediments.

The exact nature of ocean carbon pathways is not known and is currently being investigated. In the UK much of the research in this area is being spearheaded by the BOFS (Biogeochemical Ocean Flux Study) programme which began in 1988. Over all, BOFS has two principal goals: to improve the understanding of ocean and ocean-atmosphere biogeochemical processes with particular reference to carbon and, secondly, from this to develop predictive models of man-induced changes. Emphasis is being given on quantifying carbon exchanges: across the sea surface; across the thermocline (the temperature gradient between the warmer surface, and cooler deeper, waters); and into the sediment on the ocean floor. The IPCC estimates that the oceans absorb at least some 2.0 ± 0.8 Gt of the carbon humans release into the atmosphere a year, which still leaves us with a carbon imbalance of some 1.6 ± 1.4 Gt (see Table 5.1). The BOFS programme will hopefully give us an insight as to where much of the missing carbon is going[94].

Table 5.1 Sources and destinations of anthropogenic atmospheric carbon. This table is adapted from the IPCC[20]. Note that the figures in the table are rounded and that the errors are not additive since the uncertainty is greatest regarding the way the carbon cycle 'pie' is divided as opposed to its size, hence the size of imbalance

Human induced atmospheric carbon credit	**GtC/yr**
Emissions from fossil fuels into the atmosphere	5.4 ± 0.5
Emissions from deforestation and land-use change	1.6 ± 1.0
(1) Total emissions	7.0 ± 1.5
Biogeochemical atmospheric carbon debit	
Atmospheric accumulation	3.4 ± 0.2
Current ocean uptake estimate	2.0 ± 0.8
(2) Total estimated atmospheric debit	5.4 ± 1.0
Net imbalance (1 – 2)	1.6 ± 1.4

The degree of uncertainty in the above table may seem inconsequential compared to the role the oceans play in total carbon cycling. However, the above table is an estimate of the carbon cycling of just the annual anthropogenic contribution: there are also natural reservoirs of carbon and that overall (including carbon from natural sources) the oceans absorb some 105 Gt of carbon and release a total of roughly102 Gt a year.

Other than the oceans, there is another candidate for a sink that mysteriously absorbs half of the carbon dioxide put into the air by human actions. It could be that this sink is vegetation on land. Could growing trees be better at absorbing carbon dioxide than we thought? Could we have miscounted the number of trees in the world? Could peat bogs be sucking carbon out of the air? (This last is of particular interest as some researchers feels that there is evidence for the mysterious sink to be on land in the Northern Hemisphere – which is where most fossil fuel is burnt).

In fact the first half of the 1990s saw considerable effort in trying to establish just what this mysterious carbon sink was. There is quite a body of research in this area but, unfortunately, much of it is conflicting, with conclusions differing between an oceanic and a terrestrial sink (and even combinations of the two). Though this particular facet of the greenhouse debate is of great interest to those studying the greenhouse (both natural and anthropogenic) dimension to climate change, it is so involved and, at the moment, inconclusive, that there is little merit in summarising it here.

Nonetheless, a great challenge lies in identifying how the carbon cycle functions as a whole so that the fate of human contributions can be deduced. Quantifying current carbon budgets for various parts of the cycle will be a start but, as was discussed in the previous chapter, there is the question of whether the various geophysical mechanisms that govern the cycle remain constant – for instance is the basic pattern of global circulation inherently unstable? Here the question of whether ocean currents, especially the deep ocean currents, can flip between alternate circulatory states[89] is central to the greenhouse issue. For example, as discussed in Chapter 3, cold, saline water produced at the poles sinks to enter the deep ocean circulation system, this water later rises in upwelling regions along the equator where it warms, releasing the carbon dioxide that had originally been absorbed (see Figure 3.5). What would happen if this circulation were stopped, say by more rain in a warmer World diluting the concentration of the saltiest parts of the surface portion of this system? This and the other ocean circulation questions can ultimately be addressed with the results from a second European Community research project – the World Ocean Circulation Experiment (WOCE). Soon data from both the BOFS and WOCE will be integrated[95] and, considered in the light of the many other initiatives now taking place, we will begin to understand the nature of the global greenhouse experiment that the industrial nations have unwittingly instigated.

Among these other initiatives are those involving the study of the other greenhouse gases: principally methane, nitrous oxides and the CFCs.

METHANE

Methane (CH_4) is both a chemically as well as radiatively (greenhouse) active gas in the atmosphere. Its chemical role is such that it contributes to another environmental problem – stratospheric ozone depletion. It reacts with chlorine in the atmosphere to produce hydrochloric acid which in turn, under the action of sunlight, forms free chlorine that can react with, and so destroy, ozone – conversely the hydrochloric acid so produced can be washed out of the atmosphere so allowing ozone to build up. This is not to say that the greenhouse issue is synonymous with that of ozone depletion, though this is a common public misconception (see CFCs below and Chapter 12), merely that the two are connected (the other principal connection being that CFCs are both greenhouse-active and ozone-depleting agents).

Methane, like carbon dioxide, is a natural component of the biosphere but one whose atmospheric concentration is being altered by human action. As with carbon dioxide, it is possible to ascertain the pre-industrial atmospheric levels by analysing the gases trapped in ice laid down centuries ago[96]. These show that the atmospheric concentration has risen steadily from a pre-industrial level of just under 0.8 ppm, more than doubling to a 1990 figure of 1.72 ppm. (See Figure 5.3). This is equivalent to an atmospheric reservoir of some 4.9 Gt (or 4.9 billion tonnes) and is growing at a rate of under 1% a year. The growth in atmospheric methane itself seemingly parallels that of the human population, furthermore the concentration in the Northern Hemisphere (where human activity is greatest) is 0.08 ppm higher than the Southern. Finally, as with carbon dioxide and indicative of methane's greenhouse properties, analysis of bubbles trapped in Antarctic ice for the past 160 000 years show that the atmospheric methane concentration has varied in tandem with the climate[97].

Between 400 to 600 Mt of methane enters the atmosphere each year. Over 90% of this is removed, entering two principal biosphere sinks (absorption by soils and an atmospheric chain of chemical reactions), so that the atmospheric turnover to methane increase ratio is high compared to that of carbon dioxide. This last results in methane's atmospheric lifetime, of about a decade, being shorter than carbon dioxide's 50–200 years. (The atmospheric lifetime is the atmospheric residence time of a typical molecule.) In 1990 the IPCC gave estimates as to the various sources and sinks for atmospheric methane[20]. However, while the 1990 IPCC report was being written and published data were coming to light that revealed that the rate of growth of atmospheric methane was decreasing; it was still rising, but not as fast. This, together with improved best estimates, caused the IPCC to revise its figures in 1992[93]. However, these 1992

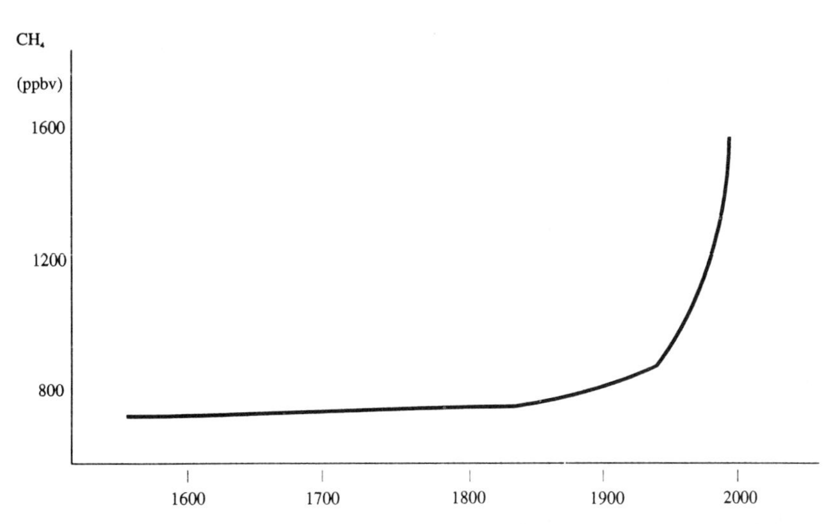

Figure 5.3 Appropriate trend of atmospheric CH_4 concentration over the past four centuries

revisions were nearly all still within the IPCC 1990 declared margins of error (see Table 5.2).

Table 5.2 Principal sources and sinks of methane. (IPCC 1990[20], 1992[93]). The revisions the IPCC 1992 reflect the decrease in atmospheric accumulation, and increased confidence and accuracy in measurements

	IPCC 1990	*IPCC 1992*	*1992 Range*
		Mt p.a.	
Source			
Natural wetlands	115	115	100–200*
Rice paddies	110	60	20–150
(Animal) fermentation	80	80	65-100*
Gas drilling & industry	45	estimates merged	25–50⁺
Coal mining/industry	35	in 1992 data below	19-50⁺
Coal mining and gas drilling	–	100	70–120
Biomass burning	40	40	20–80*
Termites	40	20	10–50*
Landfills	40	30	20–70*
Oceans	10	10	5–20*
Freshwater ecosystems	5	5	1–25*
CH_4 hydrate destabilisation	5	5	0–5
Sink			
Removal by soils	30	30	15–45*
OH atmosphere reactions	500	470	420–520
Atmospheric increase	44	32	28–37

*unchanged from the 1990 IPCC report
⁺1990 estimate ranges only as 1992 report merged contributions from coal mining and gas drilling

There are a number of methane sources and, as with carbon dioxide, the principal ones are the consequence of life metabolic processes. The principal sources arise from anaerobic metabolism as organic matter decays in wetlands, both natural and artificial. It is thought that close to half the atmospheric methane is supplied in this way, with a fifth being contributed by natural wetlands (bogs, swamps and tundra), and half this amount from rice paddies. Though these are undeniably major sources, there is considerable doubt as to the accuracy of the estimates, especially as habitat and crop data from the Third World is so unreliable. (About 90% of the World's rice comes from Asia, and about 60% of this is grown in India and China.) However, World rice production has approximately doubled since 1940 and so it seems reasonable to suggest that the methane from this source has similarly increased.

As well as wetlands and rice paddies, there is the contribution from enteric fermentation – methane given off by ruminent animals. Some 80 Mt are thought to be released this way. Principal ruminant sources include all cattle and sheep, the populations of which have grown markedly this century and which have not been offset by decreases in wild ruminant populations, including elephants and the North American bison: consequently the methane generation from enteric fermentation may have increased at least two fold over this time[98].

Two other sources are also thought to have grown markedly this century: methane from biomass burning, fossil fuel, and the natural gas industry. Methane from the former results due to the incomplete combustion of organic material primarily in tropical and sub-tropical regions. The practice of slash-and-burn in forests is of particular concern as this not only generates methane but is so extensive that it is non-sustainable and is, in recent times, thought to be a significant contributor to species extinction. This last especially applies to other ways in which tropical forest is cleared. By the end of the 1990s over 90% of Brazil's accessible coastal forest had been cleared and some 8–11% of all Brazil's Amazon basin forest was lost: these areas contain an estimated 10 000 and 30 000 animal and plant species respectively. The forests of the Northern Andes, the habitat of some 40 000 plant species have also suffered with over 90% of its area cleared[99]. As tropical forest clearance results in the release of carbon dioxide as well as methane, this practice is of great significance to the whole question of climate change. As discussed at the end of Chapter 4, here is another example where there are yet other motives – in this case species conservation – for implementing greenhouse policies.

This century's other fast-growing source of methane is the fossil fuel industry. The carbon in the methane from this source contains no ^{14}C isotopes, having been photosynthetically trapped millions of years ago which is more than enough time for all the ^{14}C to have decayed. (Indeed, the isotopic ratio of ^{14}C, ^{13}C and ^{12}C varies between the principal methane sources and so helps in their identification.) Coal gives off methane, and if seams are exposed to the air they will release it into the atmosphere. The ventilation of coal mines was thought by the IPCC in 1990 to contribute between 19–50 Mt of methane a year with a best estimate of 35 Mt. The natural gas industry is estimated to contribute, via venting, drilling and leaks in transmission, 25–50 Mt a year, with an IPCC 1990 best estimate of 45 Mt. Here leaks in transmission have stimulated much discussion in that on one hand, with methane's higher hydrogen to carbon ratio, natural gas provides more energy per molecule of carbon dioxide generated on combustion than either coal or oil. Conversely, in terms of greenhouse policy, this benefit might be lost if gas leakage is sufficiently great i.e. over 2.8%. In 1990 an Earth Resources Research study[100] provided a least-case estimate for leaks from the British Gas network of 1.9% but a worse case scenario of 5.3–10.8% ayear. British Gas replied by claiming the figure to be closer to 1%[101]. One could retort that 2.8%, although closer to 1% than 5.3%, natural gas is still more 'greenhouse friendly' than burning oil or coal. The debate is further complicated if one takes into account methane's low atmospheric residence time compared to carbon dioxide and that there is less environmental damage from acid rain with natural gas compared to coal[102]. However, it would be unwise to assume that just because estimates of methane leakage from gas supplies in the UK and other OECD countries may be around 1%, that methane leakage in other less-developed countries or the former Soviet bloc countries is as low. If leakage is greater in those places then these countries might be unwise (soley on greenhouse grounds) to further use natural gas energy. This last even applies with a lower estimate[103] as to methane's global (climate) warming potential than those in the IPCC 1990 report[20]. Such discussion on climate and the human use of fossil (natural gas) methane and the current growth in natural gas consumption is likely to be short-lived (of concern for only a few decades) due to the low level of natural gas reserves worldwide[104]. In terms of policy making, the other non-energy-use sources of atmospheric methane are arguably of longer-term concern.

The second largest natural source of atmospheric methane, but providing less than the 115 Mt a year from wetlands is that from termites: the IPCC 1990 place termites' contribution in a wide range between 10 and 100 Mt a year with a best estimate[20] for this of 40 Mt a year, though by 1992 it had revised this estimate downwards to 20 Mt a year (just within the original range)[93]. While laboratory experiments on small sample populations provide valuable data, it is difficult to translate this into a figure for the global termite contribution. Here, the estimates for their World population remain uncertain, as do the figures for their biomass consumption in the wild; both are critical factors in determining the total termite methane contribution. Another unknown is how the world termite population will respond to climate change: will it grow or decline?

Landfill sites used to 'dispose of' domestic and industrial rubbish also generate methane as organic matter anaerobically decays. Here the IPCC 1990 best estimate is that the landfill contribution is 40 Mt of methane, but revised this downward to 30 Mt in 1992. Their greatest and least estimate contributions remained unchanged between the 1990 and 1992 reports at 70 and 20 Mt of methane a year. The uncertainty largely arises because the exact organic content of landfill sites is unknown, as are the exact nature and extent of various landfill practices.

So what can be done to reduce this contribution to the atmosphere? There are techniques being developed. For example, methane can be trapped for use as a fuel. As the landfill site fills up, a layer consisting of a network of porous pipes is laid before more rubbish is accepted – sometimes clay or some non-porous material is used to cap the pipe network and separate the next layer. After many layers and a full site, the whole thing is capped with clay leaving only one main pipe which enters the landfill and connects with each layered network of porous pipes. This whole assembly channels the methane to the single outlet where it may be burnt to provide power. Such schemes reduce anthropogenic greenhouse contributions in two ways: by preventing methane from entering the atmosphere, and by displacing traditional energy sources.

Altogether these methane sources are largely, though currently not completely, offset by the loss of methane to two principal sinks (see Table 5.2). The largest of these is the reaction with atmospheric OH radicals and entering a complex chemical cycle which includes the formation of formaldehyde. Combining two sets of data, of ice core methane and formaldehyde content, it is possible to begin to establish a history of atmospheric methane and its oxidation[105]. The atmospheric OH sink is thought to account for some 470 ± 50 Mt. It mops up the majority of methane emitted each year.

The second lesser sink, accounting for only 30 ± 15 Mt of methane a year, consists of the removal of methane by soils (an estimate unchanged by the second IPCC report in 1992[93]). Here the future performance of this sink may well be affected by land use. There is evidence to suggest that applying fertiliser and cultivation of grasslands might decrease the absorption of methane by soils, and increase the production of nitrous oxide, itself a greenhouse gas[106].

The difference in methane fluxes from and to sources and sinks accounts for any growth and decline in atmospheric concentration which can, of course, be directly measured. In 1990 some 44 ± 4 Mt of methane were accumulating in the atmosphere annually. However, currently there has been a substantial slowing of the global atmospheric accumulation rate since the early 1980s and, *if* this continues, atmospheric methane concentration will peak shortly after the beginning of the twenty-first century[107]. Subsequent work confirms that in the early 1990s the rate of annual methane accumulation was declining. As to the cause, this could be due to a change in one or more of methane's sinks or sources (only a few per cent would be required in either or both). The exact situation remains unclear[108]. What is clear is that the principal sources of methane are located in the Northern Hemisphere, which is not surprising as that is where most coal mines, natural gas transmission, termites, landfills, rice paddies, and natural wetlands exist.

Although methane isotopic signatures are fairly distinctive, as noted above there is considerable uncertainty in the estimates for sources and sinks. Furthermore, it may be that if the errors were mainly overestimates then there may be another major source of atmospheric methane. One suggestion has been that methane, in the form of $^{14}CH_4$, is formed by nuclear fission power stations – principally Pressurized Water Reactors (PWRs)[109]. Methane output from PWRs has been measured in New York State in the USA. It arises from the radio-conversion of water's oxygen to carbon and its subsequent reaction with hydrogen (the cover gas in PWRs). If the magnitude of PWR methane production is as large as is suggested then it would rival, if not dwarf, the principal sources currently estimated by the IPCC. Such speculation aside, the near 1% growth per annum in atmospheric methane continues, albeit currently declining, and is currently equivalent to an annual increase (albeit decreasing) of 11.5 ± 0.5 ppbv a year[108].

HALOCARBONS

The third most important contribution to greenhouse warming comes from a chemical family of gases – the halocarbons. Within this family are a subgroup, the CFCs (chlorofluorocarbons). With the exception of CH_3Cl and CH_3Br, these are all manmade compounds. Their uses vary: CFCs–11 and –12 are used as foam blowing agents (that is putting the 'bubbles' in types of plastics); CFCs–12 and –114 and HCFC 22 as refrigerants; CFC–113, CH_3CCl_3 (methyl chloroform) and CCl_4 (carbon tetrachloride) solvents; and halons 1211 and 1301 as fire retardants.

Some halocarbons do occur naturally. Methyl chloride (CH_3Cl) is released from the oceans, but there is an anthropogenic contribution from biomass burning. Similarly methyl bromide (CH_3Br) is also released from the oceans, and again there is also an anthropogenic contribution.

Despite early doubts as to whether it was possible to measure concentrations of the level of a few parts per trillion (ppt), the atmospheric concentration of CFCs was first measured away from human habitation in 1971 by James Lovelock. He and his colleague Robert Maggs showed that CFCs were persistent and long-lived in the Earth's atmosphere and that two other halocarbon gases, carbon tetrachloride and methyl iodide (CH_3I) were also to be found in both hemispheres. At the time Lovelock believed that atmospheric CFCs were of no meaningful environmental consequence. Since then, concentrations of some CFCs have increased by 500%, while their roles as ozone depleting and greenhouse climate forcing agents are now better understood, and Lovelock has joined those who would regulate their emissions[53].

That CFCs are involved in a series of polar stratospheric reactions that destroy ozone and that they are greenhouse gases, has caused some confusion among the lay public who muddle ozone depletion with the greenhouse effect. As we shall see in Chapter 12, when we shall look at how the greenhouse issue is perceived, there is evidence for this public ozone-greenhouse confusion both from the USA as well as the UK.

Such confusion aside, scientists, and now politicians, do recognise the environmental damage resulting from the release of CFCs. In March 1985 the Vienna Convention for the Protection of the Ozone Layer was signed, and subsequently implemented in September 1988. It recognised 'the possibility' that CFCs and other chlorine-containing substances could damage the ozone layer. It also led politicians to other negotiations.

On September 16th, 1987, an initial 27 nations signed a treaty in Montreal – the Montreal Protocol[110] – to reduce the release of CFCs by 50% of the 1986 levels by the end of the twentieth century. Two years later, in May 1989 at the first meeting of the Protocol parties in Helsinki, 81 nations agreed (albeit in principle) to ban eight ozone-damaging halocarbons. A declaration was signed calling for CFCs and halons to be phased out by the year 2000. Then in July 1990 at the second meeting of the Protocol parties (in London) there was agreement to actually phase out CFCs, halons and carbon tetrachloride by the year 2000 and methyl chloroform by 2005. In 1991 the European Community directive 594/91 was implemented with European CFC use to cease by 1997. The following year the fourth meeting of the Montreal Protocol parties (in Copenhagen) agreed to bring forward the phasing-out schedule.

By 1992 it was apparent that the Montreal Protocol was working as the rate of growth in atmospheric concentration of CFC–11 and CFC–12 was slowing, even though concentrations were still rising! If the early 1990s trends continue through the decade then the atmospheric concentrations of these two CFCs should peak before the year 2000. However, this does not mean that the concentrations of other CFCs will follow suit, so damage to the ozone layer around the poles is likely to worsen each year into the next century. Nonetheless these international policies are a welcome sign.

However, there is a problem in that replacement chemicals tend to be more expensive than the original CFCs so that less-developed countries who want their populations to have access to industrial products (such as refrigerators) and who do not have money to spare, face an increased bill that they can ill afford. In 1991 the USA National Academy of Engineering sponsored an international conference to review the progress made in developing CFC replacement technology – it was apparent that such technology transfer from the developed to the less-developed nations was more of an ideal than a reality (though the Montreal Protocol did make an allowance for some modest financial assistance from the developed to less-developed nations). The USA National Academy conference pointed to a further difficulty in that it is not just a question of supplying CFC replacements but also the hardware – refrigerators, air conditioners, foam-making machinery and so on – as all the old CFC hardware will have to be either replaced or adapted *cum* retrofitted[111].

Another difficulty is that though some of the CFC replacements are 'ozone friendly', they are still greenhouse forcing agents.. DuPont is the World's biggest producer of the CFCs which are being phased out this decade. These CFCs are highly stable and so persist until they reach the stratosphere where they break down under the action of UV light, with the breakdown products going on to react with ozone. Many DuPont replacements are blends of less stable CFCs and will only deplete ozone by a fraction of the CFCs they replace. These blends could be considered an interim measure and many can be used in existing coolant machinery. Potential replacements for CFCs in refrigeration and foam, respectively, include hydrochlorofluorocarbons (HCFCs) 22 and 142b which have only 5% of the ozone damage capacity of the most harmful CFC – CFC–11. However, they do still retain a significant greenhouse climate forcing capability – HCFC–22 has still some 30% the potential of CFC–11 to cause greenhouse warming[86]. As to the future, DuPont estimates that by 2000 about 45% of refrigerant market will be met by alternative products in new equipment, 35% by alternatives in existing equipment, and the remainder by CFC conservation – reducing emissions by better maintenance, recycling and reclamation[112].

The current concentrations of the most abundant and greenhouse-active halocarbons are listed in Table 5.3. Even though the Montreal Protocol will be

Table 5.3 Principal CFCs and their current atmospheric characteristics

Halocarbon		*Concentration ppt*	*annual increase %*	*lifetime years*
CCl_3F	(CFC-11)	280	4	65
CCl_2F_2	(CFC-12)	484	4	130
$CClF_3$	(CFC-13)	5	–	400
$C_2Cl_3F_3$	(CFC-113)	60	10	90
$C_2Cl_2F_4$	(CFC-114)	15	–	200
C_2ClF_5	(CFC-115)	5	–	400
CCl_4	carbon tetrachloride	146	1.5	50
$CHClF_2$	(HCFC-22)	122	7	15
CH_3Cl	methyl chloride	600	–	1.5
CH_3CCl_3		158	4	7
$CBrClF_2$	(halon 1211)	1.7	12	25
$CBrF_3$	(halon 1301)	2.0	15	110
CH_3Br	methyl bromide	10.15		1.5

NOTE: CFC numbering Reading from left to right, the first figure represents the number of carbon atoms minus one (this number is left out if it is zero), the second is the number of hydrogen atoms plus one, and the third is the number of fluorine atoms attached. It is assumed that the other attachments are chlorine atoms. (Source: IPCC[20]) renegotiated, perhaps on a virtually on-going basis, because of their longevity, their contribution to ozone depletion and climate change will largely be unaltered during the first third of the next century. The IPCC[20] estimate that the atmospheric concentrations of CFC–11, –12 and –113 will still be some 30–40% of the 1990 value by the end of the twenty-first century.

NITROUS OXIDE

Nitrous oxide (N_2O) is chemically active and produced from a variety of natural sources. The current atmospheric concentration is about 310 ppb which corresponds to an atmospheric reservoir of some 1 500 million tonnes. However ice core bubble analysis shows that the pre-industrial atmospheric concentration was stable for many centuries at close to 285 ppb. Furthermore, ice core data indicates that N_2O concentrations may have been lower during the Little Ice Age. Here, unfortunately, the spread of the data is such that it is difficult to confirm the extent of this dip, though it does suggest that N_2O is a significant contributor to the natural greenhouse effect and, as with CO_2 and CH_4, to natural variations in global climate.

Worldwide, soils are the largest contributor, contributing approximately two thirds of the N_2O released into the atmospheric reservoir. Of this, the majority arises from tropical forest regions, amounting to over twice that from temperate forests. In both tropical and temperate regions N_2O is thought to arise from denitrification in aerobic soils, although nitrification under anaerobic conditions might be another mechanism. It is difficult to make an accurate calculation due to the tremendous variations in conditions within such ecosystems.

Not to be confused with N_2O arising from forest soils, N_2O is known to be released by soils treated with nitrogenous fertiliser. Here, the range of the IPCC estimated contribution is the largest of all the estimates for the various sources they identify[20] – see Table 5.4. Recent work[106] on wheat-growing prairie soils suggests that fertilising cultivated grasslands increases N_2O emissions by a factor of 2–3 and that this might be a significant source of the post-industrial increase in the atmospheric reservoir. Importantly to the greenhouse issue, fertilising cultivated grassland also reduces the soil's ability to remove atmospheric methane.

After soils, the oceans are the next largest source of N_2O, particularly in upwelling areas where deep water rises to the surface – there ocean water has been known to be 40% super-saturated. While it is possible to measure N_2O emissions accurately from various ocean sites, as well as to measure the N_2O content at various depths (which suggests a comparable ocean reservoir to the atmosphere), the exact mechanism governing N_2O generation has not been established.

The burning of fossil fuels was once thought to be one major source of

Table 5.4 Estimated sources and sinks of atmospheric nitrous oxide

	Range (million tonnes N per year)
Source	
Oceans	1.4–2.6
Soils (tropical forests)	2.2–3.7
(temperate forests)	0.7–1.5
Fossil fuel combustion	0.1–0.3
Biomass burning	0.02–0.2
Fertiliser	0.0–2.2
TOTAL	4.4–10.5
Sink	
Removal by soils	0.4?–?
Decomposition in the atmosphere	7.0–13.0
TOTAL	7.4?–?
Observed atmospheric increase	3.0–4.5

atmospheric N_2O. The evidence for this was found to be somewhat erroneous due to experimental artefacts arising from the way the gas was collected. The current view is that only about 2.5% of atmospheric N_2O originates from fossil fuel burning. Similarly, due to the same experimental artefact, biomass burning was also thought to be a significant source. The current view is that biomass burning produces less N_2O than that from fossil fuels.

Against these sources, N_2O is primarily removed from the atmosphere by oxidation in the upper atmosphere: to form nitric oxide (NO); and to form nitrogen (N_2) and oxygen (O_2). NO is involved in ozone destruction but because N_2O has many natural sources, much of its ozone depletion capacity cannot be attributed to human action.

There are other sinks, such as aquatic systems and soils but these are thought to be minor, though no accurate estimation of their magnitude has ever been made.

The IPCC estimate of the atmosphere's N_2O budget (see Table 5.4) is uncertain. It could be that there is an additional key source of atmospheric N_2O so far unaccounted for. One recent suggestion[113] is that the use of nitric acid (HNO_3) as an oxidant in the manufacture of nylon may account for about half of the 0.2 ppb a year unexplained difference in the increase in atmospheric N_2O between the Northern and Southern Hemispheres. N_2O has an atmospheric lifetime of some 150 years and, as commercial nylon production began in 1939, a significant proportion of the current atmospheric N_2O reservoir may be attributable to nylon production. If subsequent research bears this out then there will be a strong case for changing the way nylon is made.

OZONE

Though carbon dioxide, CFCs, methane and nitrous oxide are the principal anthropogenic greenhouse gases, they are not the only ones. Of the remainder, ozone is arguably the most significant but its atmospheric chemistry and that of its precursors presents researchers with many problems.

As we have seen, the question of stratospheric ozone (O_3) depletion is linked to the greenhouse issue in that some of the principal greenhouse gases – CFCs and nitrous oxide – are involved in the destruction of stratospheric ozone. If ozone depletion in the stratosphere is a problem, then at first it may seem odd to consider ozone as a pollutant. Here, the key is to divide atmospheric ozone into two groups: ozone that exists in the high atmosphere (stratospheric), and ozone found close to the ground (tropospheric).

Stratospheric ozone is involved in regulating the climate and absorbs both incoming high and outgoing low energy (frequency) heat. Less stratospheric ozone allows more heat to penetrate the troposphere, on the other hand it traps less heat, being radiated back into space. Conversely, more ozone in the stratosphere prevents heat from reaching the surface but then again traps and part re-radiates heat back to the ground. The solar warming (a function of the total vertical distribution of (O_3) and the low frequency/longwave cooling (a function of the vertical distribution) are similar in magnitude. Whether ozone serves to warm or cool therefore depends not only on how much ozone is present, but where it is found in the atmospheric column. As there is less ozone over the poles – due to the nature of atmospheric chemistry at low temperatures – its effect on climate varies with latitude. Stratospheric ozone depletion around the poles is currently increasing.

Tropospheric ozone is not involved as such in providing an ultra-violet shield and today tropospheric ozone's increased concentration (by a factor of 2–3 on average over pre-industrial levels) is the result of human activity. Unlike stratosphere ozone, tropospheric ozone itself acts as a greenhouse, climate-forcing gas.

Near surface ozone monthly mean concentrations are typically 30–50 ppb. They are highest in the mid-northern latitudes in the spring and summer, and are lower in the tropics. Monthly mean concentrations aside, peaks of 70 ppb in tropospheric ozone have been observed downwind of UK cities, and in the exceptional heat wave of 1976 ozone concentrations reached 212 ppb in London and 258 ppb in Oxfordshire[114]. Since the mid-1970s tropospheric ozone has increased at remote sites (such as Alaska and Hawaii) by 0.8% per annum and between 1% and 2% a year from the mid-1950s to the early 1980s in Europe.

Tropospheric ozone's contribution to global warming is complicated by its complex interactions with the atmospheric chemistry of other greenhouse gases – specifically the latter's interactions with ozone's precursors and, importantly, reactions involving nitrogen oxides (NO_x). High concentrations of NO_x (NO +

NO_2), above about 20 ppt combined with high concentrations of methane, lead to an increase in ozone (hence the aforementioned ozone peaks downwind of cities). Conversely, low concentrations of NO_x combined with increases in methane lead to a decrease in tropospheric ozone. Tropospheric concentrations of NO_x vary greatly from place to place as well as in time, the latter being due to the short NO_x atmospheric lifetime of about one day. Typically NO_x concentrations range from 0.001 ppb in remote areas to as high as 10 ppb in North-eastern USA and Europe, outside of urban areas. Fortunately, ozone has a comparatively short atmospheric life of several weeks so that, unlike long-lived CFCs, changes in policy that affect ozone chemistry will have a rapid affect.

Not only is ozone's atmospheric history extremely complex but there is great uncertainty as to whether it is a major greenhouse gas in the troposphere. The IPCC report[20] lists tropospheric ozone as a possible major contributor and says that, compared to the principal greenhouse gases, ozone's climate changing contribution 'may also be significant, but cannot be quantified at present'. Indeed, not all greenhouse analysts include ozone in their calculations (those that do – such as UNEP's Global Environmental Monitoring System – put tropospheric ozone's contribution at around about 10% of the warming experienced). The second IPCC report in 1992 did not attempt to quantify possible warming from tropospheric ozone. Rather it noted that computer models of how tropospheric ozone might be formed from its chemical precursors differed considerably from each other. Increases in tropospheric ozone from NO_x compounds differed from model to model by as much as a factor of three, while those modelling ozone from methane differed by 50%[93]. The climate forcing effects from ozone remain unclear, other than that ozone is not one of the principal contributors to climate change. Whether or not it plays a significant, albeit minor, role has yet to be determined.

RELATIVE CONTRIBUTIONS OF GREENHOUSE GASES TO WARMING

As we have seen, each of the principal greenhouse gases exists in the atmosphere at a different concentration and – because of their varying sources and sinks – have different atmospheric lifetimes. Greenhouse gases also differ in their ability to absorb and re-radiate heat. Consequently, releasing a unit – be it a molecule, kilogram or whatever – of carbon dioxide will have a different effect on the climate to that of methane, nitrous oxide or a particular CFC. The atmospheric concentration of these gases will not by itself reveal how the climate is likely to be affected by emissions. In order to appraise a greenhouse gas' contribution to climate change – to ascertain the degree of climatc forcing imposed by a greenhouse gas – indices characterising the relative forcing of greenhouse gases have been developed. These include:

(1) **Relative molecular forcing**. This provides a relative indication of the climatic forcing imposed by a gas on a molecule-to-molecule basis compared to a base standard. The base standard is usually taken as carbon dioxide, since its atmospheric concentration is relatively large and, although not the strongest greenhouse gas from a molecular perspective, it is the gas with the greatest anthropogenic effect. The relative molecular forcing values are particularly useful in that they directly relate to the forcing from equal volumes of the gas.

(2) **Relative mass forcing** is similar to the above but is calculated on a kilogram-per-kilogram basis. It is calculated from the relative molecular forcing of a gas using its molecular weight. It is of particular use when contemplating the atmospheric release of 'mass' of gas where the gas is a commercial product in its own right (such as CFCs) as these are marketed by mass.

(3) **Contribution from past, present and likely future changes in gas concentration**. This can be taken on either a relative or an absolute basis. The merit of this measure is that it takes into account the

Table 5.5: The relative molecular forcing, and the relative mass forcing, compared to carbon dioxide see also Appendix 4 for 1995 GWP. (Adapted from the IPCC 1990[20])

Greenhouse gas	*Relative molecular forcing compared to CO_2*	*Relative mass forcing compared to CO_2*
CO_2	1	1
CH_4	21	58
N_2O	206	206
CFC-11	12 400	3970
CFC-12	15 800	5750
CFC-113	15 800	3710
CFC-114	18 300	4710
CFC-115	14 500	4130
HCFC-22	10 700	5440
CCl_4	5 720	1640
CH_3CCl_3	2 730	900
CF_3Br	16 000	4730
Possible CFC substitutes		
HCFC-123	9 940	2860
HCFC-124	10 800	3480
HFC-125	13 400	4920
HFC-134a	9 570	4130
HCFC-141b	7 710	2900
HCFC-142b	10 200	4470
HFC-143a	7 830	4100
HFC-152a	6 590	4390

atmospheric lifetime of the greenhouse gas and its abundance in the atmosphere. Carbon dioxide, for instance, is a comparatively weak greenhouse gas on a molecule by molecule or kilogram basis, but not when one considers that carbon dioxide concentrations over pre-industrial levels are between two and four orders of magnitude above other greenhouse gases such as methane or nitrous oxide. It is important to ascertain the time basis being used which is typically the contribution since pre-industrial times, the contribution over a century, or a 50 year period.

(4) **Global Warming Potential (GWP).** The GWP differs from the above indices in that the above either characterise the individual gas or are based on concentration changes in the atmosphere, whereas the GWP strictly relates to greenhouse gas emissions. Not only does the GWP take into account a gas' climate forcing potential on a molecule basis, calculating GWPs include a time factor – the atmospheric lifetime of the gas – but it can also include the indirect greenhouse effects due to

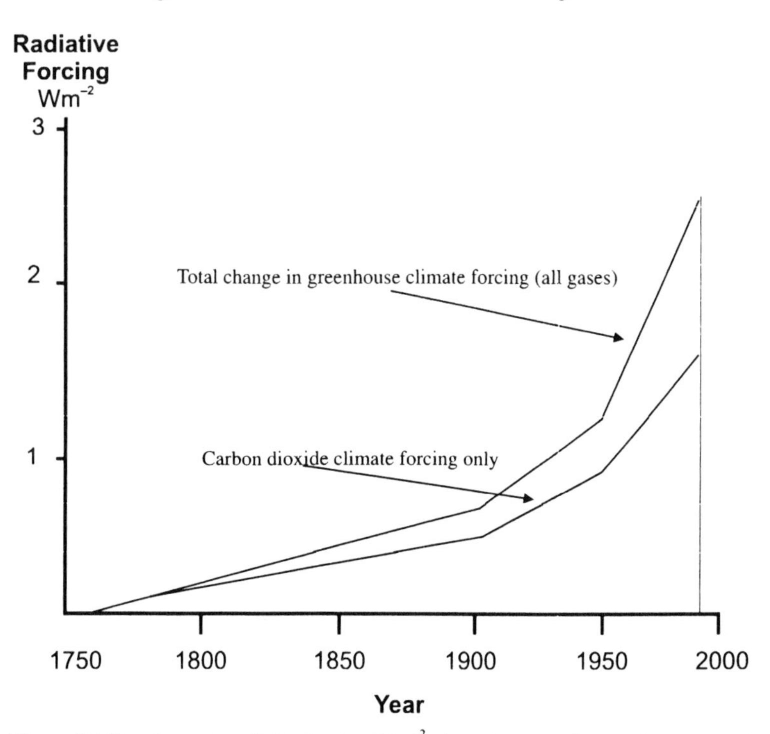

Figure 5.4 The change in radiative forcing (Wm^{-2}) due to increases in greenhouse gases between 1765 and 1990. The values given are changes in climate forcing from 1765 concentrations. The top line represents the total change in forcing from changes in all greenhouse gases. The bottom line represents the changes due to the climate forcing change from atmospheric carbon dioxide. (Diagram simplified from the IPCC)

its atmospheric chemistry and the resulting changes in forcing that may arise.

Algebraically the GWP over 'n' years for a kilogram of a greenhouse gas 'g' released, where 'f_g' is the instantaneous forcing due to the release at the time 't', and 'c_g' is its concentration compared with warming potential resulting from the release of a kilogram of carbon dioxide. It is represented as:

$$\mathrm{GWP} = \frac{\int_o^n f_g c_g \, dt}{\int_o^n f_{CO_2} \, c_{CO_2} \, dt}$$

Though GWPs are arguably the most useful of the above indices, at least in terms of identifying the likely climatic impact from the release of a particular greenhouse gas, there are problems in their calculation. These include:

- The estimation of the various atmospheric lifetimes of greenhouse gases and the likely variations in the future under a different (in part anthropogenic greenhouse-induced) biosphere regimen. In particular, an accurate estimation of carbon dioxide's atmospheric lifetime has yet to be determined, consequently, it must be borne in mind that current GWP calculations use an approximation in determining the reference warming potential.
- The gas' greenhouse forcing of climate is affected by the number of gases and their concentrations which may have overlapping spectral absorption bands.
- It is difficult to determine the true extent of the indirect effects of an additional unit of greenhouse gas. Methane, as discussed earlier is thought to affect the concentration of other greenhouse agents; primarily stratospheric water vapour, tropospheric ozone, and carbon dioxide.
- It is difficult to estimate how much higher concentrations of carbon dioxide absorbed into the oceans will affect the oceans' subsequent ability to absorb more carbon dioxide.
- It is difficult to decide the best time period over which to calculate the GWP integration. A short timescale may be helpful to a policy maker in determining the best policy that will have a more marked effect in the near future, yet this would be less useful for a long-term strategy. Again, methane is 21-times more active a climatic forcing agent than CO_2 on a molecule-for-molecule basis, but is short-lived by comparison. Consequently, its GWP relative to CO_2 is 56 and higher using an integration timescale of two decades but lower with

a GWP of 21 when calculated on a 100 year basis. When calculated for over 500 years, methane's GWP is only 6.5 (see Appendix 4 for 1995 calculated GWP).

Although the exact magnitude of many of these uncertainties is unknown, general estimates have been made. One calculation that takes into account the two major effects from increasing spectral absorption band saturation and ocean absorption of carbon dioxide – each of which can affect the climate forcing from a pulse of additional carbon dioxide by a factor of three – shows that the effects are opposite and largely cancel out even in different carbon dioxide emission scenarios. The researchers conclude that the concept of GWP still remains useful[115].

The one thing that most indices have in common is that the standard they all frequently work from is that of carbon dioxide. There is no denying that there are very good reasons for making carbon dioxide the reference point, but that does not mean that other standards do not have merit. Indeed, because of the uncertainty in the biospheric cycling of carbon and that carbon dioxide's radiative forcing behaviour is different from that of other greenhouse gases, another reference may sometimes be required. Climatic modellers working on computer global circulation models (GCMs) have complained as to the inadequacy of using CO_2 as a proxy in simulating the climatic forcing effect of other greenhouse gases.

When looking at the atmosphere's changing chemical mix, carbon dioxide may not be the best representative or proxy for representing the effect of other greenhouse gases. Researchers at the Atmospheric Science Research Center, Albany, and the National Center for Atmospheric Research, Boulder, USA, have found that their models show carbon dioxide affecting the 'long-wave radiative heating' of the atmosphere in a different way to that of other greenhouse trace gases, both with height and latitude. The warming effect from doubling carbon dioxide is spread more equally with latitude and is in the main beneath 8 km, whereas the warming from the increase in trace gases reaches greater altitudes and is found throughout the atmospheric column. Looking at their model's annual and global mean vertical distributions of the radiative heating rate highlights the differences. Carbon dioxide appears to show peak warming near the Earth's surface, a small amount of cooling at 8–16 km, and considerable cooling above 20 km. By contrast, the trace gases provide two warming peaks of similar magnitude – one near the surface and one near the troposphere/stratosphere boundary with less cooling above 22 km. Additionally, the secondary effects of increased trace gases (such as an ozone) compound the problem of using carbon dioxide as a standard for expressing concentrations of other greenhouse gases[116].

Such difficulties in creating a 'warming index' does not mean that one cannot estimate each greenhouse gas' contribution to warming, only that 'estimates' are all we can come up with at the moment. Nonetheless, there is value in such estimations and no doubt these will improve with time and more work. Recognising the difficulties illustrated above provides a feel for the accuracy and the need to note exactly what, and over which time period, is being examined. Differences in researchers' calculations as to the contribution to warming from the various gases are therefore only to be expected. However, that such estimates do follow a broad pattern does make them of use to those involved in determining policy.

A key paper proposing a GWP index, by Daniel Lashof and Dilip Ahuja[117] does highlight these difficulties and is contrasted here (see Table 5.5) with two other estimates of the contribution from each of the anthropogenic greenhouse gases. To illustrate the variations in GWP estimates and how these effect the contribution of each of the principal greenhouse gases to anthropogenic warming, three contribution estimates are listed in the Table. The latter two were based on GWPs calculated on a 500 year basis for each gas. Again, both were published in 1990, one being from the IPCC[20] report itself, and the other a paper by Hansen, Lacis and Prather[118], both for emissions between 1980 and 1990. The Lashof and Ahuja estimate was calculated differently, it looks at the gases' contribution to climate over a 1000 year period and considers the emissions from only one year – 1985. Furthermore, it attempts to take into account that carbon dioxide's warming effect is non-linear: that an injection into the atmosphere of an additional unit of carbon dioxide does not have the same warming impact as did the injection of the previous unit due to the saturation and overlap of absorption bands with other greenhouse gases.

It has to be said that the Lashof and Ahuja estimate differs more markedly

Table 5.5 Relative contributions of greenhouse emissions to anthropogenic global warming, given as percentages

Greenhouse gas	*IPCC*[20]	*Hansen, Lacis & Prather*[118]	*Lashof & Ahuja*[117]
Carbon dioxide	55	57	71.5
CFCs-11 & 12	17	25	9.5
Other CFCs	7		
Methane	15	12	9.2
Nitrous oxide	6	6	3.1
Carbon monoxide	–	–	6.6

The first two estimates of the relative contributions of anthropogenic greenhouse emissions to global warming were from 1980–1990 calculated from global warming potentials integrated over 500 years. The climatic forcing contribution from ozone is not shown separately. The Lashof and Ahuja estimate is of the relative contribution to global warming from emissions in 1985 only, and it is based on a substantial revision of carbon dioxide's climatic forcing including integrating the forcing effect over a 1000 year period from many of the other estimates which the IPCC's 1990 report include. It also has to be remembered that the 'accuracy' of their calculation (not to mention those of others) is based on incomplete knowledge of carbon cycling: such estimates will no doubt change. Even so, with current calculations, the absolute ranking of greenhouse gases remains the same and this is unlikely to change as such estimations are refined. Furthermore, no matter which estimate turns out to be the most accurate the dominance, hence importance, of carbon dioxide remains.

RECONCILING UNCERTAINTY

No apology is needed for introducing uncertainty into any discussion about the future, for the future is inherently uncertain and to ignore that debases any forecast's value. That this, and the previous, chapter has been a warts-and-all perspective of current thinking (albeit a summary) as to the extent and the causal factors governing anthropogenic global warming means that you the reader have been presented with a spectrum of views; although common themes are present. Scientists' views are not 100% identical. This is because scientists cannot see into, and obtain data from, the future. If they claimed that they could then others, such as policy-makers, should be really sceptical of their predictions! Yet, and with special regard to policy-makers, it would be easy to use uncertainty as a reason for a *laissez-faire* approach to human-induced climate change. Equally, it would be as easy to employ uncertainty as a mainstay for those wishing to maintain the *status quo*: there is no doubt that those who currently benefit from a business-as-usual scenario are tempted to do just that. However, we could just as easily turn uncertainty on its head and talk about certainty. After all the one thing we do know for certain is that there is a future whatever that may be, we also know how far away any particular future year is, and we know about the past, not to mention the present – the starting point for any forecast.

The discussion on the future extent and causes of global warming are as - certain (or uncertain) as the future of many day-to-day occurrences. To illustrate this I predict with the *utmost certainty* that I will need to eat after next week if I am to live next year. (Of course I could be killed, run over by a car tomorrow, but then that makes any discussion as to my future health redundant in the same way that the sudden outbreak of, for example, global thermonuclear war, would make greenhouse discussion superfluous.) I also predict with a *fair degree of certainty* that I will have supper a week from today. Of course I may forgo supper because my neighbours could suddenly surprise me with a late afternoon high tea, but though they may do me some kindnesses next week this particular unannounced kindness is unlikely. On the other hand if I were to predict the exact menu for my supper next week than I *would* be taking a gamble, for here certainty ceases to dominate and uncertainty clouds the issue. By now the parallels with the greenhouse debate will be apparent.

To summarise:

- The scientific community *is* certain that human emissions of greenhouse gases are increasing the atmospheric concentration, and that this will enhance the greenhouse effect leading to global warming. The scientific community calculates 'with confidence' that different greenhouse gases have different potency, with carbon dioxide dominating.
- The scientific community is less certain as to the exact breakdown and nature of the sources (and sinks) of the anthropogenic greenhouse emissions, and in calculating the likely climate change resulting with some precision. Nonetheless, the certainty remains in that Mankind is making the Earth warmer than it would otherwise be.

This leaves us with the question as to where we are currently going? Here the IPCC 1990 business-as-usual scenario[20] points the way (Figure 5.5). Throughout the remainder of this text we will refer back to the IPCC 1990 business-as-usual scenario, which is the scenario used by many of the other works cited here. (The IPCC's 1992 follow-up report gives analogous scenarios which gradually differ with time, but only forecast 0.5°C less warming by the year 2100 from that of 1990: just over 2.5°C compared to just over 3.0°C warming. The 1995 report (published in 1996) uses the 1992 scenarios but allows for a minor cooling effect due to short-term sulphur pollution aerosols in some parts of the World.) Furthermore, the 'best-estimate' for the forecast temperature rise is within the 'upper' and 'lower' estimates predicted in 1990.

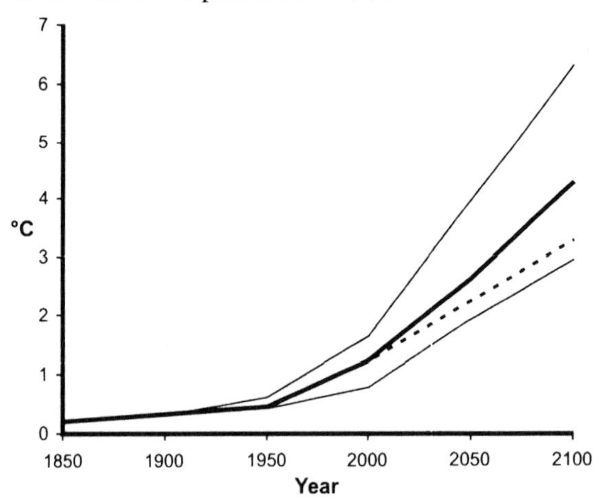

Figure 5.5 IPCC 1990 business-as usual (B-a-U) scenario simulating the anticipated increase in global mean temperature with best estimate (bold) and upper and lower estimates (light lines) showing the temperature rise above the 1765 value. (The IPCC subsequently revised the way it presented its scenarios and its forecasts in 1992 and adhered to these for its 1995 review (published in 1996 albeit with a minor modification to allow for some regional short-term cooling from sulphur aerosols. The IPCC 1992 scenarios a and b and are portrayed approximately by the dotted line)

CHAPTER 6

BUSINESS-AS-USUAL

As the World warms what will be in store for us if we simply carry on the way we are; that is if we carry on with 'business-as-usual'? Examining our prospects in this light will give us a standard, albeit a hypothetical one, against which to compare other scenarios subject to different policy regimens. Fortunately for us, authoritative forecasts of basic indicators are available from organisations such as the United Nations: forecasts that are regularly used by governments, policy-makers and industry. The two basic indicators of greatest use in discussing greenhouse issues are those of population and energy consumption. Population is used as an indicator because it is people who create the demand for, and consume, energy. Energy, because carbon dioxide emission from the energy supply industries dominate the current growth in carbon dioxide atmospheric concentration, which dominates anthropogenic warming. Clearly the two are inter-related but to begin with it is worth considering each in turn.

POPULATION

The Reverend Thomas Robert Malthus, writing in eighteenth century England, noted that the human population tended to grow exponentially – 2, 4, 8, 16, 32 . . . , rather than additively – 2, 4, 6, 8, 10 . . . [13] (Figures 6.1a and 6.1b). Indeed, in the animal world there is frequently an incremental rate of recruitment to the breeding population, that is, small breeding populations tend to grow exponentially. Whether the overall population grows or declines depends on whether or not the overall recruitment into the breeding proportion of the population enables the population's birth rate to exceed its death rate. Generally, especially for unpreyed-upon populations which have not approached the carrying capacity of their environment, the population will begin by growing exponentially. Human population growth has for many years followed a similar pattern. In its purest form this growth can be expressed mathematically, with the equation:

$$N_t = N_0 e^{rt}$$

where:

N_t = the population at time t
N_0 = the population at time zero
e = the base of the natural logarithm – 2.71828 . . .

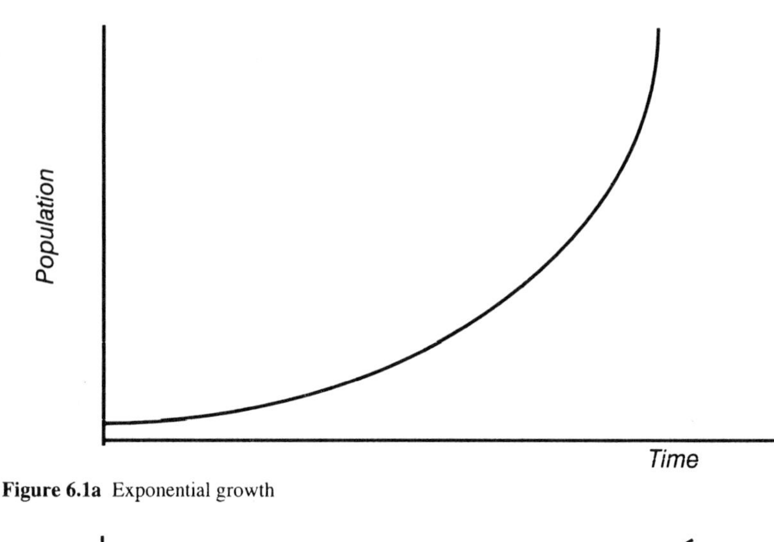

Figure 6.1a Exponential growth

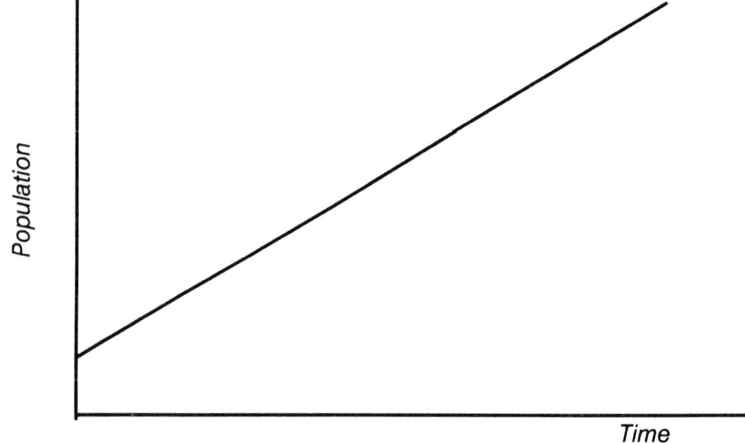

Figure 6.1b Additive growth

r = the rate of population increase
t = the time elapsed since time zero

In the real world, populations do not continue to grow exponentially indefinitely, for if they did the World would be swamped with large populations of living creatures. The oft-quoted example is that of a bacterium dividing into two every 20 minutes. If this continued unabated then it would result in the production of a colony about a foot deep over the surface of the planet in just 15 days; an hour later and the colony would be getting on for 8ft deep. Fortunately, the Earth has never seen such a bacterial colony because, as populations increase in size their food supply decreases; large numbers may attract predators and make the larger population an easier target, hence more predators can be supported;

and, with increased numbers, the population's waste products will begin to affect the very environment that sustains it, and large numbers living close together are more disease-prone. Since these limiting factors are related to the poulation's size, then as a population grows then so does the effect of such controls. Ultimately, the limiting effect of the controls equals that of the ability of the population to grow, so the population stabilizes. This type of growth is 'logistic' and can be expressed thus:

$$N_t = \frac{K}{1 + e^{rt}}$$

where K = the carrying capacity

In the real world these mathematical descriptions of population growth are ideal constructs and few populations follow such growth patterns exactly. Even so, elements of these mathematical constructs can be seen within more complex population changes. Frequently populations continue the exponential growth phase of the logistic pattern so that the population exceeds the environmental carrying capacity, with the extra population eroding the ability of the environment to sustain it. The population will then suddenly crash to beneath the theoretical carrying capacity so relieving its pressure on the environment and allowing environmental regeneration to take place. For example, a population of sheep might grow in an exponential manner exceeding the environmental carrying capacity with the sheep over-grazing the land, close cropping or trampling the grass so that the root systems are destroyed. With less grass the population plummets which in turn lowers the level of grazing allowing the grass to grow back.

However, the theoretical logistic pattern is reflected in actual human population growth, though the rate of population increase (r) was initially low. Some

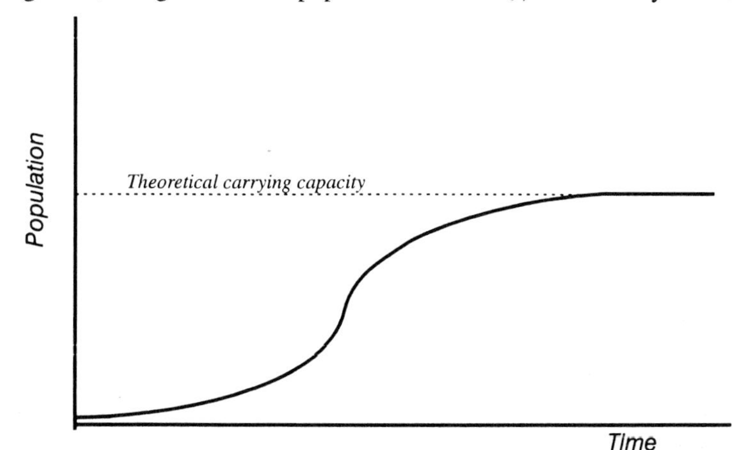

Figure 6.2 An ideal 'logistic' growth pattern. Growth begins in an exponential-like way then tails off so that the population stabilises

3000 years ago the World population was thought to have been around 0.05 billion. From there the population grew slowly to between 0.15 and 0.2 billion about the time of Christ. By the time the Normans had invaded England in the eleventh century the world population was approaching 0.3 billion. It almost doubled again over the next 600 years to the start of the industrial revolution when the exponential pattern dominated, and the population doubled again by 1850 to 1.2 billion. By 1950 the World population had reached 2.5 billion and had doubled again to 5 billion by 1988[119]. In 1990 the World population had reached 5.3 billion, and was growing at a rate of a quarter of a million people a day. Today (1997) it is nearly 6 billion.

The latter half of the twentieth century has seen the beginning of a drift away from a purely exponential pattern of growth to a more logistic one. Developed nations have seen their populations begin to stabilise: population growth in the UK has virtually halted. However, whereas in 1986 the population of the developed industrial nations was 742 million, that of the rest of the World was some five times greater. Between 1990 and the year 2000, according to The World Bank, the industrial nations will have an annual growth rate of only 0.4% though the World's as a whole will be about 1.7%[4]. One consequence of this is that while estimates of the total World population by the year 2025 will be around the 8.5 billion mark, only about a fifth will be in developed countries compared to one third in 1950[120]. Between 2025 and 2100 the World population growth will continue but the rate of growth will decrease so that the World total will probably lie between 10 to 14 billion. After this, at the end of the twenty-first century and beyond, much will depend on whether we will have overshot our carrying capacity: this will determine whether our population stabilises or partially crashes.

The early twenty-first century will, therefore, be a time of continued population growth, mainly in the less-developed countries. But exactly where will this population live? The last quarter of the twentieth century has seen human populations migrate into some previously sparsely inhabited regions: the forests of South America being one such area. Yet the principal growth area worldwide results from population movement from the countryside to urban centres which augments the indigenous growth already underway in these areas. This urban expansion has turned the late twentieth century into the age of the megacities – cities with a population exceeding 5 million. In 1950 Mexico City was the only city in the developing world to qualify for that name. Twenty years later ten more cities joined the megacity league. Ten years later still, by 1990, the number of megacities outside of the industrialised developed nations had topped twenty[121]. Indeed megacities were becoming so numerous that by 1991 the World Health Organization was unofficially, but arguably more appropriately, redefining the term by beginning to refer to megacities as those cities whose population exceed 10 million[122].

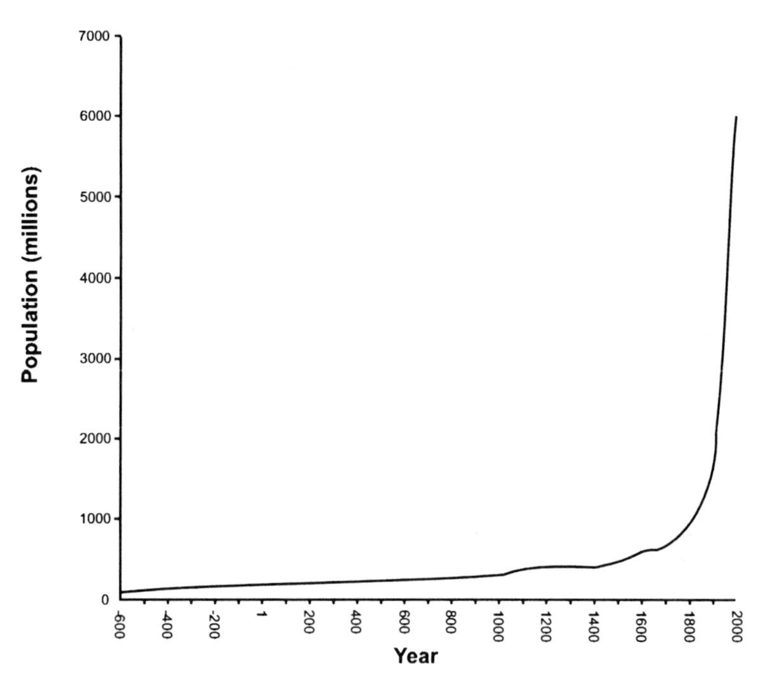

Figure 6.3a World population (millions) over the past two and a half thousand years

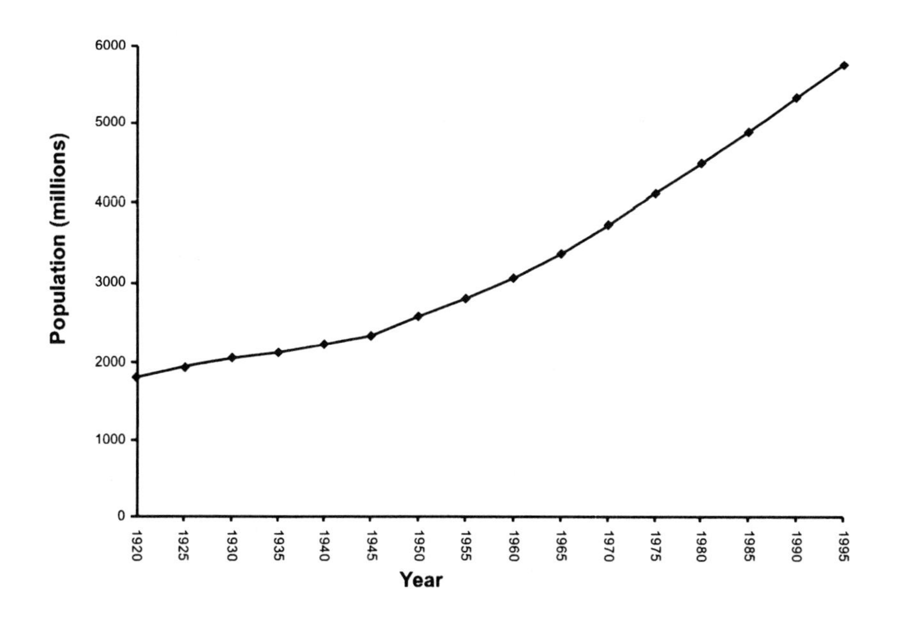

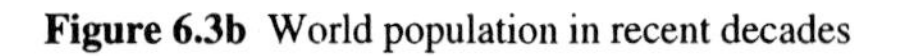

Figure 6.3b World population in recent decades

Table 6.1 Megacities in 1990[121]

No	*City*	*Country*	*Population*
1	Tokyo-Yokohama	Japan	28 700 000
2	Mexico City	Mexico	19 400 000
3	New York	USA	17 400 000
4	Sao Paulo	Brazil	17 200 000
5	Osaka-Kobe-Kyoto	Japan	16 800 000
6	Seoul	Korea	15 800 000
7	Moscow	Russia	13 200 000
8	Bombay	India	12 900 000
9	Calcutta	India	12 800 000
10	Buenos Aires-La Plata	Argentina	12 400 000
11	Los Angeles	USA	11 500 000
12	London	England	11 025 000
13	Cairo	Egypt	11 000 000
14	Rio De Janeiro	Brazil	10 975 000
15	Paris	France	10 000 000
16	Jakarta	Indonesia	9 900 000
17	New Delhi	India	9 800 000
18	Manila	Philippines	9 200 000
19	Shanghai	China	9 185 000
20	Tehran	Iran	8 100 000
21	Chicago	USA	7 900 000
22	Karachi	Pakistan	7 300 000
23	Beijing	China	7 040 000
24	Bangkok	Thailand	7 000 000
25	Istanbul	Turkey	6 500 000
26	Taipei	Taiwan	6 100 000
27	St. Petersberg	Russia	5 900 000
28	Tianjin	China	5 625 000
29	Madras	India	5 600 000
30	Lima	Peru	5 400 000
31	Philadelphia-Trenton-Winnepeg	USA	5 390 000
32	San Francisco-Oakland-San Jose	USA	5 225 000
33	Hong Kong	China/UK	5 175 000

To put the cities into the context of the total World population, in 1978 around 30% of the World's population lived in cities. By the year 2000 this proportion will in all likelihood be more than 50%, and possibly reaching two thirds (66%) by about 2025. Furthermore, population growth in the megacities of the developing world is some four times greater than in the corresponding rural areas. By the year 2000 it is thought that there will be some 60 megacities (with populations exceeding 5 million), of these 45 will be in the developing world.

One potential consequence of the World's population being concentrated in urban areas, and in megacities, is that it is easier to supply such a concentrated population with electricity and other energy imports than to distribute these resources to a widespread rural community. The potential demand for energy per capita worldwide is likely to increase with population, and so in turn is the anthropogenic contribution of atmospheric greenhouse gases. Factors that may modify this include those affecting the population's ability to grow – it will make a difference to climate change policies if we are heading for a World population by the year 2100 of 14 billion as opposed to 10 billion. Again, the degree of climate change and the nature of greenhouse energy policy will be determined by the ability of the population to create the wealth to spend on consuming greenhouse-gas-generating energy resources.

That the growth in World population may decrease during the twenty-first century is to be welcomed but at what cost? Part of this reduction in population growth can be accounted for by the lower birth rate of wealthier populations, but the significant factor is the increased death rate due to the lower health standards to be found among the poor. In 1990 the World Health Organization (WHO) announced that for the least-developed countries 'all the socio-economic indicators, and in particular the health indicators, are in the red'[123]. Life expectancy is under 50 years as against 74 years in the developed countries; 150 children out of every 1000 die in the first year of life compared to 15 in industrialized countries. In 1990 the WHO noted that the annual death toll from the 8 million cases per annum of the most widespread infectious disease, tuberculosis, was increasing and was then estimated at 2 907 000 with over 98.5% of the fatalities taking place in the non-industrialised countries[124]. With human population densities increasing, not to mention the migration into disease-harboring regions, the Malthusian check of disease can only become more pronounced. Non-fatal disease also has its impact, a population increasingly debilitated by disease will find it harder to generate the wealth to combat illness, or to implement a greenhouse policy. In addition to turberculosis, the early 1990s saw an increase in cases of malaria and cholera. In 1991 the incidence of malaria in Africa alone reached 100 million clinical cases with 5.2 million outside of Africa. Worldwide some two billion people are exposed to malaria and the malarial parasite *Plasmodium falciparum* has evolved resistance to almost all drugs in use and varying degrees of resistance to chloroquinine, by far the cheapest and previously the most useful anti-malarial drug, has been reported in virtually all the countries where the malaria parasite lives. WHO accepts that malaria is so established that it cannot be controlled by major campaigns.[125]. Other diseases are spreading. Up to 1991 cholera was virtually unknown in the Americas but the seventh World pandemic, which began in 1961, of that disease has seen it spread from previous areas of dominance in Africa and Asia. During 1990 some 1.5 million African children under the age of 5 years died from cholera. In 1991 WHO's Director-General, Hiroshi Nakajima, pointed to poverty as the principal barrier to health: 'What must be faced, however is the reality of increasing poverty and widespread underdevelopment around the World ... Cholera is but one dramatic symptom of the failure of development'[126].

Perhaps the largest single *new* health factor to affect the size of the World population and its growth during the twenty-first century is AIDS (Acquired Immune Deficiency Syndrome). This does not mean that diseases that have been with us for millennia (such as cholera) are not important, but new factors can affect us in new ways. AIDS typifies this, though none but the callous could seriously advocate disease as a method, albeit a tacit one, of population control. AIDS affects human population growth in two ways: directly and indirectly, primarily through changes in sexual behaviour.

Clinically, AIDS was unheard of prior to 1980, though cases of HIV (Human Immunodeficiency Virus) infection – the virus that causes AIDS – were subsequently discovered to have occurred prior to that date. Early in the 1980s it quickly became apparent that AIDS was a terminal condition with no known cure. Consequently the effect of the disease on any population can only be to lower it, not only by increasing the death rate but – because it chiefly alters the death rate of the reproducing cohorts – also through lowering the birth rate. This arises because AIDS is transmitted through bodily fluids from an infected person to the uninfected, primarily through sexual contact. Though the AIDS pandemic began in the USA and Western Europe within the homosexual and drug-injecting communities, AIDS worldwide is primarily a heterosexual disease and the incidence of heterosexual HIV infection in the USA and Europe is on the increase. WHO estimated that at the beginning of 1991 there were over 1.5 million cases of AIDS worldwide and that between 8–10 million HIV infections may have occurred – or 0.17% of the World's population. Of these, it is estimated that some 1.5 million adult HIV infections and a quarter of a million AIDS cases occur in Australasia, North America and Western Europe[127]. By 1994 WHO's 1991 figure of 1.5 million AIDS cases worldwide had more than doubled to four million, and the estimated number of HIV infections had risen to more than 17 million – very roughly one third of one percent of the World population[128]. WHO estimate that this will double by the year 2000.

The problem with such estimates is that they rely on incomplete data, and for every case diagnosed there are many others undiagnosed, especially in less-developed countries with little co-ordinated state-wide health care. No country has yet screened its entire population so that it is likely that many of those infected go uncounted, but there are ways of estimating these. Test screens carried out at antenatal clinics in London in 1990 revealed that one in 500 purely heterosexual women in that part of the city were infected with HIV[129]. Furthermore, from this country's health service there is evidence of HIV infection under-reporting by doctors, suggesting that current estimates as to the level of infection are, if anything, low[130]. That AIDS is one of the main wild cards in future World population estimates is exemplified by its effect on the previously established population growth of the countries in which it is most prevalent. In 1991 it was reported that the effect on Uganda's population growth by AIDS –

or 'slim' as it is locally known – was to lower the country's forecasted 2025 level of 37 million to 22 million[131]. In terms of numbers of energy consumers alone, this will have some impact on anthropogenic emissions but much depends on whether trends continue over the next two decades as they have throughout the 1980s.

This 'much depends on whether trends continue' is at the crux of the uncertainty as to AIDS' likely impact on greenhouse gas emissions. However, the impact, even should it be great, will be spread over time, and manifest itself more in the context of a different pattern of greenhouse emissions rather than any sudden change or discontinuity. With population issues such as these there are many other wild cards, and even further demographic factors relating to AIDS itself. For instance, culturally over the longer term, AIDS will inevitably alter sexual behaviour and this in turn may alter birth rates.

It is not just the increased mortality that will lessen fossil fuel consumption but also how those who remain unaffected react – arguably this could have a greater impact! Those more conscious of 'safer sex' will be less likely to contract the virus, but one spin-off will be less unwanted pregnancies and a lower birth rate through greater access to, and awareness of, safer sex contraception. (Again with people more aware of long-term consequences they may be receptive to environmental policies?) This safer sex message applies to the industrialised developed countries equally as to less-developed nations and here bear in mind the energy consumption per capita compared to less-developed nations. Small changes in developed nation's population in energy terms equate with far larger changes than in developing nations.

Although the growth in World population continues there are the beginnings of some welcome trends even if far from the prospect of real population stabilization. Never before has contraception been so widely available. At the start of the 1990s just under two thirds of the population in developing countries had access to one modern method of contraception at a cost of less than 1% of their wages. Countries such as Uganda do have access to the basic resources required to enable its sexually active population to practice safer sex – but not the ancillary education and associated health infrastructure to get the message across the psycho-cultural barriers.

The good news about HIV infection is that nobody needs to get it; the bad news that nobody knows who has got it. It is the latter that prevents greater accuracy in the forecasting of how the World's population growth will logistically move from an exponential to a more stable mode.

It is these recent health factors, such as AIDS, that impact on the population in new, so by definition quantitatively unknown, ways that lead to greater uncertainty with regard to climate change. Other examples of newly-emerging infections include: *Escherichia Coli* O157:H7, cryptosporidosis, multiple-drug-resistant *Pneumococcus*, and vanomycin-resistant *Enterococcus*. Combine the population effect of these new pathogens with the sporadic but increasing number of outbreaks of traditional diseases and strains thereof (itself due to the increased demographic opportunity for spread in a more crowded world) and the uncertainty effect on population forecasts becomes difficult to ignore. Climate and World development issues such as population are intrinsically bound together. It was for this reason that both matters were addressed in the 1992 UN Conference on Environment and Development, though such are the vagaries that we will never be able to reconcile these concerns through perfect forecasts of the future.

Yet, as we shall see, while complicating the greenhouse issue with uncertainty, such factors could well be irrelevant to the broader question of whether under the business-as-usual scenario fundamental changes to energy policies worldwide need to be adopted (for all reasons including those other than climate). In terms of energy policy priority it does not matter whether there will be 10 billion or 14 billion people consuming energy by the year 2100. It is not just that 14 billion people will tend to consume energy 40% faster than 10 billion, rather how long finite energy resources can sustain development. More appropriately, we should be asking what are our options for modifying greenhouse emissions, establishing the alternatives, and what do we really require and want? Finally, to find out whether the business-as-usual scenario is really tenable, we need to know what are the limits to our current growth in finite fossil resource (not economic) terms? Even if in many ways we decide for perfectly legitimate reasons to continue business-as-usual, it would seem that as it stands in the context of population and disease our current business-as-usual strategy is not optimal and requires at least some modification.

ENERGY

If, in the late twentieth century it seems that the prospect of a twenty-first century global transition to a more stable level of population remains unclear, the 10 to 14 billion estimates for 2100 population at least provides a tentative upper and lower boundary. But population alone will not ultimately determine energy consumption. There are other ways to estimate likely future energy demand, and supply. And, of course, we can increase the reliability of energy forecasts by forecasting closer to the present than 2100.

So what of the year 2010? By then not only will the human population be greater but the World's car population will have grown too, indeed it is forecasted to have doubled from its early-1990s size. By then today's (non-OECD) developing nations will also be consuming more food, clothing, housing, transport, lighting, heating, etc. Collectively they will be consuming more resources than the developed industrialised (OECD) nations, though not on a per capita basis. What will this do to the World's energy demand if we go down the business-as-usual route? Here, virtually by definition, one of the best answers comes from asking business or industry itself: after all industry is accustomed to making medium-term forecasts a couple of decades into the future. In relation to greenhouse issues the energy supply and power industries are the most germane.

Asea Brown Boveri (ABB) is the World's largest power engineering and mass transport group with an annual (1989/90) turnover of some $25 billion. By the end of the 1980s its chief executive and president, Percy Barnevik, was already thinking how to supply the extra three billion people alive after 1987 that will contribute to the estimated World population of eight billion by the year 2025. He recognised the relevance to his business of the 'embryonic' new megacities. When people move into cities from underdeveloped rural areas energy demand goes up by about an order of magnitude. He also foresaw an associated pollution problem, and not just of greenhouse gases. But where Barnevik saw global problems he also saw business opportunities. His organisation, or some 40% of it, is affected by environmental issues and he is placing more emphasis on developing systems that are cleaner and more efficient[32].

That energy consumption is related to standard of living has long been recognised. By the end of the 80s USA primary energy consumption was just under two billion tonnes of oil equivalent (Btoe) and its GDP was a little over $700 billion. (Primary energy being the energy from the raw fuel consumed, some of which goes to make secondary energy such as electricity from coal -electricity from nuclear power and hydroelectricity counts as primary energy.) The UK primary energy consumption was a tenth of this at 0.2039 Btoe and it had a GDP of $88 billion[4, 1]. This gives an energy consumption of GDP ratio of 2.8×10^{-3} and 2.3×10^{-3} \$ Btoe^{-1} for these countries, respectively. These figures are quite close and, surprisingly, the ratio is broadly true for the less-developed nations who consume less energy but equally produce less wealth.

Instead of looking at (cash) productivity per unit of energy consumed, we could consider energy consumption per unit productivity. Here, we could see how a nations' energy consumption relates to its Gross National Product (GNP), which includes a nation's overseas earnings from investment and the 'invisible' earnings such as from the money brought in by tourism. Here too there is a broad consistency between the nation's energy consumption and its wealth in terms of GNP. In 1984 out of 22 former non-Iron Curtain European countries over three quarters – some 17 nations – consumed between 10 and 30 megajoules of energy per $US of GNP (standardised to the 1975 $ by The World Bank). Twenty-one African countries out of 49 fell within this energy/wealth band. No country consumed less energy per dollar GNP than Chad (3426 kilojoules per dollar) and no more than Egypt (35983 kilojoules per dollar) with the single exception of South Africa (63014 kilojoules per dollar)[133].

All of this ties in with how much energy the average person in each country consumes. Grouping together countries geographically and economically serves to emphasise the connection between wealth or economic activity and energy consumption. By the end of the 1980s the average person:

- in the USA was consuming 8 tonnes of oil equivalent a year (toe/yr);
- in the USSR and Eastern Europe 4.4 toe/yr;
- in Western Europe 3.2 toe/yr;
- in Japan 3.4 toe/yr;
- and in the less-developed nations a person consumed 0.6 toe/yr.

The exception to the rule being the USSR and Eastern Europe whose high energy consumption is not so much due to their economic wealth (if anything the contrary) rather the ready availability of indigenous energy resources. In 1989 these amounted to 42.5% of the World's proven reserves of natural gas, 30.4% of coal, and 5.9% oil.

For many years those policy-makers concerned with energy in developed countries held firm to the above tenet that energy consumption was directly related to economic performance. Indeed, not only was there the international evidence of nations with higher GDPs tending to consume more energy than those with lower, but there was the supporting documentation within many countries' own history. As these countries grew and their GDP rose then so did their energy consumption (see Figure 6.4).

Combine the above broad relationship between energy consumption and wealth with the detailed knowledge of a country's past energy performance and its economic history, and it is possible to begin to forecast its energy future. Amalgamating such national forecasts worldwide provides an estimate of the likely global energy demand for a number of years, not to mention global energy flows. This highlights the current pattern of national inequalities in that the World's most developed countries, those in the OECD (Organisation for Economic Co-operation and Development), representing about 15% of the World's population, together consume half the world's energy. However, of critical importance is the USA which, with just under 5% of the World's population, consumes nearly a quarter of its energy! In other words, while in the mid-1990s the average Western European (European OECD citizen) consumed a little over 3 mtoe a year, compared to an average for all world citizens of 1.4 mtoe, the average USA citizen consumed over 7 mtoe! If (as the less-developed countries wish and are beginning to do) the rest of the World develops in an attempt to attain a 'Western' standard of living then, subject to availability, the World's energy consumption would increase some six-or seven-fold. But availability is of course the constraining factor. There simply are not the fossil fuel resources in existence to consume energy at six or seven times the rate we currently do for more than a decade or so.

In 1980, worldwide some 6729 million tonnes of oil-equivalent of primary

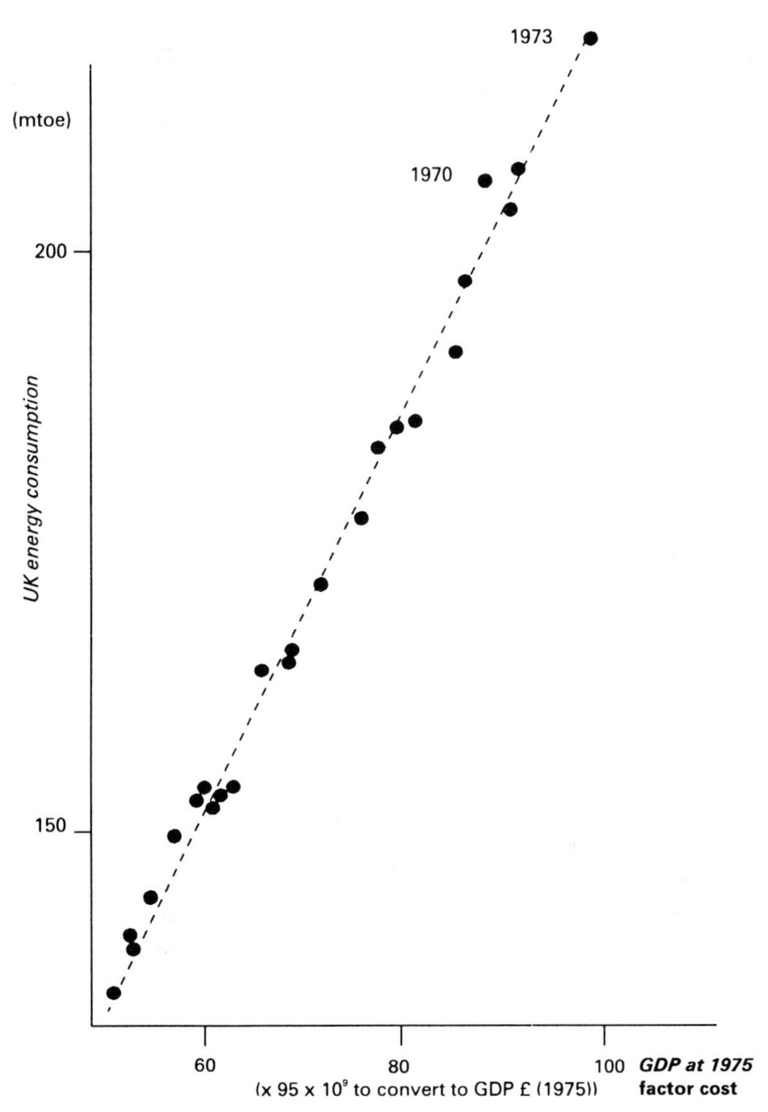

Figure 6.4 Up to the oil crisis of 1973, the energy policies of many developed nations were governed by the belief that energy consumption was directly linked to a nation's economic activity. It is not difficult to see why when looking at a plot of UK energy consumption against its economic Gross Domestic Product at 1975 = 100 factor cost. There appears to be a near straight-line relationship

energy was consumed. By 1990 this had risen by over 19% to 8033 million tonnes[134] and stayed within 4% of that figure for the next four years[135]. The total energy consumption was larger – this figure only includes commercially traded fuels, and therefore does not include fuels such as wood, peat and animal wastes which are important sources of energy in some parts of the less developed world. Of the 1990 primary energy demand, some 87% was met by the consumption of the finite, or fund, fossil fuel energy resources. Geological surveys provide industry and governments with a working estimate as to how much of these resources are left. Of course there are undoubtedly reserves yet to be discovered, such has been the extensive nature of the energy industries' exploration the yet-to-be-discovered resources are unlikely to be greater than the now-proven reserves. For example, in 1991 BP's new Colombian oil and gas find of 2–3 billion barrels of oil equivalent[136] (then a tenth of that company's reserves) amounted to less than 5% of the World's primary energy demand for that single year. There will probably be other finds, and some may be as big, or bigger than the 1991 BP find, but it is unlikely that there remains very much more than is currently known.

So, assuming on one hand that the present-known fossil fuel reserves are indeed a realistic working estimate of about half of all that remains, and on the other that demand is likely to increase, then resource lifetimes can be crudely calculated. Of course, doubling known reserves is merely increasing them by an arbitrary proportion. Industry is not concerned with idle speculation, so it calculates its estimated remaining lifetime of resources by simply dividing the proven reserves by rate of consumption (see Table 6.2). These 'lifetimes' are known as the 'resource to production' ratio. These are quite useful, though of course they are not true representations as the ratio calculation does not include the other vaguaries that could alter. (In the real world fossil fuel demand does vary and will in the medium term increase, but then again in terms of resource lifetimes this will be in part off-set by newly discovered reserves, new extracting technology and different economic considerations. If the price of the resource goes up then more can be spent on the extraction of difficult-to-get-at reserves.) Current indications of fossil fuel resource lifetimes as provided by reserve-to-production ratios are listed in Table 6.2. However it must be remembered that these R/P ratios are a World average and the R/P ratios for the developed (OECD) countries are lower: these nations will be hard pressed to supply their industries with indigenous oil and gas at the 1990 rate after the year 2010 and will increasingly be relying on non-OECD supplies.

Table 6.2 Current (1994) global annual demand, proven reserves, and steady-consumption lifetime or reserve to production ratio (R/P)

Fuel	*Annual demand*	*Proven reserves*	*R/P*
	(billion tonnes oil equivalent)		*(decades)*
Oil	3.2	136	4
Natural gas	1.9	102	6
Coal	2.2	572	23*

* The calculation of the R/P for coal includes a consideration for the varying energy content of different types of coal

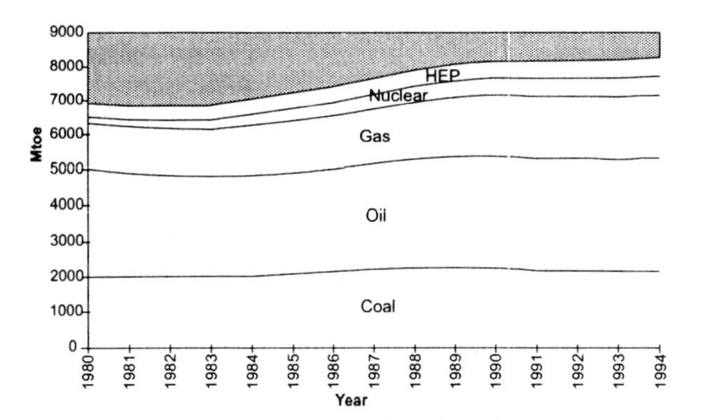

Figure 6.5a World primary energy consumption by fuel 1980–1994

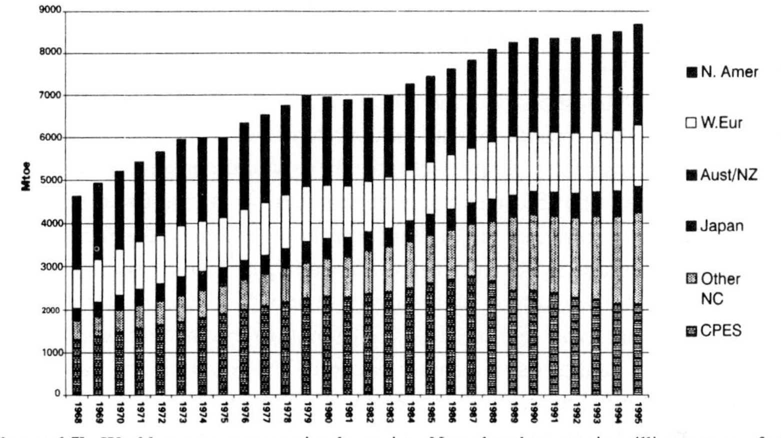

Figure 6.5b World energy consumption by region. Note that these are in million tonnes of oil equivalent for potential fossil fuel dependency comparison purposes. To determine potential fossil fuel dependency all the energy values for the various geographic regions are converted into million tonnes of oil equivalent (mtoe). The energy output from HEP (hydroelectric power) stations and nuclear power stations assumes that its oil-fired equivalent would be operating at a thermal efficiency of about 33%

To put the above R/P figures into context it is worth remembering that the R/P ratio does not mean that oil will suddenly run out some four decades after the mid-1990s. What they do indicate is that oil will not be extracted at the same real-term price as in the mid-1990s, that by that time we will be extracting the harder-to-get-at reserves and paying far more in real terms for a barrel of crude. The knock-on from this will be felt in the way we use a barrel of crude in the year 2030, in that other energy resources will become more competitive, that energy over all will probably become a more expensive commodity with all the industrial and political implications that entails. These changes will not suddenly appear in the year 2030, but will be gradual. Meanwhile, despite being a finite resource, dependence on fossil fuels continues.

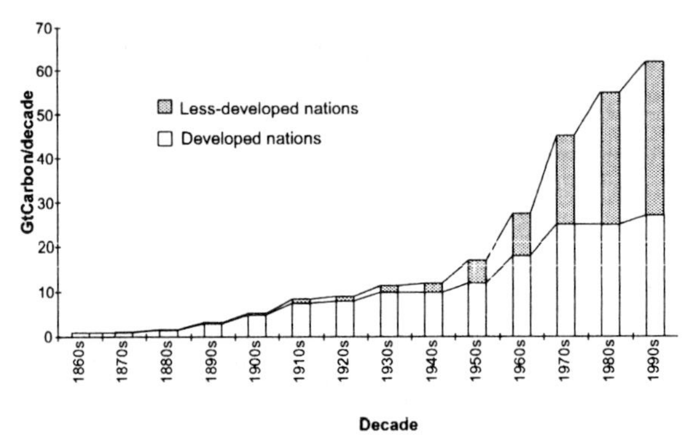

Figure 6.5c Human generation of carbon dioxide through fossil fuel burning by decade broken down into developed (OECD) nations and less-developed nations. The estimate for the 1990s is based upon figures for the first half of the decade with 1.5% growth *per* annum assumed for the second half of the decade

Whereas in 1990 some 87% of the World's primary energy consumption came from the finite, fund resource fossil fuels, our dependence has decreased from a 1980 figure of over 92%. Conversely, the 1990 world consumption of fossil fuels exceeds that of 1980, such has been the increase in total (both fossil and non-fossil) energy consumption. In short, whilst proportionally the World is less reliant on fossil fuels at the end of the 1980s compared to the beginning, this independence has been offset by the overall increase in energy demand (see Figure 6.5). While over this period oil consumption rose by less than 5%, the consumption of natural gas increased by about 35%: both oil and natural gas have low R/P ratios, hence expected resource lifetimes. The 1990s is therefore seeing a greater anthropogenic contribution to the atmosphere's reservoir of carbon dioxide than any other decade in human and geological history. This greenhouse contribution will in all probability increase at the very least in the short term up to the early twenty-first century.

BUSINESS-AS-USUAL CONSEQUENCES

Based on present population and energy trends there is little doubt that the human race is heading for a demographic crisis. This may seem an extreme statement. It has been argued that such a view is scare-mongering, and that some new development, source of energy, food, or whatever will arrive and all will be well. Of course, if this did happen then it would be a new factor causing a deviation from the current trends (some possible promising examples will be discussed later). Again, even if a new factor did enter the equation there would be the problem of its dislocation of the *status quo*, be it economic, technological and/or sociological, not to mention the question of impact timescale – how long would it take for some currently unforeseen development to significantly affect matters? What can be counted on is that we will not be able to sustain current trends indefinitely – a business-as-usual way of life with no regard for the future.

In 1988 the renowned agriculturalist Sir Kenneth Blaxter gave a keynote address[137] on the occasion of his retirement as President of the UK Institute of Biology (the UK equivalent to the American Institute for Biological Sciences). It was built on an earlier series of 'R M Jones' lectures he gave at Belfast University and which focused on the inability to sustain the human race under its present resource-use regimen. He drew special attention to agriculture's ability to feed our planet's steadily growing population[138]. He pointed out that the Malthusian dilemma, despite Malthus' critics, is still very much with us today; that in November 1974 the heads, or senior ministers of 127 states, at the UN's World Food Conference proclaimed that 'Within one decade no child will go to bed hungry, no family will fear for its next day's bread and no human beings future and capabilities will be stunted by malnutrition'. Anyone who has seen the pictures from Ethiopia in the 1980s will realise how empty that proclamation was. Furthermore, 'our best farms are already approaching the thermodynamic limits to photosynthesis' productivity – wheat, maize and rice grown under ideal conditions already give yields of some 90% of the theoretical maxima'. Finally, Sir Kenneth noted that environmental problems, including global warming, are as much the concern of the developed Western countries as those of the Third World. We face, he said, 'perhaps the greatest biological challenge that Man has ever encountered'.

Relating Blaxter's biological challenge to that of anthropogenic climate change, we are in effect talking about adding an additional pressure to a system already Malthusianly stressed. True, climate change has naturally occurred in the past, and will do so again (albeit in the distant future with Milankovitch changes in incoming solar radiation) but, anthropogenic climate change is something we can begin to address. If we do not, as discussed in Chapter 4, then the effects can only serve to increase human suffering. According to the Daily and Ehrlich, Stanford University team[82] in terms of the World food supply, global warming could double the death rate from starvation to 400 million over two decades (see page 81).

Others have linked anthropogenic climate change to resource use. Esso's General Manager of Corporate Affairs in Australia, Des Mackney, is aware that as we enter the twenty-first century not only will petroleum imports severely impact that country's balance of trade (the same applies to other energy-resource-dependent industrialised nations) but its economy as a whole. - Conversely, energy resource conservation, as part of a package to keep Australia 85% energy self-sufficient in the 1990s, will not only result in less carbon dioxide being generated but in all likelihood increase employment, provide economic growth and reduce the national debt by over 3% of GDP[139]. Mackney speaking from within the energy industry is disturbed by the business-as-usual scenario. Energy forecasts in the early 1980s were inaccurate not least due to the lack of policy stability emerging (hence price) from the OPEC oil cartel – in the main due to internal changes in Iran followed by the Iran-Iraq war. Premium gasoline prices varied from US$150 per tonne in 1978 to over US$370 in 1979/80, down to US$155 in 1986/7, before peaking again in September 1990 at around US$450. Forecasts for the 1990s were equally varied and there is greater disagreement on future liquidity levels in the World trade system[140]. Uncertainty is not confined to greenhouse issues: there are as great, if not greater, 'gambles' for industrialists under the business-as-usual scenario, even in the short term.

However, there are a couple of predictions we can make with a fair degree of confidence regarding World energy for the decade up to the end of the century. Qualitatively, the current trend in diversification of primary energy resources harnessed will increase from being oil- and coal-dominated in 1960 to, in the year 2000 being more equally split between coal, oil, natural gas, and the combined resources of hydroelectric, other renewables and nuclear power. Quantitatively World demand – in-keeping with the business-as-usual scenario – will be higher in the year 2000 probably by at least 10% over the 1990 figure of 8000 million tonnes of oil equivalent (not including the non-world market fuels such as firewood). Malthusian pressure forecasts on the business-as-usual scenario, though noticeable, could be made relatively insignificant (by national/international energy policy or equally its absence) at least up to the middle of the twenty-first century's first quartile. After that time it would take great courage and a brave forecaster to predict the exact consequences of the 'business-as-usual' scenario. We can broadly, as the IPCC did[20], predict that by the year 2100 the atmospheric CO_2 concentration might reach some 825 ppm and anthropogenic warming of 3°C over the 1990, or 4°C over the pre-industrial, level with the consequences that that will have as discussed in Chapter 4. It is also likely that by 2100, worldwide the number of malnourished and destitute will increase and not just in absolute terms but as a proportion of the human race. Whether or not the average (both mode and mean) 'standard of living' in Western Europe and North America will be as high as today's is currently unanswerable, though it is not inconceivable that under the business-as-usual scenario their economies could crash through lack of the resources that fuel them. In short, whereas the business-as-usual scenario offers possible gains in the short term, these would be at the expense of greater economic security over a lengthier time period.

Given that by definition our starting point is the current business-as-usual trend, what are the most relevant factors to the greenhouse debate? Looking at carbon dioxide, it being the single most important greenhouse gas, and how it is generated, it can be readily seen that worldwide the electricity supply industries (ESIs) are not only a major source of CO_2 but electricity supply is increasing; again not just in absolute terms but as a proportion of the total World energy market which itself is increasing. Here the USA dominates, with only 5% of the World's population it contributes approximately a quarter of the 5 billion tonnes of carbon entering the atmosphere annually – one quarter of this in turn comes from the USA ESIs which means that the USA ESIs are responsible for about 7% of the major global warming gas.

At the other end of the 'development' scale, the Third World will see a greater twenty-first century growth in electricity demand (given that it is starting from a lower base). Proportionally, growth in the less-developed nations' electricity demand is curretly perhaps twice that of the USA's whose own governmental growth forecasts are of 2.0–2.6% per annum for the early twenty-first century. By then there will be an extra billion people worldwide, most in the less-developed countries needing energy supplies. According to the USA Agency for International Development, Third World demand for electricity has been growing at over 7% per annum over the past twenty years. Even a conservative annual growth forecast of 4.5% p.a. will have cost $125 billion a year throughout the 1990s in additional generating capacity[141]. Currently, a fifth of the anthropogenic CO_2 released arises out of electricity generation worldwide, and in the high energy use per capita nations (such as the USA and UK) this is higher – in 1987 over a third of UK CO_2 emissions (37%) came from electricity generating power stations. Should these current fuel consumption trends continue the ESI CO_2 contribution can only increase both in absolute terms as well as proportionally in relation to the other human activities that increase atmospheric CO_2,

Table 6.3 A compilation of key parameters to the IPCC 1990 'business-as-usual' forecast for the years 2030 and 2100

	2030	*2100*
CO_2 concentration	450 ppm (27% up on 1990)	840 ppm (138% up on 1990)
Methane	2550 ppb (48% up on 1990)	4000 ppb (131% up on 1990)
Global temp increase above 1990		
– high estimate	2°C	5.3°C
– best estimate	1°C	3.3°C
– low estimate	0.5°C	2.0°C
Sea-level rise above 1990		
– high estimate	29 cm	1.1 m
– best estimate	18 cm	0.66 m
– low estimate	8 cm	0.31 m

Note: In the light of data gathered in the early 1990s, it may be that the rate of accumulation of methane may have been over-estimated. However, a few years deviation from a long-term trend does not mean that the trend has necessarily changed such as deforestation or non-electricity generating fossil fuel burning. Consequently ESI energy policy, or lack of, is of central import to anthropogenic global warming.

The IPCC's 'business-as-usual' scenario[20] assumes that the World's energy supply will become increasingly coal-intensive and on the consumption side only modest efficiency increases will be achieved. Carbon dioxide controls will be modest, and deforestation will continue until the tropical rain forests are effectively depleted and agricultural emissions of methane and nitrous oxide are uncontrolled. For CFCs the Montreal agreement is implemented but with only partial international participation. Table 6.3 summarises the IPCC climatic predictions from Chapter 5 (Figure 5.5) business-as-usual scenario, the likley effects of which were discussed in Chapter 4.

As discussed earlier, in Chapters 3 and 4, it is worth noting that the change in the mean global temperature between now and the last ice age lies between the IPCC best and high estimate forecasts for the temperature change predicted under their 'business-as-usual' scenario through to the end of the next century (Figure 5.5). This is quite a temperature change and one might be forgiven for wondering why effects, such as the predicted sea-level rise under this scenario, are not more pronounced? Here we have to remember that the IPCC prediction is based on a dynamic scenario: throughout the IPCC 'business-as-usual' scenario concentrations of greenhouse gases increase. In reality we might eventually expect human greenhouse emissions to stablilise and for the atmosphere to come in to a new series of steady-state rate of exchanges with the various greenhouse gases. This will inevitably take many decades but it is useful for those making computer meteorological models to study such equilibrium states and at the time of writing of the IPCC report there were a score of equilibrium response models to a doubling of atmospheric CO_2. The difference between the dynamic and equilibrium scenarios underlines the point that even if greenhouse gas emissions are stabilised, or even began to decline after the end of the twenty-first century, then global warming would continue for some time (decades) to come. Indeed, even if we begin to reduce our rate of emissions of carbon dioxide and methane (IPCC 1990 scenarios 'C' and 'D') we would still be committed to both global warming and sea-level rise through to the end of the twenty-first century – obviously not as marked as under their 'business-as-usual' scenario.

The limitation in discussing the 'business-as-usual' scenario itself is that it depends on the current pattern of 'business' and the 'business' trends at the time of writing. A similar discussion taking place in the lower-than-now-energy-efficient early 1980s, or 1970s prior to the oil crisis of 1973, would come to a slightly different conclusion from the current analyses. Indeed economic forecasters are beginning to consider greenhouse issues. The Policy Studies Institute 1991 forecast for the year 2010 predicts that for the UK at least there will be:

- some limited promotion of energy efficiency in all sectors of the economy;
- higher prices for energy and private transport;
- increased investment in public transport;
- a switch from coal to natural gas in power stations;
- increased development and use of renewable energy[120].

Even though 'business' does change with time and will undoubtedly respond in some (currently unknown) way to greenhouse issues, 'business-as-usual' forecasts do provide a baseline, a starting point, from which we can identify future options and ways to proceed.

CHAPTER 7

ALTERNATIVES

If we do nothing, in other words follow the business-as-usual scenario, then according to the consensus provided by the IPCC (both in 1990 and in its later reports) we will see not just a continuation of global warming but its acceleration. To prevent this likelihood the 'business-as-usual' trends must be abandoned – fossil fuels cannot continue to be used in a profligate way and alternatives must be sought. Fortunately, there is a suite of readily identifiable alternatives. Each have their own costs and benefits, but some combination of these alternatives could provide a workable strategy that would slow the rate, and the final magnitude, of anthropogenic warming.

These options can be divided into two groups. The first is to decrease our dependence on fossil fuels by:

- Increasing our reliance on existing non-greenhouse energy resources and developing these further. Here, the principal resource available is nuclear fission that might even be developed further with fast-breeder technology. Long-term nuclear developments might also include nuclear fusion.
- Increase our reliance on renewable energy, principally hydroelectric and wind power, and to develop these technologies further by the extensive harnessing of solar, wave, tidal and geothermal resources.

The second group of options allows us to continue, at least while reserves last, with our fossil fuel-dominated energy supply by:

- Increasing the efficiency with which we use energy resources. In other words to increase the amount of 'useful' energy obtained per unit from the burning of fossil fuels.
- Increasing the size of the biosphere's carbon sinks, such as through reforestation.

Realistically, these options do not provide a complete solution and there are problems associated with each. However, they all have something to offer as we can see by examining them in turn.

NUCLEAR FISSION

Nuclear fission is the process by which the binding energy of very heavy elements is released as they are split into lighter fragments, closer in atomic number to the most stable (in nuclear terms) element of all, iron. This element has the greatest binding energy per nucleon. Uranium is the favoured heavy element used and a comparatively cheap fuel compared to coal or oil.

In terms of World energy, nuclear power has grown from its beginnings in the mid-1950s to providing over 573 million tonnes of oil equivalent in the mid-1990s, 7% of the World total demand for primary energy in 1990. Of this, 36% was produced in Western Europe and 35% in North America: together the developed OECD countries accounts for over 87% of the World's nuclear energy output which itself grew by between 3% and 6% per year during the first half of the 1990s[135].

Nuclear power is a growing business and one whose energy-producing reaction creates no carbon dioxide. In the UK there are 40 nuclear reactor units (civil nuclear power sites each have more than one unit) with two more under construction. Nuclear power provided 55.5 TWh (terawatt hours), 19.3% of Britain's electricity in 1988; electricity that would have otherwise been supplied by CO_2-producing coal or oil combustion. In 1991 the UK Energy Secretary, John Wakeman, reported that 'Britain would emit around 15.5 million tonnes more carbon if the electricity currently provided by [the then] existing nuclear power stations were to be generated by coal, increasing total UK emissions by nearly 10%'. He also gave assurance that the then proposed 1994 review of UK nuclear power would take such greenhouse considerations into account[142]. Using this proportion as a basis for calculating the carbon saved by nuclear power in the USA gives us a figure of some 170 million tonnes of carbon: equivalent to between 3–4% of the estimated carbon released worldwide. Worldwide nuclear power broadly displaces close to a tenth of the carbon released through fossil fuel combustion. Consequently, in terms of greenhouse criteria, whatever our nuclear plans for the future it would be unwise to abandon our existing investment in operational nuclear plant.

Carbon displacement arguments are powerful and are being cited by the nuclear industry to exemplify its value: Christopher Harding, chairman of British Nuclear Fuels, argues that the nuclear sector's ability to displace carbon emissions is a sign of increasing acceptance for the industry[143]. If we are to follow this proposal to its conclusion we need to see how much nuclear power can contribute toward meeting World energy demand? This very question was dealt with in a presentation[144] to Imperial College by Brian Eyre, Deputy Chairman and Chief Executive of the UK Atomic Energy Authority (UKAEA), who provided estimates as to the lifetime of uranium resources (as 'burned' in the main late-twentieth century reactor types) from the year 2030, based on both the currently proven uranium reserves as well as a speculative figure for reserves yet-to-be-found. These two figures formed the basis of calculation for three scenarios:

(1) That by 2030 electricity is 20% of the World total energy consumption outside of the (1990) centrally planned economies;
(2) 30% share as per above;
(3) 40% share as per above (i.e. about 3.5 times the current annual demand for uranium).

The resulting uranium resource-lifetime estimates are portrayed in the Table 7.1 below.

Table 7.1 Lifetime of uranium resources using 'burner' ('thermal') fission reactors only

Rate of uranium demand	*Known reserves*	*Speculative reserves*
Current	81 years from 1990	333 years from 1990
2030 (20% ESI nuclear)	14 years from 2030	159 years from 2030
2030 (30% ESI nuclear)	5 years from 2030	101 years from 2030
2030 (40% ESI nuclear)	0.5 years from 2030	71 years from 2030

In terms of the estimates for the lifetimes of other energy resources, the low 20% supply known reserve scenario puts nuclear power broadly on a comparable footing with natural gas – i.e. close to depletion by the middle of the next century. However, the speculative, yet-to-be-discovered, uranium resources could easily last until well beyond the end of the next century – but not as long as coal. In short, as used in 'conventional' thermal or 'burner' reactors, the estimated lifetime of uranium is less than that of fossil fuels: nuclear (burner) fission power is a comparatively scarce finite resource. There is another more efficient way of using uranium and that is by employing fast breeder reactors instead of the burner, or thermal, type reactors.

FAST FISSION: BREEDER REACTORS

As an energy resource, 'burner' or 'thermal' reactors use just a small proportion of the uranium available, only the ^{235}U isotope; the more plentiful ^{238}U isotope has no central role in energy production. Burner reactors are comparatively easy to construct and operate, with the nuclear reaction relying on low energy (less than 100 eV) thermal neutrons and a moderator used to slow the neutrons down – the moderator being graphite in the UK first and second generation, Magnox and AGR (Advanced Gas-cooled Reactor) reactor types, and water in the US Pressurised Water Reactors (PWRs). On the other hand, 'breeder' reactors use fast neutrons (with an energy in excess of 20 keV) which react with the plentiful ^{238}U to produce ('breed') plutonium which can then be used as a fuel. It has been estimated that introducing fast breeder reactors will extend the World's uranium resource-lifetime some 50 fold[145].

The theory is fine but there is a difficulty. The problem with breeder reactors arises out of the design requirements that have to be met when using fast neutrons. As they do not sustain the fission process as easily as the lower energy thermal neutrons, about 200 times as many neutrons are required compared with conventional burner reactors. To obtain this number, a high neutron flux density is required, the core is also compact to minimise the distance neutrons have to travel; both factors result in a high energy density. In order to remove this large quantity of energy, a very efficient coolant is needed with high thermal conductivity. Metals are the best conductors of heat and so the preferred design option employs a liquid metal – molten sodium. Here there are several problems. First, since the coolant enters a region of high density, fast neutron flux it becomes transmuted to a radioactive isotope of sodium (^{24}Na) so a second 'clean' sodium coolant circuit is required to transfer the heat to a third water coolant circuit that heats the boiler that drives the turbine that generates the electricity. This brings us to the second difficulty – sodium and water react explosively on contact. Fast breeder reactors are therefore carefully designed to minimise the possibility of the clean sodium and water circuits being breached and for mixing to take place. If such an event were to take place then the resulting chemical explosion would, in all probability, result in a substantial release of radionuclides.

So far the twin difficulties of achieving a high rate of energy extraction from the core under high neutron flux with chemically reactive coolant has delayed the arrival of the commercial fast breeder, but the technology is slowly being developed. Britain has a prototype fast breeder reactor (FBR) at Dounray in Scotland and the French have their FBR, the 233 MW Phenix reactor, in operation since the mid-1970s. Since then the French have constructed a 1200 MW Super Phenix with the rest of Western Europe looking on. Currently the UK is closing the Dounray fast reactor (in the mid-1990s) and is to join others in the EEC developing the European Fast Reactor (EFR): a fast breeder reactor whose capital costs are within 20% of the burner, or thermal, PWRs. Commercial fast breeder technology may not yet be with us but it will be available early in the twenty-first century, enabling fission to continue to partially offset fossil fuel demand, hence greenhouse impacts. To what extent fission will be able to do this will depend on our overcoming three problems:

(1) Long term nuclear waste disposal – nuclear waste remains radioactively dangerous for centuries.
(2) Nuclear weapon proliferation – civil nuclear programmes provide much of the nuclear capability required for weapon development. *cf.* UN concern over Iraq in 1991.
(3) Overcoming lack of social acceptance, which in part arises from the above and from accidents such as Windscale (1957), Three Mile Island (1977), and Chernobyl (1986).

The above represent nuclear power's three principal areas of uncertainty that are impeding its extensive development: it is also a reminder that uncertainty is not exclusively associated with climate change and that uncertainty in some way affects virtually all areas of human activity. Here, the degree of both actual scientific uncertainty combines with that of the uncertainty perceived by the public and from political action (human phenomena not entirely based on hard data). Long-term waste disposal is perhaps the question that concerns those in the nuclear sector the most: the other two areas being outside its areas of direct influence. The nuclear industry, barely three and a half decades old, simply has not lasted the thousands of years necessary to see a completion of the nuclear process from nuclear fuel fabrication to harmless nuclear-decayed waste. Currently, the preferred solution is to bury the waste for thousands of years in terrestrial sites of high geological stability and away from subterranean water. Sweden has been operating one such site since the 1980s and the USA has been looking at a site, Yucca Mountain, 10 miles northwest of Las Vegas, Nevada. To date, approval for the development of a high-level nuclear waste repository at Yucca Mountain has yet to be given as uncertainties remain as to the site's geological suitability over a required 10 000 year period, and to the likely environmental impact – that the site is the home to a species of desert tortoise, recently declared endangered, has not helped the Yucca appraisal programme costing some $2 billion [146,147]. Meanwhile, across the Atlantic in the UK, in 1989 the company that handles UK nuclear waste, Nirex, announced that it would be investigating two sites, Dounray and Sellafield, both owned by the UK Atomic Energy Authority (UKAEA). In July 1991 Nirex announced that Sellafield was the preferred site and asked for a Public Enquiry into its proposals. More recently (1994) Nirex has applied for permission to construct a rock characterization facility in order to properly assess the geology beneath the Sellafield site. Some basic geology-testing work has been conducted with some 20 boreholes drilled to a depth of just under 2 km. If events go according to Nirex plans, then Nirex could be in a position to apply for planning permission for a proper waste depository around the turn of the century[148]. Nonetheless, long-term high-level nuclear waste disposal remains a stumbling block for the nuclear industry.

Nuclear proliferation, or the threat, provides another restriction for the development of this energy resource. Nations which do not have nuclear weapons could develop them if they had a fully operational civil nuclear programme. The isotope enrichment and waste reprocessing needed for civil nuclear power can also be used to manufacture weapons-grade fissile material with all its international implications. In the 1980s Israel and South Africa jointly detonated a warhead that they themselves had developed using resources from their civil nuclear programmes. Similarly, in 1991 following the Gulf War there was much concern that Iraq was continuing to develop its own military nuclear capability. During the war itself the allies bombed Iraq's two civil reactors in an attempt to halt the military programme, and after the war the UN sent three teams from the International Atomic Energy Agency to Iraq to verify that the UN directive banning Iraq from developing nuclear weapons was being adhered to. Whereas the disposal of high-level waste may ultimately be solved, the threat of nuclear proliferation will remain and prevent fission from being contemplated as a greenhouse option for many countries.

Finally, extensive use of fission will only come about once the public have accepted it. There is considerable public concern principally on two fronts: the day-to-day workings of the industry, and the (remote) possibility of an accident, be it an inadvertent release of radioactive material or a major malfunction resulting in a core fire or meltdown. There is also some public concern over long-term waste disposal discussed above.

Much of the public's concern relates to the intangible low level radiation: it cannot be smelled, seen or touched; the exact opposite of, for example an oil well blow-out. Because those harnessing conventional energy resources run 'conventional' risks this does not mean that these risks are any less real than those of the nuclear industry, but perhaps because these risks are readily perceived so, frequently, are the solutions and from this the beginnings of public acceptability. For instance extracting coal, a long-term primary energy mainstay, has its environmental and human costs. In 1960 about 14% of UK miners suffered from pneumoconiosis, but by 1980 improvement in mining conditions and the increase in automation dropped this figure to around 4% [149]. Not surprisingly the nuclear industry points to the double standard imposed by its critics. It views its safety record as impeccable when compared to the many deaths occurring in the other energy industries or the considerable public risk from incidents such as Flixborough (the chemical works explosion) and Seveso (the dioxin release disaster)[150]. The risk from nuclear power to the public from ionising radiation is also low. The total British death rate from cancer is about 130 000 a year, of which about 20 000 result from smoking and 100 000 from unknown causes. The carcinogenic effect of these unknown causes has been estimated to be roughly equivalent to that resulting from a radiation exposure of 20 rem (200 mSv) per person[151]. The total annual radiation burden imposed by nuclear power on a member of the public is roughly 3 μSv (0.003 mSv). So it would seem that perhaps about 99% of our cancers arise from something other than the nuclear power industry.

Conversely, it has been possible to not only measure radioactive emissions from nuclear sites, principally Sellafield (the nuclear fuel and waste processing plant) in the UK but to identify whether the radiation arises out of airborne or liquid discharges. A Sellafield study even identified a mechanism for the transportation of plutonium from the sea (into which it was originally discharged) back to the land[152]. The amounts were vanishingly small (10-14 Curies per cm^2) but of concern, considering that the area around the plant sees a higher incidence of cancers, particularly leukaemia. As a result, an inquiry was set up headed by Sir Douglas Black. In 1984 it confirmed that the incidence of leukaemia around the plant was significantly higher than average but could find no causal relationship with the plant so gave a 'qualified' reassurance but called for further studies[153]. One of those published in 1990 did find a possible connection: many of the fathers of those children with leukaemia had worked at Sellafield which suggested that perhaps there was radiological impact taking place prior to conception and that this was being passed on to offspring of those exposed[154]. A year later another study into a similar cancer cluster around the Dounray nuclear installation concluded that there was no obvious connection between the cluster and the nuclear facility, though there was an 'apparent association with use of beaches within 25 km of Dounray[155].' (One of the currently preferred alternative theories is that it is not nuclear installations that cause cancer clusters, rather large construction projects that attract a new population into the locality, and that this somehow increases the 'infection' of leukaemia and non-Hodgkin's lymphoma[156]).

But it is not just those near nuclear installations that might be at risk. The Chernobyl incident in 1986 demonstrated that radiation from nuclear installations could be released, albeit accidentally, and affect the lives of those some considerable distance away. Not only were some upland farmers in the UK and reindeer herders in Scandinavia faced with restrictions on what they could sell, but four years later an estimated 19 excess cases of leukaemia a year, over a base line rate of 2293 in Europe, were thought to be caused by the accident[157]. Shortly after the accident the UK radiation watchdog organisation, the National Radiological Protection Board (NRPB), predicted that on top of the average 300 000 thyroid cancers expected in EEC countries over a 50-year period, there would be an excess of 2000 as a consequence of the Chernobyl incident[158]. Such figures are small – less than a 1% increase and a result of an accident, not the normal working of a civil plant. However, an accident is not a prerequisite for determining if there are ways for detectable amounts of radiological isotopes to reach humans from the nuclear industry. Unfortunately for fission's public credibility, radiological routes back to Man have been reported in the academic literature. Returning to the UK Sellafield reprocessing plant, a study in 1991 found that those eating local lamb and drinking local milk in the Scottish Western Isles do have excess body concentrations of ^{137}caesium along with ^{134}caesium, though not above recommended safety limits, which suggests that nuclear reprocessing (most likely from Sellafield) is the source[159].

Individually, *few* of the above concerns warrant a major policy revision of the nuclear industry; together they are disquietening and the nuclear industry needs to seriously address them if it sees civil fission programmes as a growth area.

FUSION

Fusion is a potential energy resource for the future. Unfortunately, no practical and economic reactor has yet been built, let alone energy generated commercially. However, research reactors have been built and slow, but expensive, progress is being made.

Just as fission makes use of the release of nuclear-binding energy from splitting very heavy elements into those whose weight is closer to iron (the element with the most stable nucleus), so fusion makes use of the energy release from combining light elements to those whose weight is closer to iron. Fission is based on the splitting of atomic nuclei, and fusion on the fusing together of nuclei. This is not easy as atomic nuclei are positively charged and so repel each other, which is why in nature it is only those systems with enough energy to overcome this repulsion, such as those within the huge gravitational wells of stars, that are capable of sustaining fusion. It is the fusion processes within our nearest star, the Sun, that keeps our planet warm and allows life to flourish.

The good news about fusion is that it uses commonly-found light elements, such as hydrogen, as a fuel. The fuel itself is not turned into long-lived radioactive waste – only the inner reactor casings are transmuted by fast neutrons to radioactive elements – reducing the level of radioactive waste by perhaps half compared to fission. The theoretical return of energy per unit mass of fuel undergoing fusion is higher than that for fission. Fusion will not add to fission's ability to facilitate the proliferation of nuclear weapons. As with fission (excluding fission as a possible source of nitrous oxide – see Chapter 5), fusion will have a minimal greenhouse impact. But, as with fast-breeder fission, there are problems anticipated with some commercial fusion designs that propose using lithium as a coolant. This would not only react vigorously with water but also with air. If a fusion reactor's lithium did catch fire then there would probably be a release of radioactive tritium (^{3}H). Such a risk is of a comparable order to that associated with fast breeders although technical obstacles are not insurmountable.

Current research towards an economically viable plasma (matter's high-energy state) for fusion is promising, albeit expensive and slow. Fusion faces the difficulty that not only does it require considerable energy to start reaction, but once established it generates so much heat – in excess of 100 million °K – that no physical chamber can hold the plasma. At the moment there are several options for containing the plasma, the preferred being the use of a 'magnetic bottle'. This is the method employed by leading researchers including the UK based Joint European Torus (JET). Back in 1960 the World's most advanced machines were a factor of one million away from the conditions needed for commercial fusion. Today JET has managed to attain a plasma of the sufficient temperature, density, and length of time required for more energy to be generated by the plasma than is consumed in its creation: the difficulty has been bringing these three criteria together at the same time[160]. Meanwhile, in the USA, experiments at Princeton University's Tokamak fusion reactor have managed to push the energy output to two thirds of that consumed.

So successful are the current programmes that the Princetown team are hoping to see close to energy break-even soon, and there are now proposals to extend the JET project's lifetime. Meanwhile, there are plans for the next generation of experimental reactors. Europe, by itself, is planning the Next European Torus (NET), and in conjunction with Japan, the USSR, and USA (as well as Switzerland and Sweden) is working towards designing an International Thermonuclear Experimental Reactor (ITER)[161]. Sites for three ITER centres, San Diego in the USA, Garching in Germany, and Naka in Japan were decided on in 1991. The decision whether to extend the agreement to include construction is planned to take place before 1998.

We are still a long way away from the commercial reactor. To begin with, plasma energy break-even is different from system break-even: one being a theoretical figure and the other a practical reality. Then again, a greater energy output is still required for true commercial viability and all this once the experimental designs have been scaled up. We have to reconcile the fact that there are many layers to the 'feasibility' of fusion; including scientific feasibility, engineering feasibility, and socio-economic feasibility[162]. Fusion is unlikely to provide energy on a scale beginning to rival fission until well into the next century, and as such can only be regarded as a potential long-term greenhouse-friendly energy resource, one that is free of the traditional (fossil fuel and uranium) supply constraints and one that does not rely on the large spaces required for collecting low-energy density renewable energy, be it solar or wind.

RENEWABLE (FLOW) ENERGY

Other than the nuclear energy resources offered by nuclear fission and fusion, renewable energy – HEP, solar, wind, wave, tidal or geothermal – is the only resource group that does not result in large-scale release of greenhouse gases. These resources are often referred to as 'flow' resources as they each make use of a continual energy flux, as opposed to 'fund' resources which rely on a buried 'fund' be it of fossil fuel or uranium. In the main, this energy 'flow' is driven by solar radiation. This may be obvious in the instance of solar power but it is as true for hydroelectric power (which is dependent on the water cycle and in turn the solar-powered evaporation of the oceans); wind power (using wind arising from the solar radiation warming the atmosphere); and wave power (which is dependent on waves formed by the wind). The energy from tidal and geothermal power has different origins. The former is the harnessing of the Earth's angular (rotational) momentum which is so vast as to make our energy demands seem insignificant over the millennia. The latter is again on a planetary scale and depends on successfully tapping the Earth's internally- generated heat (which primarily arises from the radioactive decay of heavy metals in the Earth's interior) – as we shall see although not truly renewable, geothermal energy can still be exploited in a flow-like way. Of the renewable energies hydroelectricity has been the most extensively developed.

Hydroelectricity

Whereas in 1990 nuclear power contributed some 5.7% to the World commercial primary energy demand, 6.7% – over a sixth again – was met by hydroelectricity. In 1990 the USA produced 72 million tonnes of oil equivalent (mtoe) worth of hydroelectricity power (HEP), or 3.6% of its commercial primary energy demand. Combined with Canada, which relied that year on HEP for over a quarter of its 238.9 mtoe primary energy demand, North America as a whole almost mirrored the world average figure by obtaining over 6.1% of its commercial primary energy via hydroelectricity. In Western Europe nearly 7% of primary energy consumption is derived from HEP, primarily from the mountainous countries of Norway, Sweden, Austria, France and Italy (in decreasing order of HEP importance). The UK, with few regions capable of providing a sufficient head of water to make HEP a useful energy resource, relies on it for less than 1% (some 1.5 mtoe) of its primary energy requirements[162].

As with all the renewable options, HEP is greenhouse-friendly with no greenhouse gases being emitted. Another attraction is that, whereas electricity generation from fossil fuels is around 30–40% efficient, HEP is 80–90% efficient. Nonetheless, this does not mean that HEP schemes have no environmental impact. They require a head of water, as the power output from HEP is directly proportional to both the mass of water falling as well as the height through which it falls. Consequently, HEP schemes frequently require the flooding of a valley and the disruption of a river's flow. This last can often be minimised, but with some large schemes there have been unforeseen impacts. HEP dams can silt up and become difficult to clear so making the resource effectively non-renewable. The large Aswan dam (Egypt) was supposed to serve the dual purpose of supplying electricity and controlling the Nile floods, but has had the secondary effect of trapping the nutrients that were normally swept down to the delta. Consequently, the Nile valley farmers have had to use artificial fertilisers[163]. Even so, such problems can be avoided if the schemes are constructed with proper planning and at the appropriate scale.

As to the potential contribution that HEP might make worldwide, it is conceivable that, at most, HEP output might triple from 540 mtoe to 1.5 billion tonnes of oil equivalent by the middle of the next century[164, 165]. Certainly, HEP is currently the only renewable energy resource being harnessed on a meaningful scale and so will probably lead other sources for some decades to come while these alternatives are developed.

Solar power

All life on Earth ultimately relies on solar energy for warmth and food – photosynthesis harnesses about 1% of the solar radiation and some of this energy may become trapped as future fossil fuels. Equally, it is the Sun's energy disturbing the Earth's atmosphere that causes the wind and, in turn waves, so that energy from these sources is just another way of utilising solar power. However, such a definition is too broad to be useful, so here solar power will be referred to in the context of the direct, artificial harnessing of solar radiation.

The solar energy the Earth receives is considerable, even compared to the scale of human energy use. A disc the size of the Earth and in the Earth's orbit at right angles to the incoming solar radiation would receive 1.37 kW m^{-2}. However, allowing for day and night, atmospheric absorption, and excluding areas within the polar circles, the typical solar input at the Earth's surface is roughly about 240 W m^{-2}. So that if the World's population as a whole was to consume energy at a similar rate to that of a member of the European Economic Community, then only 0.01% of the incident solar flux would be required[163]. The energy *is* there for human use, it is a question of harnessing it.

In the main, harnessing the Sun's rays requires it to deliver thermal energy, either directly to where the heat is required, using building design to act as a greenhouse (passive solar energy), or to use the Sun's rays to warm water flowing through simple flat plate collectors. The latter low-grade heat collected can be made to warm a surprisingly large volume of water – a swimming pool or a tank to act as a thermal store – or it can raise the temperature of water in a domestic hot water tank several degrees before further conventional heating is needed. However, both systems only work well if incorporated into building at the design stage: little can be done in a house with few windows and a poor midday Sun-facing aspect which lacks a way to circulate and manage the heat internally.

Commercial solar power has yet to come into its own and has largely been considered as a resource very much in its experimental and prototype phases of exploitation. By the mid-1980s solar power was providing some 400 000 tonnes of oil equivalent a year or roughly about 0.001% of the World's oil consumption, or less than half this again of the World's primary energy consumption[166]. Much of this production came, not from Europe or the USA, but Israel and Japan. With regard to the UK, it has been estimated that solar power could meet 0.1% to 0.15% of the nation's demand[167]. While this may not seem much, in 1990 cost terms when fuel oil fetches £76 and coal £45 a tonne[168], this corresponds very roughly to a potential annual saving on that nation's fuel bill of some £600 million.

A more difficult, and expensive, but useful way of harnessing solar power is to convert the Sun's rays directly into electricity via photovoltaic cells. These work with an efficiency currently of about 6% (some research prototypes are around 35% efficient) by allowing light to fall onto two sandwiched wafers (a 'p' and an 'n' type) of a silicon semiconductor. If a cascade of semiconductors is used to trap light of different wavelengths then, in principle, over two thirds of solar energy could be converted. The incident photons excite the electrons within the 'p' type semiconductor which then migrate to the 'n' type and so create a potential difference – a voltage – across the sandwich; an effect first observed in 1954. Unfortunately, the large expanse of semiconductor material required to generate a useful current is costly, although the price is falling largely due to the growth of the silicon semiconductor industry. In 1977 the average commercial price for photovoltaic systems was about \$16 000 kW^{-1} generating capacity and fell to around \$7000 by 1979[169] – to be competitive with other conventional energy sources the costs in 1977 would have had to have been about \$800 kW^{-1}. Even so the price has continued to fall for both photovoltaic and thermal solar power systems throughout the 1980s (see Table 7.2). This has encouraged an increase in the use of photovoltaic solar power worldwide since its first megawatt capacity in 1978, through an estimated supply capacity of 43 MW in 1989, to a forecasted 1200 MW by the end of the century, with virtually all of this growth taking place in remote sites. At this level of supply photovoltaic solar power will still be very much a minor and exotic energy resource, but this could begin to change should the cost-falling trends continue to hold into the next century. Nonetheless, solar photovoltaic could begin to play a key role in reducing emissions in the latter half of the twenty-first century.

Table 7.2 Cost of solar power generation in 1988 cents kWh[170]

Energy resource/Year	*1980*	*1988*	*2000 (forecast)*
		(1988 cents kWh^{-1})	
Solar thermal	24	8	6
Solar photovoltaic	339	30	10

How fast costs will fall and how soon photovoltaic solar power begins to take on a significant amount of the fossil fuel energy burden largely depends on political will. At the 1982 Versailles G7 World leaders' meeting solar photovoltaic power was identified as one of the major renewable energy sources for the twenty-first century. By 1991 the governments of all the industrial nations, with the notable exception of the UK, had mounted photovoltaic R&D programmes. In May 1991 the UK rectified this by launching a research programme, with the government citing the greenhouse implications as a policy rationale. The UK Minister of State for Energy said: 'This is a particularly important source of electricity generation, with no emissions of pollutants'. To date, the fruits of many nations' research is encouraging enough for some to forecast the beginnings of commercial photovoltaic power – at a price competitive with coal – by early in the next century[171].

Wave power

Solar power is only one way of harnessing the Sun's energy but heat from the Sun stirs the atmosphere which in turn generates ocean waves from which energy can be extracted making wave power an indirect form of solar energy. Continental waters exposed to coastward prevailing winds have the most continuous high level of wave energy. By the time the prevailing winds have reached these areas they have travelled thousands of kilometres over the ocean in which to impart their energy, build a swell. Not only has the ocean swell - considerable energy but the decay time generally exceeds 12 h which helps even out the energy available in coastal waters on a day-to-day basis even if there is no wind blowing locally. Off the UK some 75 kW on average is contained in a metre of wave frontage per second. From this it can be estimated that within UK territorial waters the potential is there of some 120000 MW bound up in the waves: twice England and Wales' combined electricity generating capacity[172] but, commercially, perhaps only a quarter of England and Wales' electricity needs could be met this way. An additional attraction is that there is a seasonal peak that coincides with that of demand. Looking beyond the UK, one estimate is that there is a potential 110 GW of wave power available off Europe's coastlines which is equivalent to about 85% of the present European Community electricity demand (though the only European country with a serious, albeit small, wave programme is Norway, a non-EC nation)[173].

There have been several designs proposed for wave power machines, one of the most favoured designs was put forward by Stephen Salter of Edinburgh University in 1974 – the Salter Nodding Duck[174]. The Duck is shaped so that waves will cause rocking, thus allowing a turbine to be turned and generate power. The Duck rocks in such a way that the water behind it is not disturbed: if it were disturbed then energy from waves in front of it would be transmitted to create waves behind it, and not be extracted. Prototype tests on Salter's Ducks have shown that in ideal conditions some 90% of the waves' energy can be extracted; though in more realistic conditions only half the energy can be harnessed. A full size Duck would be about 38 m long so that 400 would perhaps generate 1000 MW. The key to any wave power design is that the devices have to be sensitive enough to be able to extract energy from comparatively small waves, yet robust enough to function in storm conditions. For Ducks it has been suggested that they might be designed to be able to submerge and so survive intact through the roughest of weathers.

As we shall see later, the UK has been slow to develop wave power on a large scale; indeed the UK Department of Energy publicity booklet on *Taking Power*

From The Water, while mentioning tidal power, pump storage and hydroelectric schemes, dispenses with wave power in just two paragraphs and the conclusion that: 'systems for large-scale offshore electricity generation are not now expected to be economic, but small shoreline systems, might provide economic power sources'[175]. This view was to be severely criticised in 1990. Up to that time only the smaller shore-based machines were being planned in the UK but, by 1991, there were two operational shoreline power plants elsewhere in the World: one in Norway and the other in Sakata, Japan.

The first fruits of the UK shoreline programme were inaugurated in July 1991 when the UK Energy Minister, Colin Moynihan, officially opened the modest 75 kW system on the Scottish island of Islay. The Energy Minister took the opportunity to comment on the potential of renewables to produce power cleanly, claiming that: '[The scheme] demonstrates the importance the [UK] government attaches to electricity generation from environmentally friendly renewable resources'[176]. However, such small-scale, shoreline schemes will hardly make an impact on displacing CO_2-emitting fossil fuel consumption during the next century even if vigorously developed. The real potential, as always subject to cost, probably lies in long offshore barrage schemes of a scale that can literally harness wave power in mega- and gigawatts.

Tidal power

Unlike wave power, tidal power relies on tapping the rotational energy inherent in the Earth's angular momentum. As the Earth rotates beneath the Moon a 'bulge' of water is formed on either side of our planet as a result of the lunar gravitational pull (and the resulting centrifugal reaction from the rotating Earth–Moon system). Although there are complicating factors (in that the seas and oceans are bounded by land), in essence the arrival and passing of this bulge that creates tides. Furthermore, the movement of the tidal bulge is not frictionless, and energy has to be conserved, so that gradually the Earth's rotation is being slowed with days becoming longer: 100 million years ago the Cretaceous day was only 23 h 16 min compared to 24 h at present[177]. However, the passing of the tidal bulge can be tapped for energy. When the tide is high a reservoir, or basin, can be filled then emptied later at low tide allowing the water to run out through electricity-generating turbines. In this way the Earth's huge rotational (or its angular momentum) can be harnessed.

Tidal power stations are not new, one was built at La Rance in France in 1967 (then) for £50 million: it has an installed capacity of 240 MW and there is another, but only for turbine trials, in the Annapolis Basin in Canada. In the UK the greatest interest in tidal schemes have focused on the Severn estuary which has the distinction of the highest tidal range in Europe. This fortunate characteristic is due to the estuary's large funnel shape – extended by the Welsh coastline to the north, and the northern coastline of England's southwestern peninsula to the south – and that the funnel's length is the same order of magnitude as a quarter of the tidal wavelength which results in a tidal resonance effect. The estuary's large tidal range is of key importance to prospective tidal station engineers as the scheme's power capacity is proportional to the square of the tidal range.

A great deal of work has gone into the feasibility and costings of a Severn tidal scheme since the first Governmental study[178] in 1981. The report considered a number of proposals on the basis of £2.3 million worth of studies; three were identified as worthy of further appraisal. In 1989 a second report[179] put forward a preferred option, a scheme enclosing a basin of 480 km^2, with a total installed capacity of 8640 MW producing an average annual energy output of 17 TWh (7% of England and Wales' electricity consumption) which would, in turn, result in an annual saving over its design life of 120 years of some 8 million tonnes of coal (7.5% of UK 1990 coal consumption). The scheme's basic capital cost was estimated at £8.28 billion (in April 1988 monetary terms) which excludes the £200 million required if a 16 km dual carriageway road was constructed along the barrage's length. The cost of electricity (April 1988 money) at the barrage boundary would range from 1.7 pence/kWh at 2% discount rate to 7.2 pence/kWh at 10% discount rate[179]. (For comparison's sake by the end of 1988 medium-sized industrial consumers bought their electricity at 4.6 pence/kWh though a proportion of this last includes distribution costs[180]). The size of the scheme is so large, and its benefits so diverse, that it is unlikely to be successful as a private enterprise and government support will be required.

The UK has a number of other suitable sites for tidal schemes and is currently leading the World in research into this form of renewable energy. After the Severn, the next favoured proposal is one for the Mersey with a barrage linking Liverpool with the Wirral. This scheme would have a rated capacity of approximately 700 MW with a net power output of 1.5 TWh and with a construction cost of £800 million (at 1989 prices) and annual operating costs of £10 million per annum (which includes allowances for dredging and lock operation[181]).

Tidal schemes do not produce continuous power but, nonetheless, can offset coal and gas consumption used for electricity generation and so reduce national greenhouse emissions. There are, as with *any* form of power generation, other environmental costs, such as those on estuarine water quality, but principally there is a loss of estuarine intertidal mudflat habitat. Mudflats provide a good source of invertebrates for other animals higher up the food chain and estuaries provide a habitat for many species of fish and birds. Britain's estuaries support an average peak of 1.5 million wading birds (approximately 40% of the East Atlantic Flyway population) as well as half a million wildfowl. In the UK twelve species of wader and six of estuarine wildfowl are in sufficient numbers and under threat that they are included in the 'British Red Data Birds'.

Unfortunately, the very reason why many UK estuaries are suitable for tidal schemes, the large tidal range which in turn enables there to be a large area of inter-tidal mudflats, is the reason why birds are attracted to them. The Severn alone, for part of the year, is the home to some 51 000 waders and 26 000 wildfowl, and the Mersey to 31 000 waders and 32 000 wildfowl[80]. There are many pressures on estuaries, including land reclamation and leisure uses, so tidal schemes cannot be considered in isolation. One particular dilemma directly related to the greenhouse effect in that sea-level changes of only a few metres (as mentioned earlier in Chapter 4) will have a profound effect on the estuarine environment, yet 'greenhouse friendly' tidal schemes would play a part in offsetting this last impact. These twin sets of circumstances paint a bleak picture for the future of UK estuarine habitats[182] unless the various impacts of tidal schemes (and energy strategy options) can be meaningfully costed and set against the benefits – including those from beyond the market place. In the case of tidal power, such a cost-benefit approach might allow for funding to safeguard other major feeding grounds against a rising sea-level. (Such provisions might seem wishful but there are precedents: the Welsh Dinorwic pump-storage scheme had a considerable sum spent to enable it to blend into its National Park setting, and as we shall see in the USA one coal power station had its construction tied to a tropical re-forestation programme.)

Worldwide tidal power programmes are not pursued as vigorously as in the UK. In the USA sites such as Minas–Cobequid, Shepody and Passamoddy might all support tidal schemes with a greater generating capacity than the proposed Severn scheme. Similarly in South America there is San Jose; in France, Mont St Michel; and in the former USSR the White Sea. Together these and other smaller schemes might generate in the order of 500 TWh a year – perhaps equivalent to 1% or 2% of the World's oil consumption. This may not appear much but tidal power is only one of several sources of renewable energy available: tidal energy has a contribution to make.

Wind power

About 370 TW, 0.2% of the solar energy received by the Earth each second drives winds (and in turn part creates ocean waves)[163]. To put this into a human perspective, the energy in winds is roughly 50 to 60 times that 'artificially' generated by Man. Harnessing these winds is in theory quite simple. The wind blows to turn a turbine that generates electricity. The practice is a bit more difficult in that windmills must be strong enough to survive the roughest of conditions yet sensitive enough to extract energy from low wind speeds, and to function as continuously as possible (which means overcoming metal fatigue). With so much wind energy available in quantity, early exploitation of this resource was possible and there are many examples of harnessing this power over many centuries: the use of wind to grind corn and for sea transportation has been extensive and well documented. In more modern times the cheap availability of more concentrated and controllable fossil fuel energy has meant that electricity generated from wind power has largely been ignored until quite recently.

The oil crisis of 1973 revitalisized interest in flow resources and in particular wind power. By the beginning of the 1980s there were several prototype designs operating. The period 1976–81 saw a range of government programmes aimed at developing larger-scale turbines both in the USA and UK. The first half of the 1980s did see the emergence of a market for small- and medium-sized turbines in the USA: that country not only passed legislation ensuring that privately-generated electricity would have to be bought at a realistic price by utilities but initial Federal and State tax incentives made wind energy positively attractive at some sites, particularly along the western Californian coast. Wind power installation in California rose from 10 MW in 1981 to an extra 60 MW in 1982, new capacity trebled again in 1983 so that by 1984 it had risen to a level of 400 MW a year. By 1988 the mean size of commercial wind farms had increased to 108 MW, with farms of 200–300 MW not being uncommon. Even so, as the decade closed over 80% of the world's wind-generated electricity was produced in California. Improvements in machine performance followed and capital costs fell from an average of US$3 100/kW in 1981 to about US$1 250/kW in 1986. Currently, efforts are being made to produce commercial machines of 1–2 MW with capital costs around £800/kW (1988 prices) which equates to an electricity cost of around 3 pence/kWh (assuming a 25 year turbine lifetime and 5% discount rate, and 20% loss from maintenance unavailability etc[182]). This closely approximates the 1983 estimated cost of electricity from the then forthcoming Sizewell 'B' nuclear reactor of £131 per kW per annum[183] or 1.5 pence per kWh (allowing for the almost doubling effect of inflation up to 1988 brings us near the 3 pence per kWh target). In short, wind power technology is already close to being competitive and would be if enlarged to exercise economies of scale.

Regarding the desirability of wind power, this is largely governed by site constraints. The low-cost option is to use land-based wind farms: the costs for offshore farms are much higher and currently are not being seriously considered. In the USA siting is less of a problem than in the UK with the latter's higher population density. There have been worries in the UK over public reaction to the construction of wind farms in rural areas (a social cost of externality). In Cornwall, prior to the construction and operation of Britain's first wind farm at Delabole, 300 people living within a 2 km radius of the site were polled for their views. Of these, some 31% objected to the wind farm, principally on the grounds that the farm might spoil the view, and 40% approved. However, after 6 months of operation, 81% said that the farm made no difference to their lives and 85% approved or strongly approved of wind power[184].

As to the contribution wind power might make towards offsetting electricity generated from fossil fuels, in 1987 the UK wind power lobby claimed that 20% of all current electricity generation could be met by wind power for around 2 pence per kWh (1986 money) without causing undue land-use conflict[185]. Other studies[182] support this view and although the UK Government did not agree with this estimate that wind power had a role to play in supplying Britain's energy, it received ministerial endorsement in 1988 when the UK Government agreed to back a £30 million programme to ascertain likely wind farm sites[186]. Indeed, the potential for wind power is so attractive that in 1990 the then UK Minister for Renewable Energy was openly expressing the view that in the UK, by 2025, up to 30 TWh (or about 10% of 1990 UK electricity supply) would be provided by wind power[187*]: only a year later the Energy Secretary, John Wakeham, said that this forecast 'may turn out to be over cautious[189].' A year later, in 1992, the UK Government's Renewable Energy Advisory Group (REAG) agreed with the wind power lobby's 1987 claim in predicting that as much as 20% of UK's (then) electricity demand could be met by wind power[190].

In the USA with its lower population density and so fewer constraints on siting, the potential contribution to energy supply is greater. Given the broad favourable consensus afforded wind energy, it would seem likely that this form of renewable energy has the potential to be one of the first to begin to offset traditional fossil fuel consumption in developed nations.

Geothermal energy

The Earth contains a great deal of energy arising largely from the decay of radioactive elements within the planet (which makes geothermal power one of the few renewable energy resources divorced from solar origins). This energy is constantly being transferred as heat away from the Earth's interior. The rate of transference varies, largely with the thickness of the Earth's crust, but this heat can also be tapped where it has built up in a particular body of rock.

The first geothermal power station was opened in 1904 at Larerello in Northern Italy: currently some 370 MW of electricity are generated there. The first electricity generating plant opened in the USA at the Geysers field, California in 1960 with a capacity of 12.5 MW. One of the latest (at the time of writing) geothermal plants is the second 35 MW unit at Momotombo, Nicaragua, which came on line in 1989. Altogether this site provided Nicaragua with 40% of its electricity at a capital expenditure of just $80 million and displaced $10 million a year in oil consumption (1989 prices)[191]. US geothermal resources have also developed so that by the mid-1980s some 1000 MW was generated in this way, by 1991 The Geysers' site was producing 1500 MW accounting for 75% of US geothermal power and 6% of California's electricity. Worldwide, about 4000

* For an overview of the UK government's proposals on renewables see ref 188

MW are generated geothermally and it is generally competitive with traditional electricity generation[192].

All the above sites exploit geothermal power where the energy flow is particularly intense close to the surface, but heat is escaping from the interior everywhere. Typically, UK thermal gradients are around 25°C per km, although a Gas Council bore found a temperature of only 59°C at 3 km depth – a gradient of only 19°C per km. Worldwide gradients are of the order of 30°C per km giving an average surface energy flow of about 60 MW per m^3 [192].

Geothermal fields can take several different forms, with three general classifications: hydrothermal convective systems (the most important worldwide); low grade aquifers; and hot dry rocks (which have the most potential globally by possibly increasing the current resource use by an order of magnitude). All function by circulating liquid within the system and so extract the heat. The liquid, in the case of hot dry rocks, may have to be introduced to the site. Here, it should be noted that as the heat in such power systems is generally being removed from the rock faster than would otherwise occur and, importantly, faster than it is replaced, so that geothermal power is not really a renewable energy source; heat mining would in most instances be a more appropriate term. The Momotombo, Nicaragua has seen a decrease of steam output of over 5% in the last half of the 1980s[191]. The US Geysers geothermal field, the largest exploited field in the World, is also running out of steam so that, despite increasing development throughout the 1980s, power production peaked in 1987. Steam delivery from that field was (in the early 1990s) dropping at over 10% a year[193]. Geothermal power, though an energy alternative, should be treated as a non-renewable resource.

Prospects for further development of this resource appear to be good. It is believed that large geothermal deposits occur in the Far East, Turkey, and along the African Rift Valley as well as the west coast of the American continent from Alaska to Chile[192]. In the UK the size of economically recoverable geothermal power is thought to lie around 1000 TWh for electricity[194] (i.e. if 10 TWh could be generated geothermally per annum then perhaps 3–4% of UK electricity demand could be met this way for 100 years, or a smaller proportion for longer). This energy contribution could be augmented by using geothermal energy for district heating systems (current practice in Iceland); an exploratory borehold in Marchwood, UK, showed that water at 1.6 km depth was warmed to 73°C[167]. Currently the UK is considering the possibility of working with other European countries on an EC geothermal research programme. Both in Europe and worldwide geothermal power is unlikely to have a major effect on offsetting fossil fuel consumption, however, if by the second half of the twenty-first century geothermal power is providing around 1% of energy needs then that is 1% less that we need to worry about.

Biomass conversion

Natural systems have evolved specifically to trap energy and 40 TW of solar energy is trapped each and every second by planet-wide photosynthesis[163] and held in the form of biomass. This will continue long after the last oil field has run dry or coal bed been depleted. It would therefore seem a trifle presumptuous for us to write off biomass conversion without seriously considering tapping the energy flow that has been powering life on Earth since its earliest days.

Photosynthesis is the process by which plants combine water and carbon dioxide to form carbohydrates. From here other metabolic processes result in a range of fuels from plant oils, to ethyl alcohol and even methane. These fuels can then be burned to provide light, heat and electricity in much the same way as traditional fuels.

From a greenhouse perspective, biomass conversion has many advantages. Unlike fossil fuel consumption which releases buried carbon into the atmosphere, biomass conversion depends on plants taking carbon from the air so that when it is returned once the fuel is burned there is no net increase in atmospheric carbon dioxide. In common with the other alternative energy resources discussed, biomass conversion is greenhouse neutral as well as being able to reduce our dependence on finite fossil fuels.

To date, the country that has most benefitted from biomass conversion has been Brazil. Its government launched the country's National Alcohol Programme in 1975 to encourage ethanol production from the fermentation of sugar cane. In that year Brazil was producing 600 000 m^3 of ethanol of which a third was dehydrated for gasoline blending (the remainder being used in the chemical and pharmaceutical industries). The alcohol (20%–petrol blend known as gasohol) soon played an important part in the Brazilian economy so that by 1985 10.5 billion litres was being produced for motor fuel with a further 2 billion litres for other purposes, altogether created 500 000 new jobs in agriculture as well as industry. Savings on Brazilian oil imports were estimated at around $8 billion annually for the late 80s[195].

The Brazilian experience reveals the benefits of national or regional biomass conversion programmes including:

- the aforementioned greenhouse benefits;
- foreign exchange savings (on oil imports);
- security of energy supply (which would be important for Europe and the US);
- increased demand and diversification of agricultural products (in Canada 10% grain ethanol in gasoline would increase employment by 42 000[195]);
- environmental benefits as ethanol would replace lead as an octane-enhancer;

- spin-off benefits to equipment manufacturers and suppliers in forestry, agriculture and chemical industries;
- increased research in biotechnology, microbiology, agronomy, forestry, genetics, engineering, metallurgy and chemistry.

There are many other benefits largely peculiar to individual countries. It has been calculated, for example, that if most of the highly erodible land in the USA were planted with short-rotation, intensive-culture trees, then the biomass from these would produce energy equivalent to 60% of all the power generated from coal in that country in 1987[196]. However, this might take up as much as 8.5% of total US land area. Another estimate from the US Department of Energy puts the realistic potential for short-rotation-coppicing (SRC) by the year 2010 at a generating capacity of 25 000 MW. While the US Electric Power Research Institute doubles this, putting the total potential realizable at 50 000 MW, which is equivalent to 7% of current US electricity capacity. The crop area for this would take up 1% of the total US land area[197].

In the UK in the late 1980s the Department of Energy has estimated that some 800 000 ha of land might be used for forest energy plantations – with the wood either being burned for energy (again, providing the forest was replanted this would be greenhouse neutral), or used as a feedstock for biotechnology industries to convert to alcohol. However preliminary polls among farmers indicates that they would only allow perhaps 250 000 ha of their farmland to be used. Willow and poplar are favoured coppice species that could be cropped (using non-manual methods) on a 3–5 year cycle[198]: coppicing has the advantage that plant growth is concentrated aboveground and the root system remains intact across cycles. It was thought that some 3.8 mtce (2.53 mtoe) could be economically supplied, or 13.4 mtce (8.9 mtoe) if energy prices doubled over their 1986/7 value[199].

In 1992 the opportunities for UK and other European Community countries to obtain energy from biomass dramatically increased through reform of the EC Common Agricultural Policy (CAP) designed to reduce agricultural surpluses. (Contrary to popular belief, for all its faults the CAP has been successful in several ways: boosting productivity per hectare, and strategically increasing Europe's food security.) To reduce food surpluses the EC implemented the policy of 'set aside' whereby farmers were paid £320 per ha to leave 15% of their arable land unproductive. A few years later the UK Department of Trade and Industry's (DTI) Energy Technology Support Unit (ETSU) estimated that the UK could have one million hectares of surplus agricultural land by the year 2000, and up to five million by the year 2025. If all this land was used for energy crops, the total electricity that might be generated at a 1994 cost of 10 pence/kWh would be 194 TWh/year[197] – enough to meet over three quarters of England and Wales' electricity needs. By 1994 estimated costs of biomass energy was between about 6 pence kWh and 7.5 pence/kWh. However, the cost outlay of biomass power is high with a payback delay of several years in the case of SRC. CAP reforms allow a one-off set-aside payment equal to five years worth of subsidy, but the UK Treasury does not wish to fund what would be such a very large grant[197].

There are many other ways of utilising biomass. In the UK alone around 14 million tonnes of straw are produced annually. Almost half of this is lost (through ploughing back) yet, if used as a fuel it could provide 1% of the nation's total energy requirement[200]. In the USA, the bread-basket of the World, the potential for energy from straw is much greater where 24 million hectares of maize produces 7.5 tonnes ha of crop residues a year.

Worldwide, the theoretical potential of energy from biomass to displace fossil fuels is considerable. A study from Princetown University[201] estimates that, combined with the IPCC strategy to increase energy efficiency and reforestation, biomass could help offset emissions by 5.4 Gt carbon per year in the middle of the next century: an amount equivalent to the 1990 atmospheric input of carbon from fossil fuels, which would reduce the estimated business-as-usual 2050 carbon emissions to half the 1985 level. In Europe over 15 million hectares of cropland would have to be taken out of production to control agricultural surpluses, and in the USA some 30 million hectares are left idle to reduce production or to conserve land. This could offset emissions by some 170–225 million tonnes of carbon without bringing new land into production. Bearing in mind that there are some 800 million hectares of tropical land potentially available for reforestation which (conservatively assuming temperate coppice productivities of 7.5–10 tonnes per hectare dry biomass a year) would provide 3–4 Gt carbon.

Problems with attaining this level of biomass production include difficulties of ensuring proper nutrient cycling (which might be offset by planting a suitable species mix), ecological impacts (which do exist but are arguably far less than those of intensive monoculture farming), and the opportunity cost of using land that might be required for food production as the World population increases. Notwithstanding these, the ability of biomass conversion to provide greenhouse-neutral fuel is significant and will no doubt receive greater attention in the years to come.

Worldwide renewable energy potential

Each of the renewable energy sources may be harnessed to offset fossil fuel use. The degree to which these are employed is another matter: one of the most important factors in terms of anthropogenic climatic change. It is impossible to predict with any confidence the exact contribution these will make in the future. However, we can identify theoretical maximums but, clearly, unless investment in developing these technologies, and policy encouraging their implementation is provided it is unlikely that these potentials will be fully realised. Policies do and will vary from country to country both in principle and their implementation in practice. Already there is a spread between countries of differing proportions of national electricity supply being met by renewables: in 1989 some 2% of UK electricity came from renewable sources; 5% in Germany; 9% in the USA; 13% in France; 11% in Australia; 80% in New Zealand; and 100% for Norway and Iceland.

Aside from regional variation there is a time factor: it takes time to realise an investment, and time for markets to change. Even if we start making the investments and establishing the policy framework now it would arguably take as long for renewables as nuclear power did to mirror its current penetration (461 mtoe or 5.7%) into the World energy supply market – some three and a half decades. Such a lead time would take us well into the end of the twenty-first century's first quarter. By then a significant contribution from renewables might realistically be expected, for not only will the various renewable technologies be more advanced and the political motivation in all likelihood more developed, but the cost of fossil fuel will become that much higher as demand begins to outstrip supply.

It is important to recognise that such crystal ball gazing cannot provide accurate figures of future production and demand. On the other hand, we do have a factual starting point in the present as well as knowledge of current trends and theoretical potentials. These trends will no doubt continue in the immediate short term and only change (increasingly?) as time passes. While long-term predictions are of little value, it would not be completely out of hand to make short to medium term, high and low, estimates as to realistically achievable renewable energy goals. Identifying these will at least give us a rough handle on the order of magnitude that might, not unrealistically, be expected of the renewables' contribution to greenhouse policy, so enabling us to ascertain whether or not we should at least begin on a particular strategy.

In short, a number of broad target windows for renewable energy production can be roughly estimated based on real thermodynamic limitations as well as current estimates of: actual theoretical potential; rates of current implementation (under various economic and policy regimens); and cost trends. Such attainable energy generation target windows (high and low estimates reflecting possible policy differences) for the year 2025 are listed in Table 7.3 in the form of the fossil fuel (oil energy equivalents) that they could displace.

These broad estimates of potential contribution for the year 2025 range from the conservative 695 mtoe estimate to optimistic values of about 1360 mtoe (which compares to typical annual contributions of 190 mtoe to 200 mtoe in the first half of the 1990s). This high estimate is not absolute or for that matter necessarily realisable in the unlikely event that renewable energy became the

Table 7.3 Possible annual contribution to off-setting fossil fuels by the year 2025

	Estimates of attainable global renewable fossil fuel displacement contributions for 2025 given encouraging energy policies	
Resource/estimate	*Low* (*mtoe*)	*High*
New HEP	300	500
Photovoltaic	5	45
Wave	20	220
Tidal	50	125
Wind	50	90
Geothermal	20	40
Biomass	250	340
Totals	**695**	**1360**

Note: Assumptions made in these estimates include that coal power stations have an efficiency of about 33% and that while plants with greater efficiencies of 40% or more are beginning to come on line that they will not have a major impact on the efficiency of energy supply worldwide until well into the next century. It is also assumed that the greenhouse perspective will become broadly adopted by policy-makers (even if it is not realised) so that, with the exception of biomass production (as pressure may be there for land to provide food rather than energy), renewables will be used to offset coal consumption. The energy displacements are expressed in oil equivalents for standardisation purposes with 1 million tonnes of oil providing some 12 TWh, or 4 TWh of electricity after being burnt in a power station. These estimates exclude non-commercially traded fuels such as wood, peat or animal waste which, though important in many countries, have historically been unreliably documented. Finally, these estimates are provided by the author for discussion purposes only; they indicate what could be provided with current, and current drawingboard, technology

central pillar of international energy policy. If by the year 2025 some 1360 mtoe was being displaced by contributions from the new renewables, how would this relate to the global anthropogenic energy budget?

In 1990 worldwide primary energy consumption temporarily peaked at 8033.3 mtoe[134]. By the year 2010, allowing for the growing World population, World energy consumption may well top 12 000 mtoe[202] and increase to about 15 000 mtoe by 2025. In other words by 2025 some 9% of total energy demand could be met by new renewables (as compared to 2%–3% in the first half of the 1990s).

David Pimental (best known for his work on food, energy and agriculture) and colleagues from Cornell University have estimated the potential optimistic contribution from renewables for the USA for the year 2050. Their forecast is based on broadly equal contributions from biomass, HEP, solar thermal, photovoltaics, passive solar and wind. They estimate that in 2050 the USA could obtain about 37 quads (785 mtoe). This is equivalent to 35% of its early 1990s primary energy demand, or 17% of the 2050 total estimated US demand *assuming* business-as-usual and population growth[203].

In terms of greenhouse considerations we can add the new (greenhouse friendly) renewables to the existing non-fossil supply – today not all of total World energy demand is met by burning fossil fuel. By 1990 we already had a base of 540 mtoe of hydroelectricity and 461 mtoe from nuclear power (or 5.7% of total world consumption). Adding the 1990 HEP base and assuming (modestly) that nuclear power maintains its 5.7% contribution (861 mtoe out of an estimated 15 000 mtoe 2025 World demand), then we might expect some 18% (2700 mtoe) of World energy in 2025 to be supplied without burning fossil fuels, as opposed to 12.5% today. In short, it is perfectly possible for renewables to not only provide an increasing quantitative energy contribution but, if prioritised in energy policy, this addition would help provide an increasing proportion of a growing World energy demand to be met in a greenhouse friendly way: this is the good news.

The bad news is that even with this high renewable estimate, and *with all other things being equal*, more fossil fuels will be consumed in the early twenty-first century than today (with coal being of significant medium and long-term importance by 2025). Even with 18% of an estimated 15 000 mtoe being supplied in 2025 in a non-fossil way, some 12 300 mtoe of fossil fuels will be consumed annually, so pumping even more carbon dioxide into the air than today.

Of course, all other things are rarely equal. A greater proportion of World energy demand could be met by nuclear power which would mean more than doubling the present nuclear capacity. Alternatively the renewable contribution could be increased beyond the above estimates, although that might entail bringing on-line underdeveloped (hence less profitable) renewable technology or the displacement of food production with that of biomass for energy. Yet again energy efficiencies, both of production and consumption, could be increased. This last will be discussed in the next chapter, for it may be possible to trim the estimated 15 000 mtoe World primary energy demand for the year 2025.

CHAPTER 8

EFFICIENCY – PLUGGING LEAKS

One routine oft put on by circus clowns is the hole in the bucket skit. One of the clowns has to carry some water to put out a fire or some such. The trouble is that his bucket of water has a hole in it so that by the time the clown has reached the fire, the water has drained, his bucket is empty, and back he goes for a refill. Such antics are ludicrous, but it is from such absurdity that the crowd derives its mirth: such things do not happen in real life, or do they? Well they do and the way we exploit our energy resources is a case in point. Clearly if someone had the job of continually carrying water from 'A' to 'B' you might occasionally expect them to spill a drop. But equally you would not deliberately give that person a bucket with a hole in it unless, at least, an attempt had been made to plug it (lest humour is the exercise's purpose). Yet this situation is exactly analogous to energy use. Some energy, as it is converted from its raw state (be it chemical in fossil fuel, nuclear, solar or whatever), is lost. Furthermore, energy losses take place at every stage from resource refinement through to end use. All these losses do not benefit the end user, and so in one sense these losses in energy terms (though not financial terms) are irrelevant to the said end-user. This last may seem a bit illogical but look at it this way. People, you and I, do not want energy! The bottom line really is that energy is of little importance to us. Before you disagree just consider what you *do* want?

Nobody really thinks, 'Winter is coming up and I really must expect to consume 'x' kWh more.' What people think is more along the lines of, 'Winter is coming up and I can expect a larger heating bill.' People do not want energy. They went services. They want: heat, light, sounds, hot water, and the ability to cook. When you sit down to a meal in a restaurant, wondering about the oven's thermal efficiency is hardly utmost in your mind. Whether your meal is properly cooked is of far greater concern.

This notion that people are not interested in energy, only in services, comes from those advocating low-energy strategies. That is, good housekeeping and getting the most from what you use; to maximise usage while minimising consumption. Something, which in the instance of energy, is a surprisingly foreign thought to governments and even much of industry, yet is at the heart of all good business management practices.

If there is merit in the low energy strategists' argument, there needs to be sufficient slack in the system with enough wastage already taking place, for it to be worth investing in energy savings. For, if the potential savings are small, then only minimal investments in energy conservation will be warranted. Conversely, if the energy savings are large then not only will energy conservation investment pay dividends but such investment might take priority over other investment in energy supply: it would be better to plug holes in a leaking bucket than to invest in a brand new water outlet if the water is to go to waste. The key question is, therefore the magnitude of wastage taking place in energy resource-exploitation systems at various stages.

EFFICIENCY OF SUPPLY

All human activity is ultimately governed by the laws of the Universe and it is to these laws that we must turn in order to identify the theoretical limits to our abilities. When it comes to energy generation and use, the laws of thermodynamics and electrodynamics are the most pertinent. It is not the energy in itself that matters, rather the work that energy can provide, be it work in the form of the work done to: rotate a record deck, run a food mixer, excite the electrons in a light bulb, etc. The second law of thermodynamics states that heat cannot be converted into work unless some of the heat is transferred from a higher to a lower temperature, in other words that heat cannot be transferred from a colder to a hotter body without an external influence (work) taking place. This may seem intuitive: cups of coffee left to stand do not spontaneously heat up or become cooler than their surroundings – unless work is done be it via a heating element or a fridge. Although this may be obvious from an energy 'user' perspective (one that we find ourselves in daily), it is as true from the 'supplier' point of view.

The French engineer Sadi Carnot, in 1824, made a fundamental contribution to thermodynamics with his theoretical consideration of an ideal heat engine. His work rested on the principle that the proportion of heat which can be converted into work out of the total input of heat is dependent only on the two temperatures between which the engine works. This can be expressed in the following equation:

$$\frac{\text{Heat converted into work}}{\text{Initial heat intake}} = \frac{T2 - T1}{T2}$$

T2 = Heat input in degrees Kelvin
T1 = Heat discarded in degrees Kelvin

T2, the heat input in electricity generation, be it coal, oil or nuclear, is the temperature of the boiler before the steam enters the turbine, and T1 the temperature at the other end which equals that of the coolant or of the waste gas in the flue.

Around the turn of the century the average efficiency of electricity generation was around 5%, so that around 95% of the energy was wasted. Since 1900 there have been dramatic improvements in thermal efficiency which now typically exceeds 30%–35%. Such 'high' efficiencies have been achieved through increasing the temperature of the steam leaving the boiler and lowering that exhausted through improved cooling systems – which is why inland power stations are frequently dominated by huge cooling towers; coastal power stations can use the sea for coolant. Clearly there are limitations to the efficiency of boiler-driven turbines. Even with high pressures of 150 atmospheres (even higher pressures pose increasing engineering problems) boiler temperatures (T2) achieved are only around 550°C (or 823°K). As coolant temperatures (T1) tend to be around the 10°C–20°C (or 283°K–293°K), the absolute theoretical Carnot efficiency is about 65%, nearly twice that of current power stations.

Of course there are other practical considerations which hinder us from doubling the efficiency of boiler-based power stations. But we can begin to see the type of efficiencies that are practically obtainable by looking at current generating plant. This is made up of comparatively inefficient old plant, such as the Brighton (UK) coal-fired station (cooled by the sea) with its efficiency of close to 25%, and more modern and efficient plant such as the Drax (UK) coal-fired (estuary and cooling tower cooled) with a thermal efficiency in excess of 37%.

This 25%–37% range of current thermal efficiencies provides us with a base from which we might set a target for future efficiencies. Assuming that we only increase thermal efficiencies by 8% to 45% by the year 2025 (still well below theoretical efficiency of 65%) and close down the older less efficient plant, then efficiencies will rise. In the 1987/8 financial year the UK Central Electricity Generating Board's coal-fired stations together operated with an average efficiency of 35.5%[204]. Given the above, it might not be unreasonable to expect US and European average thermal efficiencies for electricity generation from coal to modestly increase 5% to 40% over the next couple of decades. This would in effect provide more than 12% extra electricity per unit of coal burnt. In greenhouse terms we would gain an extra 12% of electricity without a corresponding increase in carbon dioxide generation.

Such an increased efficiency benefit can accrue to all electrical generation based on the Carnot cycle; which includes all generation relying on boiling water to generate steam to drive a turbine – even nuclear power, and an efficiency improvement here not only helps displace fossil fuel generation but (not withstanding other factors) helps makes nuclear power economically competitive. Naturally, electricity generation not based on the Carnot cycle is not subject to such thermodynamic considerations.

Increasingly, and with an eye to a future with more renewable sources exploited, electrodynamic electricity generation is becoming available – that is electricity from mechanical rather than thermal energy. Wind turbines, hydroelectricity, and tidal energy all rely on electricity from mechanical energy with the efficiency of conversion approximately 90%[163]. At first this might seem good news for the renewables, unfortunately their existing high efficiency of conversion means that there is little room for future improvement. Wave power is the exception as it has so far proved difficult to 'trap' the mechanical energy efficiently prior to electrical conversion.

Electricity generation is, admittedly, only one of the ways fossil fuels are used. Nonetheless, in terms of carbon dioxide generation, the electricity utilities rivals transportation for generating greenhouse gases. The USA, which produces close to a quarter of the World's greenhouse gases, illustrates this well with some 30% of that country's fossil fuel consumption being used to generate electricity – about 2000 TWh[205]. It is exactly because so much energy is wasted in electricity generation which means that it is the dominant factor in carbon dioxide generation.

The next most important sector for fossil fuel consumption is transportation. The USA in 1985 used just under 30% of its fossil fuel consumption for transportation[205]. Again, car engines are Carnot-limited with regard to their efficiencies; which are typically a little over 25% with much of the energy being lost as heat, be it through the exhaust or the radiator. Savings from transportation can be considered not so much from improving the way the energy is generated but in the way energy is used. It is here, at the energy-use end for all forms of energy consumption (not just transport) that major increases in energy efficiency can be made and more rational ways of utilising energy found.

EFFICIENCY OF USE

Improving the efficiency of energy use has all too often been a low priority both with policy-makers and the energy supply industries themselves. The reasons for this are largely to do with the human obsession with short-term profits – why invest in improving the efficiency of consumption with a payback time of several years when the benefit from investing in consuming cheap energy today is realised in the short term? Politicians and the energy industries have also argued that investment in end-use energy efficiency (or energy conservation), as opposed to investing in energy production, is a dead-end policy. We might have the most efficient energy-consuming industrial sector possible, they say, but this would be worthless if there was no energy to feed into the system in the first place. Indeed the low-energy strategist favours balance but points out that energy investment and policy has been, and is, weighted heavily towards supply rather than consumption. Investment is concentrated in resource exploration, extraction, refinement, conversion and transportation/delivery; there is far less invested in ensuring efficient consumption.

David Rose, writing during the 1973 oil crisis, illustrated in a rough and ready way the imbalance in US investment and policy between supply and consumption with, what he called, a taxonomic breakdown of US energy resource-use, sector by sector[206]. His taxonomic breakdown consists of a straightforward count of organisations and bodies concerned with energy resource exploitation. When Rose wrote his article US energy research and development investment was some $1.5 billion, about a fifth of this was devoted to just one energy producing mechanism – the fast breeder reactor. A similar breakdown of principal energy concerns within the UK today (see Figure 8.1) again shows that energy supply occasions the attention of more large organisations (with their manpower and capital) than that of energy consumption. The UK House of Commons Select Committee on Energy, looking back on the 1980s, reported that disappointing progress had been made in promoting higher standards of energy efficiency: this over a decade after a major study by the International Institute on Environment and Development (IIED) first revealed the extent of the potential energy savings in the UK[207]. But this was not the first time the Commons Select Committee on Energy had been critical of the policy balance (and understanding) between investment in energy supply and how it is consumed. In 1981 the Select Committee were 'dismayed' to find that seven years after the 1973 oil crisis the UK Department of Energy had still no clear idea as to whether an investment of over a billion pounds on a new (nuclear) power station would be as cost-effective as spending a similar sum on energy conservation[208].

The 1979 IIED *Low energy strategy for the UK*[207] marked a watershed in thinking. It was part inspired by the new train of thought emerging at that time and which in no small part included that of Chateau and Lapillonne[209]. The idea was that rather than attempt to estimate a nation's ability to produce energy to meet demand (an explorative approach to planning), the various end uses of energy were qualitatively and quantitatively identified and aggregated to establish demand (a normative approach). The IIED team, headed by Gerald Leach,

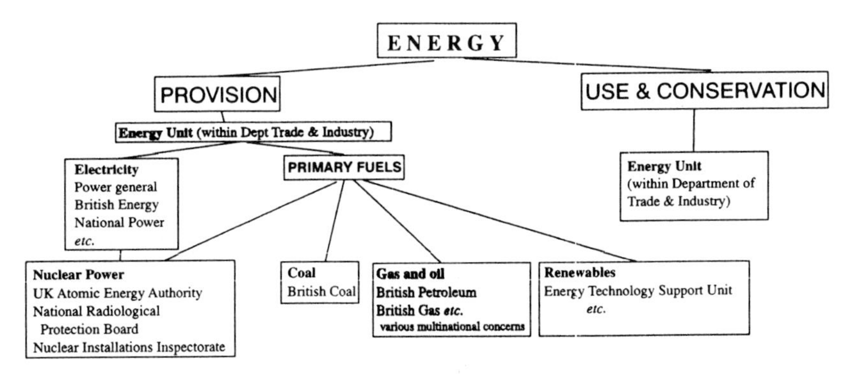

Figure 8.1 'Energy mobile' for the UK showing the principal bodies and organizations involved in energy supply and the policies relating to energy use and energy conservation. (This diagram is loosely based on a taxonomic breakdown of various energy sectors that Rose (1974) used when looking at energy policy in the USA. See reference number 206)

did just this in attempting to identify the minimum amount of energy needed to maintain Britain's economy and the British standard of living. They broke down energy use into four sectors: industry, domestic, transport, and commerce and institutional, to see how the energy was currently used before seeing how much energy could be saved using the traditional energy-saving measures then available (conservatively assuming that no development of these techniques would take place in the future). They then made an allowance for economic growth (both high and low cases) based on the Government Department of Energy's own forecasting scenarios, through to the year 2025, and calculated the demand so created for each of the four sectors.

By introducing into their calculations energy-user efficiency measures known in the late 1970s, Gerald Leach and his team concluded that under the low, 2.5% to 0.5%, economic growth scenario UK primary energy consumption could actually decline by as much as 7% in the year 2025 below that for 1976. In making this 1979 forecast Leach was affirming a trend – one of uncoupling of economic activity and energy consumption – that had begun six years earlier in the wake of the oil crisis (see Figure 8.2).

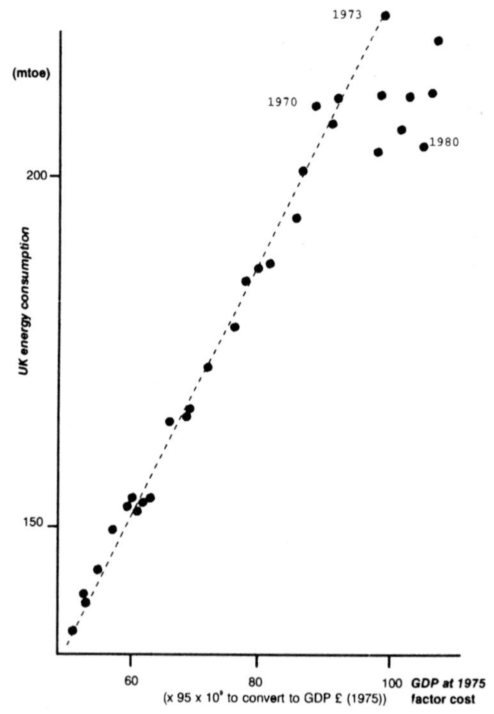

Figure 8.2 UK energy consumption (mtoe) *vs* Gross Domestic product (GDP). (100 = UK GDP in 1975). Prior to the 1973 oil crisis policy makers assumed that there was a close (near linear) relationship between the energy consumption and the economy. Following the 1973 oil crisis the economy continued to grow but consumed less energy. Compare with Figure 6.4

Since the end of World War II and up to the 1973 oil crisis UK primary energy consumption was so closely related to the Gross Domestic Product that policy-makers treated one as a barometer of the other (*cf.* Chapter 6). Energy conservation was seen by some as a real threat to the economy: a view that was to remain in some quarters through to the decade's end. This was one of the arguments used by those at UK Atomic Energy Authority in response to the IIED low-energy strategy. Aside from concern over the strategy's resource base and that the investment in conservation might exceed that of the economic return, they said that some desirable energy-consuming activities were omitted from the study. In other words they did not believe that there was the slack in the system to enable economic growth to continue as predicted, yet with a decrease in energy consumption[210].

UK energy forecasts made in the late 1970s and 1980s for energy demand in 2000/2005 were high. A projection made by the Advisory Council on Research and Development (ACORD) was particularly high, forecasting growth of some 3.5% a year (see Figure 8.3). As the economy adjusted to the post-oil crisis energy prices, so forecasted demand dropped. Today (in the mid-1990s) we are closer to 2000 and we can see that of these forecasts, that of the independent IIED Leach team were the closest[104].

Ironically, if planners had looked at energy consumption and GDP differently they would have seen a second trend: that as time passed, with every extra unit increase in economic growth accounted for proportionally more economic growth. Because this was nearly proportional a straightline relationship between energy consumption and GDP appears but from time to time in actuality energy

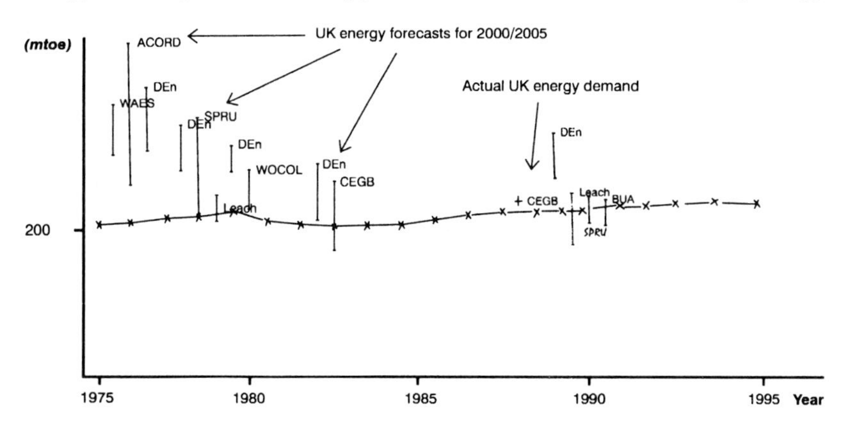

Figure 8.3 Energy forecasts of UK primary energy consumption for the years 2000/2005 (vertical line brackets) against the year the forecast was made, as well as the actual UK primary energy consumption over the years 1975–1995. By the 1990s, with less than 10 years to go, it became apparent that UK energy demand was unlikely to rise significantly. (Note especially the CEGB (Central Electricity Generating Board), and DEn (Department of Energy) forecasts). (Figure based on and updated from reference 104)

was already gradually being used more efficiently. If US and UK planners had looked at economic energy intensity (dividing energy consumption by GDP) against time, rather than economic activity against energy, then they would have seen that the energy efficiency of their economies had increased in fits and starts for nearly all of the twentieth century. If anything this suggests that an economy that uses energy efficiently is more successful than one that does not.

The value of the low-energy strategists' approach to developed countries' energy consumption is that current energy-use is so profligate that considerable savings can easily be made. Examples of such inefficient energy use abound in all the principal energy consumption sectors in most developed countries, both in Europe and North America.

Transport

Relate an era to a mode of transport and, before the eighteenth century you will probably think of the horse; in the nineteenth century, the steam locomotive. On the other hand, the twentieth century might be remembered as the age of the automobile: certainly cars play a major part in the economic and recreational dimensions of the societies in developed nations. Yet globally the World's automobile class is proportionally relatively small at only 8% of the total population (and growing) but is still responsible for 13% of anthropogenic carbon dioxide release[211]. Car-use has other costs too, for instance it is responsible for about 100 000 deaths in the European Economic Community and the USA, and a quarter of a million traffic fatalities worldwide a year (which compares with a total figure of three million accidental deaths a year)[212].

As noted earlier, the thermodynamic (Carnot) efficiency of the internal combustion engine is only about 25% and there is undoubtedly room for improvement. In the USA the energy efficiency of cars increased steadily after the 1973 oil crisis and was given further impetus by the 1980 crisis. Yet by 1988 the weighted fuel economy for US cars had reached a peak of 28.6 miles per gallon (mpg). (US light trucks peaked at 21.6 mpg in 1987.) If every car in each of the US industry's weight categories did as well as the best model, then the average efficiency would improve by almost a quarter, to 34.4 mpg. If North Americans could be encouraged to buy smaller cars – an average 12–13% drop in car weight, interior volume and top speed – then the overall economy could be as much as 45% better[213]. From these figures it is not difficult to see that combining a change in behaviour (using smaller/lighter cars) with an increase in the efficient use of cars already in existence, then energy savings of 50% would not be impossible for the US automobile sector without a decline in the passenger miles travelled. This saving could be easily attained by 2025.

Western Europe has always had lighter cars in its car fleet, so the savings here would not be so easy to meet. However, allowing for some improvement in car technology over the next couple of decades should enable comparable savings to be made without impinging on the degree of mobility enjoyed by the population today.

There is another way to create energy savings in the transport sector that does not rely on improving car efficiencies, and that is *not* to use cars for all the passenger-miles travelled but other more efficient means of transport. Throughout this century not only has terrestrial transport accounted for much of the growth in developed nations' energy consumption but the growth has largely been in road use. In the UK between 1958 and the end of the 1970s, increased transport accounted for about half the total increase in UK energy consumption. The 1958–1976 growth in road transport itself accounts for 70% of the rise in UK inland energy consumption[207] and is largely attributable to the motor car. In the mid-1970s cars and taxis contributed to some 350 billion passenger kilometres a year, a major rise from their 60 billion passenger kilometres in 1953; by the mid-1980s this had risen to over 400 billion passenger kilometres. On the other hand, buses and coaches have declined from just over 75 billion to about 60 billion passenger kilometres over the same period. (Rail passenger travel has remained roughly constant at around the 40 billion passenger kilometres a year between 1950–1986)[214]. By the mid-1980s UK road passenger transport accounted for 80% of the distance travelled by passengers but 90% of the energy consumed.

Looking at passenger travel and energy from a rail and coach perspective we can see that, although in 1986 they accounted for some 20% of travel, they used just 10% of the energy consumed in UK passenger transportation. Commuter trains are approximately three times more efficient than cars in terms of energy consumption per passenger mile; buses are even more efficient[214]. The fact remains that of the principal modes of transport used today only airplane transport, both light and international jet, are on average more profligate in their use of energy.

Another symptom of Western societies' marriage to cars is the difficulty in obtaining a divorce. Such has been the twentieth century's rising automobile tide that even low-energy strategists, such as the IIED team, consider it unlikely that the necessary investment in motorcycles, buses and rail will be made to significantly dent the prestigeous energy squandered by the car – around 1000 pentajoules a year for the UK in the late 1980s[214]. Leach's *Low Energy Strategy for the UK* cites the position in 1976 (which is still applicable today). Then, if all bicycle and motorcycle traffic had been trebled and bus and rail traffic doubled, energy use for UK passenger traffic would have been only 11% lower than it was[207].

Yet in major conurbations of the developed world the demand for public transport is frequently over-saturated implying that there is a market for expansion. The Japanese employ staff to squeeze passengers into Tokyo's subway trains. In London commuters travelling by rail, both from within the suburbs and the nearby counties, journey in trains loaded to capacity: this is more the rule than the exception. (Commuting accounts for about a fifth – the largest single reason – of passenger journeys in the UK and just under a half of all commuting is done by car.) With the rail tracks running to capacity it was necessary in 1992 to extend the length of the platforms on all the stations in London so as to run longer trains. Good news you might think. Yet the same year, due to a marginal decline in commuter traffic (hence revenue), trains were cut from the schedule and not all trains during rush hours are long enough to use the extra platform length. Not surprisingly car use continues to rise: it is cheaper (under the current economic regimen) to use energy-inefficient transport than to adopt resource-efficient methods.

The UK swell in car travel has been mirrored in the way surface freight is transported. In the early 1950s just over half of the energy used to move commercial freight was by rail: by the late 1980s over 90% was moved by road. Department of Transport figures as to the energy efficiency (MJ/tonne-kilometre) show that, while in the early 1950s the energy efficiency of road and (steam age) rail freight transport were both in the 4.5 to 5.5 MJ/tonne-km range, by the latter half of the 1980s road freight energy efficiency had only improved to about 3.5 to 4.5 MJ/tonne-km bracket whereas rail freight was using energy four times more effectively at about 1.0 MJ/tonne-km[214].

Of course the picture is complicated in that rail (or bus) travel is not direct or door-to-door be it for freight or passengers. Even so the potential savings from using rail and buses to enable the majority of urban/inter-urban journeys are quite significant. (Remembering the earlier discussion on the rise of the mega-cities and that the majority of those in the high energy consuming developed nations live in towns.) Further complications arise in taking into account the energy required to build road and rail networks in addition to that required to move freight and passengers. One study from the US Harvard Business School in 1979 looked at just this and included construction costs in their estimates. It concluded that a passenger in an average car on a motorway consumed some 250% more energy than a passenger travelling by rail[169].

What of the future? The petrol-driven automobile has come to dominate societies in the developed nations. It may seem that the investment required for alternative transport systems is huge. But petrol is a finite resource so that *ipso facto* so is the automobile age. As for investment, current investment in cars is large. Typically in the USA and Western Europe cars are bought either first- or second-hand for thousands of dollars, kept for five to ten years (during which time their value is at least halved) before being sold and a replacement bought. Meanwhile, the car has to be maintained and fed with petrol. Owning a car is often described as having a second mortgage. All this adds up to considerable personal investment. Then there is the construction and maintenance of major roads and motorways. These may be paid in the most part by some form of car or road tax (depending on which country you are in), which is more personal cost, and the remainder of the cost is met by the state through central taxation. Cars are expensive, transferring a large part of this investment into energy efficient bus and rail would help to slow, if not halt, the increased amount of carbon dioxide generation arising from the transport sector worldwide.

Calls for change in developed nations' transport policy in order to move to a more resource-equitable transport sector are not new. Such calls have generally been associated with those outside official planning bodies yet, as we shall see, there have been some governmental planners who have been registering their concern. Indeed a few of those with access to official data are beginning to press for change and this can only add credence to the lobby for change.

Lighting

Within fifty years of Michael Faraday's discovery, in 1831, of how to generate electricity mechanically, Paris was employing electricity to illuminate streets. Brussels, Madrid and St Petersburg were to follow suit and shortly, in the autumn of 1878, Londoners were seeing the 'new' electric light when lamps were installed at the Gaity Theatre. That October the 30 000-strong crowd at the Sheffield Football Association saw their team play under four of the new lights.

The lights in those days were extremely bright arc lights, too bright for domestic use, but there was no stopping the scientists and technicians in their search for ways to make use of this new form of energy. The lighting breakthrough came with the invention of the incandescent lamp, a device worked on quite independently by Joseph Swan in England and Thomas Edison in the USA. The first of these lamps, which used a carbon filament, came on the market in 1880. Swan's own house was completely lit by electricity that year: it was almost certainly the first such house in Britain and possibly the World.

It would be wrong to say that towns and dwellings were completely dark prior to electricity – there was town gas (or coal gas) and before that oil and animal fat-based lighting – but lights lacked intensity and were not so numerous. Today this is not so. Walk down any street in the centre of any developed nation town and you will have (much to the annoyance of astronomers) all the illumination you need to light your way. From space, electricity has transformed the picture of night-time Earth. On cloudless nights the entire eastern and western USA seaboards shine out, as do the towns of Western Europe. Further east, Moscow is set like a jewel in a radial web of light with the villages and towns dotted along the trans-Siberian railway a necklace of illumination pushing out towards China. So clear are electric lights from orbit that it is possible to create a mosaic of pictures of the entire planet. In addition to electric light some slash and burn farming shows up on satellite pictures in the tropics, and the gas flares from oil rigs such as in the North Sea and Middle East are also clearly visible.

In 1975 the UK domestic sector alone consumed 321 PJ of electricity of which some 38% was used for lights and appliances (the remainder being used for cooking, water and space heating[207]). In 1990 lighting was responsible for between 9% and 18% of *all* electricity[215] demand in the developed nations (both in and outside of the domestic sectors). Here the potential for savings are great. In 1989 the International Energy Agency put the potential savings from lighting (realisable from current technology) at 50%; even more with use of new technology.

Compact fluorescent lamps (CFLs) represent a major, and fairly recent, step in the provision of energy-efficient lighting. An old thermoluminescent bulb rated at 100 W can be replaced by a 20 W CFL without any loss in illumination: a major energy saving of 80%, and, over its lifetime, a saving of over 600 kg of carbon dioxide – assuming that the lamp conserves electricity generated by fossil fuel.

Even if a developed nation devotes only 9% of its electricity to lighting, and that only half a country's lighting was energy efficient neon, then conversion of the other, thermoluminescent half to CFLs would give an electricity saving of 3.6% without having to turn off a single bulb. This is unashamedly a conservative estimate. Most developed nations use more than 9% of their electricity on lighting and have less than half their lighting provided efficiently. A more realistic general calculation would assume 12% electricity supply devoted to lighting and less than about 40% of existing lighting already being energy efficient.

Domestic

Other than lighting, there are many ways that energy can be saved in the domestic sector. In 1990 the UK domestic sector consumed some 40.6 mtoe which accounted for a little over 27% of the nation's total final-user energy consumption (which excludes power station and transmission losses), or close to 19% of total UK energy production/generation. As such the UK is fairly typical of the developed nations in Western Europe and North America whose various domestic sectors consume between a quarter and a half of their respective final-user energy budgets.

Within the domestic sector the largest single use of energy is for heating – or space heating to be specific. Over half the energy consumed in this sector two decades ago was used to keep the occupants/users of buildings within the domestic sector warm: in 1975 this accounted for 64% of the UK domestic sector's energy consumption[207]. By 1986/7, due to an increase in total domestic appliance use combined with some improvement in household insulation, the proportion of domestic sector energy used on heating homes decreased to about

22%[216]. (The sector's total energy consumption itself increased by some 10% over the decade up to 1987.) In short, between 1975 and 1986/7 domestic space heating energy consumption alone (i.e. excluding heating of shops and factories) fell from 995 to 717 PJ. Today improvements made in space heating efficiency still account for the greatest single potential saving and, from a national perspective, all the more so because space heating applies to non-domestic sector buildings too: shops and factories have to be kept warm. Conservation measures, where made, have already demonstrated their effectiveness but there is still tremendous scope for improvements. Energy analysts from the Lawrence Berkeley Laboratory, California, have studied[217] how energy was used in the domestic sectors of nine developed nations (all members of the OECD) over the two decades between 1960 and 1980. Their results divide the nine developed nations into three groups. The first – Sweden, USA, Canada and Denmark – have homes that consume more energy than in the other countries due, primarily, to the higher incomes within those countries. The second group, characterised by The UK and Central Europe which are not yet saturated with central heating systems, has as high a space heating energy consumption but energy use for other domestic purposes is significantly lower. The final group in the study consisted of Italy and Japan which both have comparatively warm climates and the lowest energy use both per dwelling and per capita – these two countries also have the lowest incomes per capita. The study showed that potential energy savings were greatest in the cooler countries where the savings to be made were the largest, but that the rate of consumption per household varied between the nine nations by a factor of five!

The reasons for the disparity largely centres on energy demand and how effectively this demand services the population (by, for example, keeping *y* dwellings at *x* degree-days warmer). As the OECD nations developed, so their number of dwellings with central heating grew, the occupancy of dwellings declined and dwelling size itself increased. All these factors added up to a leap in energy demand. Off-setting this is how effectively the heat is used, i.e. the

Table 8.1 Space heating demand to raise a dwelling one degree above 18°C in three OECD countries in 1960/1 and 1980. OECD countries saw increased dwelling size, more central heating installed and, shown here in brackets, a decrease in the number of occupiers per dwelling

	Heat per dwelling per degree-day (MJ) (People per dwelling)	
	1960/1	*1980*
Denmark	27.7 (3.0)	22.5 (2.4)
France	14.4 (3.3)	21.2 (2.8)
UK	23.3 (3.1)	18.6 (2.7)

energy required to keep a dwelling a degree-day above 18°C. While all the OECD countries experienced increased demand, they differed in how effectively they used their energy. It concluded that if conservation measures were installed early on when countries were beginning to industrialise, when growth in consumer demand for energy is great, then the savings would be considerable. The energy analysts felt that this was most relevant to the developing southern nations, who could bypass the lengthy energy-profligate phase that the present developed (OECD) nations have endured.

More efficiency savings in domestic sector space heating are possible. For example, in 1990 only 19% of UK homes that could benefit from cavity wall insulation actually have it. Current UK regulations demand that new houses have 150 mm of loft insulation, but only 43% of existing homes have even the old, 100 mm, level of insulation. This is not unduly atypical of the situation throughout the developed (OECD) world.

Table 8.2 The proportion of UK homes lacking rudimentary conservation measures. Reported in ENDS July 1992[218]

Draught stripping to doors and windows (in households without double glazing)	35%
Insulation on hot water cylinder	6%
Roof (loft) insulation	11%
Cavity or solid wall insulation	75%
Double glazing	54%

Analysts from the Lawrence Berkeley Laboratory have continued to monitor the domestic sector energy consumption in a number of developed countries[219]. They conclude that there are three main factors and broad trends acting simultaneously that have affected how national domestic sectors change with time. First, there is energy efficiency and that most household electricity-using technologies were significantly more efficient in 1991 than in 1973 (the year of the oil crisis). The degree of individual use of energy-consuming devices is increasing and (assisted by cheap energy for many of the intervening years) this outweighed the effect of improved energy efficiency so that generally many OECD domestic sectors are consuming more electricity. Finally, technology is continuing to improve so that it is possible to prevent more energy wastage cost-effectively than before. They also noted that the dwelling or family saturation of energy-using technology was now more uniformly high across OECD nations. (That is, that most developed nation households now have central heating, colour TVs, washing machines, etc.) This has resulted in more uniform household/dwelling energy demand across the OECD.

The increase in the efficiency of OECD domestic appliances throughout the

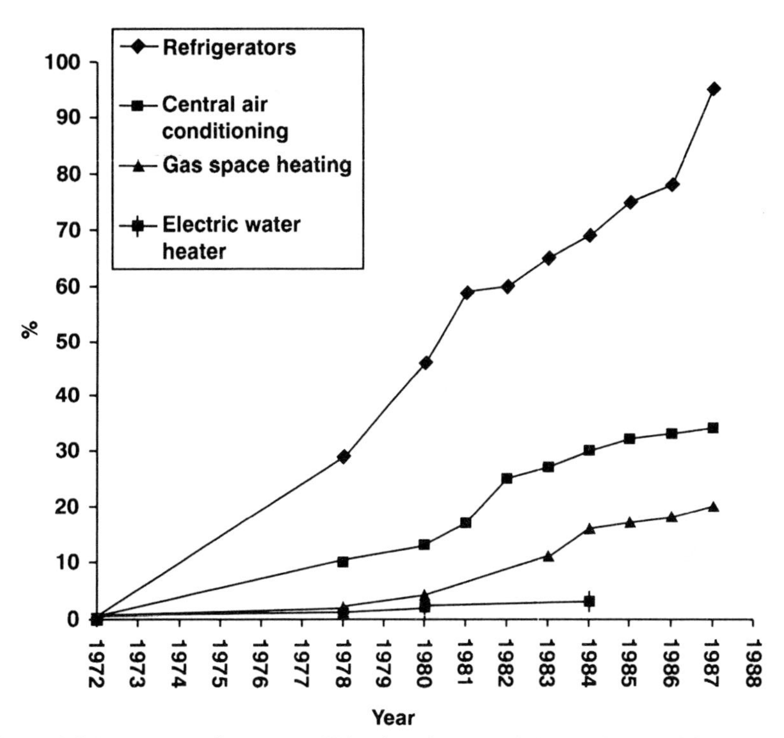

Figure 8.4 Improvement in energy efficiencies of some US household electricity-powered equipment in one and a half decades since 1972. Note, that this does not mean, for instance, that by 1987 refrigerators were 90% efficient, rather that their typical efficiency had improved by 90% over their efficiency in 1972. (Data based on reference 219)

1970s and 1980s are illustrated by the improvements in major US residential appliances, see Figure 8.4. Refrigerators and freezers have nearly halved their energy consumption without impairing their ability to cool, US room air-conditioners have improved their efficiency by nearly a third, and oil and gas heating by a tenth and a fifth, respectively. As has been mentioned, the ultimate energy efficiency of any system is determined by the laws of thermodynamics and for many domestic appliances we are nowhere near this limit. Generally speaking then, the overall result, in terms of total energy use of an appliance's energy consumption over the years, is for energy use to grow, peak, and then slowly decline to a steady state. Total energy consumption is, then, the sum of all these individual uses such as, for example television use.

In 1955 in countries throughout the OECD television sets were still a comparative luxury and not all homes had them. At that time within the UK domestic sector, TV sets consumed about 830 GWh. By 1970 more households had TV and UK energy consumption for this use had increased to around 5000 GWh but by then semiconductor technology was helping improve the energy consumption of individual televisions. By 1980 the introduction of colour TVs offset the improved efficiencies of sets and instead helped further boost consumption to over 5500 GWh. By 1984, despite colour TV becoming almost universal, total UK energy consumption through TV use had decreased almost to the 1970 level[216]. In short there are two trends. First, one of increased energy consumption as the appliance is used by more and more people. Second, there is one of decreased energy use as the device is improved and its energy efficiency increases. These twin trends apply to nearly all areas of energy consumption, though where we are on the timescale varies with each use: in terms of greenhouse considerations the key thing is to encourage efficiency in order to minimise energy consumption yet maximise the benefit from energy use. The question therefore is whether in the 1990s we can still further increase the energy efficiencies of the appliances we use?

In 1990 a report prepared for the (then) UK Department of Energy[220] looked at efficiencies of UK domestic appliances. Its conclusion indicated that substantial savings could still be made, even allowing for increased appliance use through greater future ownership levels. The report noted that aside from space heating, 21% of UK electricity sales went to powering the domestic sector and that this related to 6.5% of total UK CO_2 from energy sources (some 40 million tonnes CO_2 or 11 million tonnes of carbon). This electricity principally went to power refrigerators, freezers, cookers, home laundry, lighting and television, but while dishwashers do individually consume a considerable amount of energy, their present level of household ownership saturation is low. Having examined the efficiency of each appliance already in use and comparing this with the best new appliances, they showed that a saving of 40% can be achieved if the existing stock could be replaced with the most efficient models available in the UK; an improvement of 45% could be obtained by replacing models with the most efficient in the World; and that future developments might bring the total savings to 60%. The report also looked at the economics of realising these improvements. They estimated that the *annual* electricity savings to the consumer would be about £1.5 billion (1988 prices) and that the extra capital required to replace UK average with World best appliances would be in the region of £6 billion (i.e. four years worth of savings). Subsequent supply of energy efficient appliances would not cost appreciably more than supplying current average appliances once the industry had made the initial investment to supply the former.

Leaving such academic projections for a moment, let us take a brief look at the reality facing the consumer. Washing machines in 1986/7 accounted for about 4% of UK domestic sector energy consumption[220] yet how easily can the consumer find a machine that falls within the (1991) Department of Trade proposals for European Community standards on 'environmentally friendly' labelling? These proposals deem a washing machine as suitable for an 'environmental friendly' or 'eco' label only if it uses no more than 2.5 kW an hour and

84 litres of water for each 5 kg load. Yet a consumer looking for a brand name as a guide to efficiency will be disappointed as no manufacturer in 1991 met the minimum standards with all its models. Only 21 brands out of 60 examined in a survey by the Council for the Protection of Rural England would meet the Department of Trade standards. What is worse is the misleading way that the manufacturers themselves define which machine is 'environmentally friendly'; nearly half the machines the manufacturers mark as environmentally friendly fail to meet the Department of Trade standards[221]. It is regrettable that such problems of efficiency, and sustainable resource-use labelling (eco-labelling as it is known) are common throughout industry and business, but in greenhouse terms this again provides room for improvement, enabling us to decrease energy consumption (and greenhouse gas emissions) without affecting the benefit derived from business and industrial goods.

If we assume that by 2010 *all* households will have colour TVs, washing machines, etc, improving energy efficiency of these devices does not necessarily mean a reduction in overall energy consumption. Take the household that, in the UK, up to the 1980s watched a typical two hours of TV a night (I pluck a figure from the air), you might think that subsequently lowering the device's energy rating would lower that household's TV energy consumption for the 1990s. This would be a logical conclusion but life is more complicated. For instance, the 1980s saw the introduction of breakfast TV with some (lower energy) radio sets being turned off in favour of higher energy TV, and suddenly breakfast viewing has to be included for the 1990s energy calculation as to how much TV watching affects demand. This phenomena of increased choice stimulating demand is one of our society and culture. It has more to do with ourselves as human beings than with the energy consumption (or energy intensities) of devices and their efficiency.

Also very much to do with being human is the behavioural response to technological change and the improvement of energy efficiencies. Take one example, conservation and heating. Improving the thermal performance of a dwelling does not merely enable the owner to heat the building the same number of degree-days a year for less cost; there are more choices than that. Having paid for the insulation (albeit up front) and warmed the house to the same extent as before, the owner can do one of two things. First the owner might put all, or part, of the saving towards the cost of the insulation. Even with house owners meeting the cost of energy conservation measures through savings on their fuel bill, there may (with proper budgeting) after a while come a time when the cost of the insulation has been met, which bring us on to the second choice. Having paid for home insulation the owner has a choice of what to do with the cash savings on fuel bills – there is now an opportunity to spend and, in the parlance of economists, the owner has an opportunity cost decision to make. The owner could spend the savings on non-domestic goods and services. Alternatively, the owner could spend the savings on extra fuel in order to have a warmer home. In short, encouraging energy conservation measures by themselves do not necessarily result in lower energy consumption but warmer buildings too. In addition to positively encouraging energy conservation, it is necessary to actively dissuade energy consumption, if consumption itself is in fact to be reduced.

Behavioural relationships to energy consumption are complex and are not just confined to energy conservation and heating. Studies suggest that in the Nordic countries there is a national household energy conservation potential of up to a fifth (20%) from behavioural modifications alone[222]. Though it would be difficult to change a population's behaviour to *fully* realise these behavioural conservation potentials, equally some behavioural changes would be easy to encourage: charging more for fuel would be one obvious way.

No wonder that when governments ask their energy specialists how energy conservation measures will affect their domestic sector's future energy consumption, they can only give a broad window for a reply. The number of people per household (family size), size (area and volume) of dwelling, technology, behavioural and cultural changes with time all conspire to preclude accurate forecasting. In 1989 the Energy Efficiency Office of the UK Department of Energy attempted just this, to 'project' UK domestic sector energy consumption in the year 2010. Their answer comprised of a broad window of demand with the higher figure some 23% above the lower. They predicted that the UK domestic sector might consume more energy in 2010 than the 1695 PJ (470 TWh if all the energy was electricity) in 1985, with a lower window of 1735 PJ and a higher window of 2140 PJ[220]. However, their scenarios assumed a *laissez faire* approach to energy conservation policy and an increase in material standards of living across the board (rather than in specific areas).

So where does this leave us? In discussing energy policy in the context of the greenhouse effect we are not concerned with a range of scenarios, but just one specific scenario – that of the realisable potential for energy conservation within the domestic sector. Given all of the above, and assuming greenhouse policies will actively encourage consumers to realise more of the potential savings available, then it would not be unrealistic to target for a 10%–20% decrease in domestic sectors' energy consumption across the OECD nations over the 1990 figures by the end of the twenty-first century's first quarter.

Manufacturing, commerce and services

Of the greatest energy concerns to many politicians and energy strategists is the impact of any energy policy on the country's economy. Indeed when British Nuclear Fuels (BNFL) countered[210] the IIED/Leach *Low Energy Strategy for the UK*[207] two of the principal criticisms were that the strategy would not provide an economic return, and that some economically desirable (i.e. profitable) activities would be omitted; this despite the strategy being specifically designed with economic growth in mind. From the greenhouse point of view it *is* vital that the economy be healthy in order that environmental policy can be implemented, so that the goals of the environmentalist and the long-term view of the industrialist are one and the same. As has been discussed earlier, the Leach IIED strategy noted that since the 1973 oil crisis the GDP–energy linkage was broken. Further, that the coupling of energy demand to economic activity was misleading and that a good example of this was in the non-manufacturing sector of the economy. In the UK this (in 1978) generated over 50% of its GDP yet is responsible for only 12% of total energy consumption. Of course energy and economic activity are not entirely independent of each other so that a cut in economic activity will in all likelihood result in energy demand decreasing: for example, in the depression between 1980 to 1983 North American primary energy consumption dropped from 2062.1 mtoe by 6.6%, and Western European consumption from 1330.0 mtoe by 4.3%[1] These were unplanned energy savings; anyone can save energy by shutting down a factory. The trick in designing a low energy strategy is to ensure that energy is conserved *without* lowering economic productivity.

Saving energy and maintaining economic productivity need not be mutually exclusive goals, indeed they can be mutually compatible, enhancing one another. A firm that insulates its factory, lags its steam lines, and uses processes that require less heat, will be more competitive than a firm using more energy per unit product. The only way the firm using more energy can remain in competition with one using less will be to make financial savings elsewhere, such as salaries (but then they will not attract the best staff) or on building/infrastructure renewal (but then this will lower economic activity elsewhere in the economy, specifically those parts relating to the infrastructure renewal sector). True the firm using less energy will not be providing so much income to the energy generating industries. Equally, this last could be overcome by raising energy prices but that would make the country's goods as a whole less competitive with those elsewhere using cheaper fuels. For these reasons energy pricing is an important tool in many low energy strategies, but to be truly effective it needs to be international. Outside of oil cartels such as OPEC which does fix its oil price on the international market (assuming the member countries are in agreement), the raising of energy prices has not formed a part of a coherent international energy policy, if only because there has been no such thing as a truly international energy policy. Of course, in the long run improving energy efficiency *is* beneficial to industry and over the years there have been changes in developed nations' manufacturing energy efficiencies as well as the rates of OECD nation CO_2 emissions from manufacturing.

The USA, which is the nation with the highest emissions of CO_2 (both absolutely and on a per capita basis), devoted 27% of its energy consumption in the mid 1980s to manufacturing. Energy supply, and especially electricity supply, is one of the most capital-intensive activities in its economy. During the late 1970s and early 1980s, new energy supply plant and equipment for manufacturers constituted about 10% of business investment for all purposes. The dominant energy used by US manufacturers was electricity (43% of all fuel use), followed by direct use of natural gas (27%), with direct use of coal and oil used least (15% each[223]). Up to the end of the 1950s consumption broadly grew in line with the economy so that US manufacturing energy intensity (production in dollars divided by energy consumed) increased annually. From 1958 to the oil crisis of 1973 there was a period of stability but after that date US manufacturing energy intensity began to fall – ultimately by more than a third between 1972 and 1985 so that US manufacturers were selling more dollars worth of goods per joule of energy consumed[223].

Even allowing for inflation, such analyses are over-simplifications, but even more comprehensive calculations show that US manufacturing (and across OECD countries generally) is becoming more energy-efficient. One way of tightening such calculations is to look at the value-added energy intensities rather than energy intensities themselves. That is to say, the energy manufacturing uses to increase the value of its labour and raw materials in the making of its finished products (rather than just the energy manufacturers use divided by the value of their products). An additional refinement (essential when using energy intensities to discuss greenhouse implications) is to make an allowance for energy *not* derived from fossil fuels. This last enables energy analysts to calculate a figure for the added value carbon dioxide intensity (the CO_2 release by a nation's manufacturing sector in order to increase the value of its raw materials and labour). It means that countries like Norway (with its extensive use of hydroelectric power) and France (with its nuclear fission programme) generate very little carbon dioxide per dollar added value in their respective manufacturing sectors compared to countries like Denmark (which has neither hydroelectricity nor nuclear power).

Where does this all take us? The manufacturing sectors of the most developed nations – the OECD-9 (Denmark, Norway, UK, Germany, Italy, USA, Sweden, Japan and France) – show a remarkable value added carbon dioxide reduction in intensity overall of 41.7% between 1973 and 1987. That is to say that in 1987 we were getting, in real terms, 41.7% more dollars (pounds, francs or whatever) per kilogram of carbon dioxide released than in 1973. Part of this is due to improved efficiency (21.8%) and part to the growth of (non-greenhouse) nuclear fission. Energy analysts consider that such perspectives, when considered nation by nation and by fuel price, indicate that improved energy efficiency is driven by energy costs as well as economic growth[224] – the former being the stick, if you will, and the latter providing the wherewithal.

Moving away from manufacturing and national perspectives towards those of individual 'businesses' within the commercial and service sectors, motivations for energy conservation change. Energy pricing is still a major motivating factor in getting managers to instigate energy conservation measures, but other factors come to the fore such as whether the business owns its own property or whether it rents. There is little incentive for business to improve property it does not own. The picture is complex. One survey of over 200 small US businesses that had energy audits showed that within two years of the audit over 40% of the recommended measures were adopted[225]. The other way of looking at this is to say that after two years of being audited business still had to implement over half the recommended energy saving measures. Aside from owner occupancy, other factors determining whether managers would instigate energy conservation measures centred around cost. More than half the implemented measures had zero cost (such as turning thermostats down or switching heating off in little-used rooms) and many of these recommendations were for operations and maintenance measures[225]. Energy prices therefore, while providing some motivation for at least considering an energy audit and possibly adopting their recommendations, are in themselves not the complete answer. Grants to cover the capital cost of installing insulation are as important.

Given the above, that not all energy is used in the most efficient way using current technology and indeed that energy efficiency technologies are developing, the potential for future savings within the manufacturing sectors of developed nations is not inconsiderable. The potential for improved efficiencies in less-developed nations is greater as energy is used so inefficiently in these countries. There is the real possibility of many years, if not decades, of economic growth in manufacturing without the same proportional increase in energy consumption.

As to the potential for conservation measures to save energy within the commercial and service sectors, it is almost as significant as the savings to be made in the domestic sector. Even without investing any capital in conservation, such is the profligate way in which we currently use energy that at least 4% of heat and light bills can easily be saved at no cost whatsoever. Up to the end of 1991, and the dissolution of the Soviet Union, the UK National Health Service was the largest single employer in Europe, save for the Red Army in Eastern Soviet states. UK hospitals alone (that is excluding doctors surgeries and clinics) spent over £225 million a year (1990 prices) on heating and light. The UK Department of Health recognising the saving potentials set a target of cutting energy costs by 15% over a five year period up to 1995. With UK interest rates themselves running at between 10–15% (in 1990/1) the UK Government who fund the National Health would have had to have made an investment in excess of their Department of Health's energy bill to obtain the same return as their modest savings target would achieve, yet it was decided that little one-off funding would be made available to realise this goal. In 1991, to assist in realising these goals, a survey of the way 450 hospitals consumed energy was carried out which showed that a quarter of this target saving could be realised within a year – that is the cost of the conservation measures would be recouped also within a year[226]. Such quick-return cost-effective savings are easy to accrue from existing commercial infrastructures (and of course other savings could be obtained with greater investment and allowing a longer payback time), but even greater quick-return savings could be obtained if energy conservation was to the fore in the minds of hospital designers prior to their construction. Continuing with our example of hospital energy savings, one architectural consultancy did just this. They realised that in large UK hospitals some 22% of the energy consumed was used in the kitchen. Traditional designs concentrated on keeping the kitchen staff cool by venting this heat outside, literally throwing the energy away. By recycling this heat elsewhere not only were they able to cool the kitchen but cut down on having to buy energy to warm other parts of the hospital. Heat was also extracted from the incinerator flue, generator engines and even washing up water. Greater use of natural light not only cut down on energy costs but brightened dull gloomy rooms; which also illustrates that energy conservation measures should not be confused with those of austerity[227].

Hospitals provide just one example of the cost-effectiveness of low-cost energy savings within the service sector. What of the purely industrial sector? In 1983 the UK Department of Energy looked at 19 'illustrative' industrial energy conservation schemes from: recovered heat from horticulture and sulphuric acid plants; lighting controls in buildings; improved boilers... In total the schemes added up to £130 million pounds worth (1981 prices) of energy conservation - investment, yet all had payback times of less than three years[228]. At the time these results were published the UK Department of Energy was submitting evidence to a public enquiry on the building of the then proposed Sizewell (PWR) nuclear fission power station. The Department was anxious to demonstrate a need for the power station and it supported the power utility's (the Central Electricity Generating Board's) case. The need to invest in energy conservation rather than supply was of low priority even though encouraging evidence had already been gathered as to the potential energy conservation for the UK as a whole. A restricted internal paper by the Department of Energy's internal Energy Policy Unit[229] estimated that some 12% of the UK's industrial sector energy consumption could be economically saved; part of this would come from a 30% primary energy saving from improving buildings as heating and other services were gradually replaced by the year 2000 – further savings would be available beyond that date as replacements continued. For the nation, the paper concluded that the potential for economic energy conservation over five years would save 5000 million therms (delivered) a year at a total cost of £5000 million (1981 prices) – an annual saving of about 6% of UK primary energy demand or around 12.6 million tonnes of oil equivalent (more if one considers other potential, pre-delivery, savings such as at power stations). Since the paper was published some savings have been made as old equipment and buildings are replaced. Equally, energy conservation technology has also developed so it would not be unreasonable to broadly expect that savings of a similar magnitude are still available today.

NATIONAL AND INTERNATIONAL ENERGY CONSERVATION POTENTIALS

The above low-cost, five year return saving of 6% of UK primary energy is only a start on improving energy efficiency since even greater savings could be realised by adopting higher cost, and longer payback, conservation measures. Again taking the UK as an example, and again taking official estimates, the UK Department of Energy at the end of the 1970s considered the potential practical savings (for policy purposes) to be 60 million tonnes of oil equivalent (over a quarter of UK primary energy demand[230]). More recently, in 1991 the UK Government established its Advisory Committee on Business and the Environment (ACBE) chaired by John Collins, Chief Executive of Shell UK, and then through to the mid-1990s by Derek Wanless, Chief Executive of the National Westminster Bank. This committee felt that improvements in energy efficiencies were possible in industry, and that the motor industry could reasonably improve fuel efficiencies by as much as 10% by the year 2005. With regards to efficiencies in the domestic sector, it called for the Government to set a target of 10% energy savings by the year 2000, but that to achieve this the Government should back this target with a 'co-ordinated programme of regulation and economic instruments'. One instrument the ACBE called for was a landfill levy to encourage recycling (though it thought that the Government's target of 50% of household waste being recycled by 2000 might not be the soundest environmental option). In 1993, among the legislative energy measures it called for the UK Government to make were mandatory energy statements for both new and existing commercial premises, as well as (at first) voluntary, then mandatory energy labelling of appliances. All of which would be backed up by loans and capital allowances for energy efficiency investments[231]. By the mid-1990s the UK Government had yet to deliver on these proposals, with the exception of a new landfill tax in 1996.

The most important reason for nations to turn to governmental regulation of the market for improving energy efficiency is that users are typically small to medium-sized enterprises and individual consumers, whereas energy suppliers tend to be large business concerns operating in a semi-monopolistic environment. The former require quick investment pay-back times of the order of two to three years; that is the discount on the investment in efficiency has to be about 50% to 30% per annum. On the other hand energy suppliers look at longer horizons with pay-backs of the order of over a decade which is well within a third of the lifetime of a power station; these require energy supply investment discount rates to be far less than 10%. Nonetheless, potential savings from improving energy efficiency are there.

Saving estimates for other high energy-consuming nations have also been calculated. In 1991 the US Committee on Science, Engineering and Public Policy concluded that greenhouse emissions could be cut by between 10% to 40% for little or no cost largely through energy efficiency measures. The committee noted that: 'The efficiency of practically every end-use of energy can be improved relatively inexpensively[232].

Internationally, the initial priority for energy conservation lies with existing high energy users – the developed OECD nations such as the USA and Western Europe – as not only can they afford to implement conservation measures, but a saving of even a small *proportion* of their energy consumption represents a considerable *absolute* saving. Furthermore, as the developed nations are the ones who export technological know-how, there is the potential for exporting energy conservation technology. Even so, non-OECD nations are now beginning to consume more energy and a considerable potential is emerging for energy conservation among these countries.

Brazil is an example of a nation that is developing fast and which has a rapidly growing demand for energy. In 1970 total Brazilian electricity demand stood at 38 TWh. A decade and a half later (1985) this had grown by over 360% to 175 TWh and power-sector planners forecast that electricity demand will continue to grow through to the end of the century to about 276 TWh. By the year 2000 electricity efficiency savings of 30% are considered possible, but lower savings of 20% are thought to be practically attainable through the use of more efficient motors, lighting equipment and appliances. Currently the federal government and utilities are proceeding with a national energy conservation programme, but energy prices are low (which means that cost savings from energy conservation are low), there is little investment capital available to instigate conservation programmes and little information[233].

Globally, the case for energy conservation is compulsive, regardless of greenhouse issues. Most forecasts for World energy consumption in the year 2020 lie in the range 600 Exajoules (EJ) per year to 1000 EJ (which compares with current (1996 World energy consumption of about 347 EJ) – see Figure 6.5. (1 Exajoule = 10^{18} Joules or 23.15 mtoe in intrinsic energy terms.) Assuming the lower limit, then investment in energy supply would have had to increase from its late 1980s level at a rate of about 2% per annum, and for the higher 1000 EJ estimate at a rate of 3% per annum. In the mid-1980s, for developing countries, The World Bank had estimated that to meet a 2.5% per capita growth in energy use to the mid-1990s investments in energy supply of $130 billion on average each year for the decade to 1992 would have been necessary, and that foreign exchange allocations to energy investments would have had to have grown at 15% (real cost) per annum[234]. The World Bank (correctly) assumed this to be an unrealistic expectation, and indeed the lack of developed nations' support is forcing underdeveloped countries into short-term energy perspectives. Today these countries are short of energy and frequently experience regular power cuts. Yet it is because the cost of investing in supply is so high that a considerable investment in energy conservation can have a high impact, lowering necessary future supply investment (which would in any case give a better return when energy supplied is used efficiently), as well as producing immediate fuel savings.

In 1987 a study by the World Resources Institute in Washington DC[235], indicated that if the World adopted a low-energy strategy then the developed OECD countries could not only halt their increase in overall energy consumption, but allow for a reduction in energy consumption while still enabling their economies to grow through to the year 2020. Meanwhile the developing countries could achieve considerable economic growth (to the equivalent of developed nations of the 1970s) by ensuring that their development was underpinned by using energy efficiently. Although their overall energy consumption would increase, there would be no need for an overall increase in World carbon dioxide emissions from energy consumption. Consequently with:

- current profligate energy use,
- encouraging energy sector economics,
- cost effective national low-energy strategies,
- not to mention the sheer logic of reducing energy waste given a finite fossil fuel supply...

the case for conservation is compelling. Additional greenhouse arguments for increased efficiency, lowering energy-related greenhouse gas emissions, are almost superfluous, as are consequently, greenhouse uncertainties.

CHAPTER 9

REFORESTATION – THE GARDEN VERSUS THE GREENHOUSE

So far we have only considered reducing atmospheric carbon dioxide inputs through control of fossil fuel use: be it either through introducing non-fossil fuel technology, or through increasing the efficiency of energy use hence the lowering of overall energy demand. Such controls relate purely to the sources of atmospheric carbon (all be they anthropogenic) within the carbon cycle, and yet the cycle also contains sinks that could be employed to soak up carbon dioxide. As mentioned earlier (see Chapters 4 and 5), the concentration of atmospheric carbon dioxide naturally oscillates by five parts per million (by volume) each year. This waxing and waning is due to the net photosynthetic absorption of carbon dioxide by plants in the summer, and the net release of the same in the winter through respiration. (This waxing and waning is mirrored between the Northern and Southern Hemispheres, reflecting the opposing seasons either side of the tropics.) The trick, from the greenhouse perspective, is to increase the ability of these photosynthetic sinks to absorb carbon dioxide, so removing it from the atmosphere.

PLANT LIFE AS A SINK FOR CARBON

When talking about plants absorbing carbon dioxide one tends to think of the lush leafy plants and the tall, densely packed trees of the tropical rainforest. So it is useful to remember that less than a third of the World's net photosynthetic entrapment of carbon (what ecologists call primary production) takes place in tropical forests, and that a similar amount of primary production takes place in the oceans. Carbon exchange between atmosphere and ocean is almost as great as that between the atmosphere and terrestrial life. Hence the need for studies into the marine carbon cycle (see Chapter 5) and the increased attention the oceans have received in the search for greenhouse solutions. Some of these have been quite bizarre. In the summer of 1988, John Martin of Moss Landing Marine Laboratories (half jokingly) suggested that if the Southern Ocean were fertilised with about 300 000 tonnes of iron then the resulting boom in phytoplankton (planktonic plants) could possibly absorb enough carbon dioxide to soak up the surplus pumped into the atmosphere through human activity. (Though he has subsequently stressed the importance of understanding how our biosphere and its ecosystems work before making any such attempt[236]. Furthermore, the small amount of research that has been conducted up to the mid-1990s has not been conclusive.) Other ways of manipulating the ocean's carbon budget have been alluded to, but unfortunately require at least as great a tinkering with the ocean ecosystem about which we know so little. These include concerns that predators – Antarctic birds and mammals – at the top of the food chain, are breathing out (respiring) the carbon dioxide originally photosynthetically trapped by their prey's food, and returning this to the atmosphere. It has been suggested that between a fifth and a quarter of the carbon originally taken up by Southern Ocean plants is returned to the air in this way: the Southern Ocean itself accounts for nearly 15% of the total global ocean primary production[237]. The implication is that lowering these predators' numbers would also lower the rate at which carbon dioxide is returned to the atmosphere from the Southern Ocean ecosystem.

There are many terrestrial ecosystems within a variety of biomes, but those with the greatest ability to photosynthetically capture and retain atmospheric carbon dioxide over long time periods are those dominated by trees. Tree-dominated ecosystems are so good at generating biomass that they contribute the most of all terrestrial ecosystems to the World's total long-term net primary production. Broadly speaking (because the figures are uncertain and the global carbon cycle is dynamic) worldwide, tropical, temperate, and boreal forests (predominantly Northern Hemisphere), each respectively contribute some 49.4, 14.9 and 9.6 Gt of dry vegetative matter a year, out of a total terrestrial net primary productivity of 115 Gt of dry matter, with the World total being some 170 Gt[238]. This dry matter does not just contain carbon – the carbon atoms are easily outnumbered by those of hydrogen together with oxygen – consequently of the 170 Gt less than 70 Gt are carbon and part of this is ultimately lost through other metabolic pathways, particularly through respiration. While continental or terrestrial ecosystems absorb about 102 Gt of carbon a year, 50 Gt is respired back, and upto a further 50 Gt is lost to the soil (and from there some will return to the air) and rivers (and finally seas), and some is laid down to form (ultimately) fossil fuel. In short terrestrial ecosystems see a few score of gigatonnes of carbon naturally on the move around the carbon cycle each year: whereas the annual (1994) human production of fossil fuel (oil, coal, and gas) is only a little over 7 gigatonnes of oil equivalent[135], and our atmospheric contribution from the burning of fossil fuels is a paltry (compared to nature's carbon cycling) 6 gigatonnes of carbon a year. It might therefore be possible to slightly change the way carbon moves through terrestrial ecosystems yet have a considerable negative effect, effectively offsetting the human carbon contribution to the atmosphere.

DEFORESTATION

Humans have been destroying and replacing forests for as long as they have had the ability to handle tools. As human civilisation has grown, so have the forests been cut down and replaced by arable crops. The granaries of ancient Rome were filled from crops grown on land in North Africa that was once forest: now it is largely desert.

In modern times forest has generally been replaced by grasses, be it for fodder for cattle or cereals for human consumption. In 1950 roughly 30% of the land was covered by forest, roughly half of which was tropical forest. Since then the area of temperate forest has remained fairly constant due to replanting – though there is some concern over the lack of replanting in northeast America – and elsewhere replanting has not been as thorough as it might be, resulting in thinning primary forests and many forests with only secondary growth present. By 1975 tropical forest had declined to 12% and by the year 2000 some environmental biologists think that we will be lucky to have retained 7%[119]. At the beginning of the 1990s much Central American tropical wet forest has gone, severe inroads have been made into the forests in South America and Democratic Republic of Congo, hardly any forest remains in West Africa and forest is rapidly being cut down in the East Indies and in parts of Southeast Asia. The integrity of the Madagascan forest has gone, with only largely-unconnected islands of forest left, and on the African mainland, the forest along the eastern coast is slowly, but actively, being eroded[119, 239].

Other than fire (the traditional tool for deforestation employed by humans in the tropics), today chain saws and modern roads are accelerating the rate of deforestation. Even so, in 1987 remote sensing satellites detected around 6000 fires in Amazonia during the dry season while areas of the forest were being cleared for agriculture. This way of farming, known as 'slash-and-burn', depends on nutrients being released from forest plants as they burn so that not only is the land cleared, but the soil fertilised. On a small short-term scale, slash-and-burn agriculture is quite sustainable as long as there is enough untouched forest surrounding the comparatively small farm site to ensure that the soil is not washed away, although many decades are needed for the site to recover to any kind of forest. Where slash-and-burn agriculture is carried out extensively, or the land is cleared on a large scale for modern agriculture or ranching, then the long-term viability of the site is threatened. Large-scale clearance not only removes the forest plants wholesale – plants which may have a genetic value to agriculturalists or contain pharmaceutically valuable biochemicals – but also exposes the thin tropical forest soil directly to the force of tropical rains which wash away soil and nutrients, and soil erosion follows thus preventing the forest from reclaiming that land. Each year clouds of smoke can be seen by satellites covering millions of square kilometres not only releasing CO_2 but also nitrogen oxides – 2% of the atmosphere's nitrous oxide comes from biomass burning. Coincidentally, nearly 2% of the World's tropical forest, about 180 000 km^2, that are either cleared or modified in some way (such as for agriculture or agroforestry) each year[16]. To date, human activity has destroyed approximately nearly half of the World's tropical rainforest cover. But note the word 'approximately' as estimates for tropical forest (let alone 'rain forest') cover and deforestation must be viewed with extreme caution (notwithstanding this, even the lowest deforestation estimates are of a magnitude warranting great concern). Forests, almost by virtue of definition, are remote places which makes monitoring difficult. Remote sensing pictures (air and satellite data) are also difficult to interpret accurately, and there are problems of double counting where logging has taken place once, and is now happening again[204]. Yet this should not be an excuse for complacency. From a biological conservation point of view (though not necessarily one of carbon dioxide sequestering) there are additional concerns from habitat fragmentation. Allowing for a 1 km edge effect, one estimate of Amazonian deforestation, based on satellite data between 1978 and 1988, indicates that while about 15 000 km^2 a year was deforested, the rate of forest degradation *and* habitat fragmentation was approximately 38 000 km^2 a year[241].

Tropical deforestation

Of the estimated 180 000 km^2 of tropical rain forest disturbed each year, broadly speaking, roughly half are disrupted in some way but with partial cover remaining. The other half of forest is completely disturbed, with virtually all the tree cover removed. In addition, a further 44 000 km^2 is logged each year, and then left to 'regenerate'[16]. The erosion of forest is prompted by the prospect of substantial, but short term, gain. Tropical hardwood may be sold for a considerable price to markets in the developed nations. Tropical deforestation also results in a brief increase in the agricultural land available; though this land rapidly loses its fertility as nutrients and the thin soil are washed away – soil that will then silt up streams and rivers. As such, the pattern and impact of tropical forest disruption is markedly different from that occurring in temperate and boreal forests.

In 1990 the Food and Agricultural Organization (FAO) of the United Nations reassessed the degree of tropical forest clearance to see whether the trend in deforestation was up or down? Looking at deforestation in just 62 countries, which represented 78% of the World's tropical forest in 1980, their preliminary results indicate that the annual rate of deforestation in 1990 was 16.8 million hectares, or 168 000 km^2, (an increase of some 83% over the 1980 rate)[242].

Finally, though of lesser concern with regard to greenhouse issues, from a biological conservation perspective arguably the greatest causes of concern with tropical forests are two fold. First, forest loss is proportionally greatest in the places where least forest remains, for example West Africa and Madagascar. And second, the problem of fragmentation, so that even if a species survives in the short-term its genetic diversity suffers and its long-term prospects for survival are lessened.

Table 9.1a Comparison of FAO estimated rate of deforestation in 52 countries (common to both FAO assessments of the 62 total), and their tropical forest area, in 1980 and 1990

	1980		*1990*
		(million ha)	
* Mean annual rate of deforestation	9.2		16.8
Area of remaining tropical forest	1449		1281

* Mean annual rates for the 1980 assessment cover the years 1976–1980, whereas the 1990 assessment of mean rate covers the years 1981–1990

Table 9.1b Total tropical forest area in all 62 countries in FAO assessments and annual percentage rate of change[208]

	1980		*1990*
		(million ha)	
Forest area	1450		1282
Mean rate of change in % per annum 1981–90		–1.2% p.a.	

Note: The 10 countries of the 62 not covered by Table 9.1a are comparatively small tropical islands which have experienced little deforestation

Amazonian deforestation

Brazil's tropical rain forests are the largest in the World and so, not surprisingly, they have seen the greatest losses with close to 40% of their original 2 860 000 km^2 of forest gone, with (at the beginning of the 1990s) over 2% of the forest being cleared each year.

Recently some of the worst of Amazonian deforestation has taken place in the Brazilian region of Rondonia. The area was largely intact prior to 1970: it was home to about 10 000 native Indians and a few rubber tappers and prospectors. By the mid-1970s it was becoming clear that Rondonia has a better soil than much of the rest of Amazonia: soil that could support agriculture for longer than the few seasons that can be expected elsewhere. By the early 1980s about 3% of Rondonia's forest had been cleared but, by the mid-1980s this proportion had trebled to over 10%, and the population had increased by a factor of a 100 to around 1 000 000. Ecologically, deforestation in this area is particularly damaging: the rich soils support a greater number of plant, and in turn animal, species – the area may have been one of the long-term centres of high species diversity, having provided a refuge for many species through glacial times.

In other parts of Amazonia, where the soils are, more typically, poorer, deforestation has more to do with demographic pressures and tax incentives.

Some deforestation also arises as a by-product of mineral workings and oil exploration. It is to be hoped that Brazil's policy changes of the early 1990s of reducing tax concessions to developers will reduce the rate of clearance. However, it must be noted that the development and, so-called, aid policies of developed nations (the principal greenhouse gas emitters) have not helped. Such policies that politicians and business may claim to be altruistic have more to do with ensuring a stake in Amazonia's mineral wealth and tropical hardwood, than in ensuring development through the sustainable management of natural resources.

African deforestation

The largest amount of forest on the African continent is tropical rain forest, followed by sub-tropical and mangrove swamp forest. Altogether Africa's tropical rain forests make it the second richest continent in the World in terms of both tropical forest cover and, in all likelihood, in the number of animal and plant species. Given that tropical moist and rain forests bind up the most biomass *per* unit area of all types of ecosystem, it is easy to see that disturbing Africa's tropical rain forest will have a pronounced effect on that continent's carbon cycle: indeed, as with the disturbance of the South American Amazonian forest, the consequences are of global importance.

Nearly three quarters of Africa's tropical rain forest lie in a broad belt starting in Sierra Leone stretching across the Guinea countries of Ghana and Nigeria, and into the Congo basin nations of the Central African Republic, Gabon and Democratic Republic of Congo: Democratic Republic of Congo itself has the largest area of forest. Countries that have seen the greatest deforestation as a proportion of their original cover, are Gabon and the Ivory Coast – the latter, which used to have around 160 000 km^2, has lost over 95% of its primary forest. In terms of absolute area, due to the sheer size of the country, Democratic Republic of Congo has lost the most, but equally, for the very same reason, has retained the most with inaccessibility being a significant saving factor. Again, as with South America, the trend has been in the main one of deforestation and replacement with agriculture.

Deforestation in Indonesia

Indonesia currently contains the largest remaining reserve of tropical forest in insular South-east Asia: a region that itself contains close to a fifth of the World's tropical forest. Already over 40% of the original cover has suffered deforestation, or fragmentation, with the rate increasing annually. The preliminary analysis of the 1990 FAO tropical forest cover assessment indicates that throughout the 1980s, on average 5500 km^2 were deforested each year[243]; other estimates for the late 1980s suggest that this has grown to 7420 km^2 a year[239]. More recent figures have increased this to around 12 000 km^2 a year at the beginning of the 1990s[244].

The likelihood is for the late 1990s to see even more forestry activity. Indonesia has long been an exporter of wood, but in 1987 it began an export trade in paper. By 1991 Indonesia had 41 paper mills and, reportedly[244], over the next 15 years plans to build another 56 mills. At the moment it is difficult to see how these mills could be supplied with timber from forests managed in a sustainable way: it takes time to establish sustainable forest management regimes and Indonesia intends to build pulping mills faster than trees can grow. The current Indonesian government does recognise the problem and is trying to encourage sustainable development in the face of a growing poverty stricken population. The official target is sustainable forestry management by the year 2000 – although some officials acknowledge that this is very ambitious.

Small tropical island deforestation

Although the large reserves of carbon locked in continental forests are of greatest concern in terms of potential climate forcing, the plight of some small tropical islands, with their unique ecosystems, encapsulates what can happen when regional human population growth and poverty encounter forests. Anjouan, an island of the Comores between Africa and Madagascar, provides a powerful illustration. Between 1901 and 1990 its population grew by over 1000% from an estimated 15 000 to 175 000. Eighty percent of the population depends on agriculture for a living, so that today's population places a particularly heavy burden on the forest. Between 1972 and 1987 some 87% of Anjouan's primary forest was destroyed, leaving just 1109 ha[245].

Anjouan's population growth may seem atypical in global terms, but it is not on a regional basis. Already many mainland tropical cities have large rapidly growing urban populations and high unemployment. Should these migrate into the forest – a not unrealistic scenario – then rapid forest destruction would be the most likely outcome.

Boreal deforestation – Soviet Commonwealth and Canada

Tropical forests are not the only type of disturbed forests, the coniferous and temperate deciduous forests of the north have slowly dwindled. First, it was the woodlands of Europe; these have largely been replaced over the past 1000 years by human habitations and intensive farmland. In the late nineteenth and twentieth centuries considerable losses have been seen on two continents, in both Russian and Canadian forests.

Siberia's forests cover an area the size of the USA, some 2.3 million $miles^2$

(about 6.0 million km^2). They represent 57% of the World's coniferous forest volume, and 25% of the World's inventoried wood volume. This makes them twice as large as the Amazonian rain forests[246], though their net productivity (the amount of biomass they produce per unit area per annum) is somewhere between 19% and 25% that of tropical rain forests.

The current pressure on Siberia's forest stems from that encountered earlier on with the western Soviet forests. Historically, in just two decades from 1888 over 3 million hectares of forested land was cleared from the Russian Empire. In European Russia, the total clearance over historic time until the end of World War I has been about 70 million ha[239]. Today, Russian planners are embarking on joint ventures with western multinational companies and expanding their timber operations. Some 4 000 000 ha of forest are harvested each year in Siberia alone and Siberian timber accounts for 2.6% of the Russian federation's total foreign trade. As with deforestation in the tropics, deforestation in the Russian Commonwealth is the result of non-sustainable forestry management regimes[246]. Again, as with tropical deforestation, such quick return policies (at the expense of long-term sustainable profit) are fuelled by poverty, population growth and the need for developed nations' currency so as to reap the benefits of the World market.

Just as deforestation has not been restricted to the tropical forests, neither is it confined to the economically disadvantaged nations. Greed rather than need provides as tenacious a motive for short-term, non-sustainable exploitation of natural resources. Canada is part of what the Brandt Report called, the developed North. As well as having a far healthier economy than the former Soviet nations, it does not have the absolute pressures on it to tackle internal poverty or to catch up with other nations. Yet it too is failing to manage its forest resources in a sustainable way. Its problems stem from individual (but big) businesses operating at the expense of national (not private) natural resources.

It would be unfair to say that no replanting takes place, much of Canada's boreal forests are replanted after cutting, but inadequately. This inevitably is leading to a whittling down of the nation's future potential lumber resources. For example, annually in Ontario some 1900 km^2 of forest are felled, with replanting taking place on only about a third of this, some 700 km^2. Allowing for natural regeneration, overall there is an approximate reduction of about 1000 km^2 per year out of a total productive forest area of 213 000 km^2. In addition, vast areas of forest have been so disturbed that their ecosystem mix has departed markedly from the original (post-glacial) pristine community. There are also concerns that the current non-sustainable management regimes are enhancing the ecosystem's natural production of the greenhouse gas methane[239].

Combined, the Russian–Canadian backlog of unreplanted forest is equal to roughly half the uncut area of Amazonian forest. To be added to this is the area of land partially, hence inadequately, restored which amounts to over 1.6 million km^2 – roughly half of the area of Brazil's portion of the Amazonian forest. Both the Russian–Canadian boreal and the tropical forests, have been managed non-sustainably. But though there are similarities in governmental forestry policies of the nations bequeathed with both types of forest, the degradation has proceeded further in the more-developed northern countries.

REGIONAL CLIMATIC AND ATMOSPHERIC IMPACTS OF DEFORESTATION

Tropical forests, by virtue of definition, are located in regions where there is a high solar energy input. Research conducted by the Brazilian Institute for Space Research has shown that 80% of the solar energy falling on the Amazonian rain forest is used in evapotranspiration – the combined evaporation from soil and transpiration from plants. This water vapour rises, spreads out, cools to condense out and fall again either side of the rising vapour column. About half of the Amazon basin's rainfall is recycled in this way, with the other half coming from moist air dragged in (primarily from the Atlantic) to replace the warm air rising from the forest canopy. This forest- and solar-generated cycle regulates the climate of nearly all of Brazil and many of its neighbouring countries. From

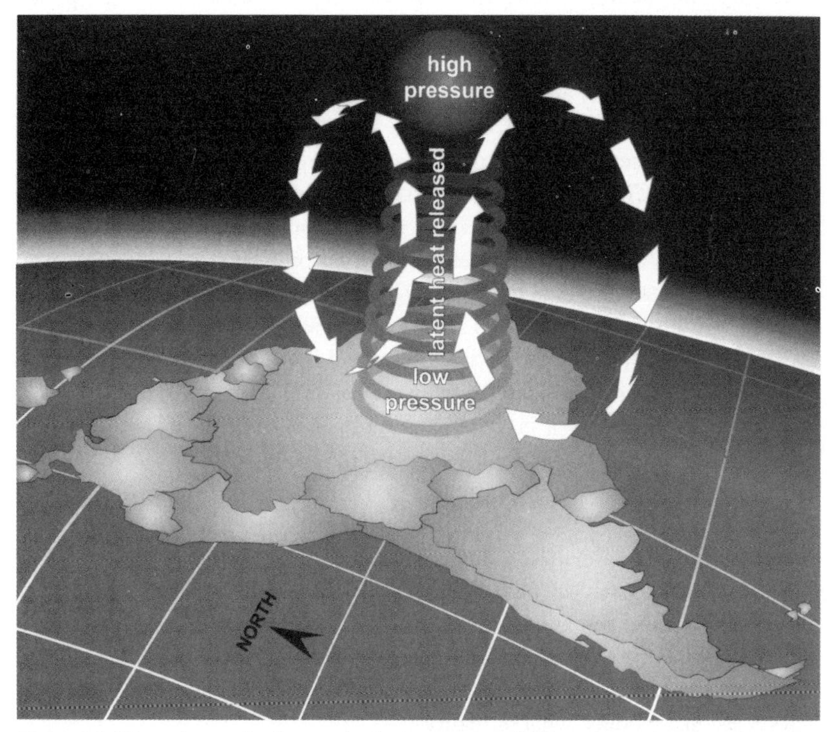

Figure 9.1 Warm air over the Amazon basin rises, cools and falls to the north and south creating twin circles of air flow

top to bottom, the height of the twin circles of air and water is of the order of 10 km (Figure 9.1). As this circular flow is powered by solar energy the area of low pressure beneath the rising column of air moves with the seasons to where the Sun most effectively evapotranspires water. When it is summer in the Northern Hemisphere, meteorological low pressures are found over northern Brazil, Columbia, and Venezuela. When the Southern Hemisphere has summer, low pressures tend to form over southern Brazil. In this way the South American tropical rain forests have a strong influence on the region's weather and climate[247].

The concern is that if the rain forest is removed, less evapotranspiration will take place which will alter not only the amount of water entering the atmosphere, but, in turn the way the heat from solar energy is distributed about the region. Of course, as with global anthropogenic climate change, no one knows for certain what would happen to the region's weather if the Amazon rain forest were removed and replaced with agriculture, but it is possible to estimate the effect. Ecologists can measure how much individuals of tropical forest tree - species will normally transpire. They can measure the humidity and temperature profiles above and within the tropical forest canopy. They have, broadly, the necessary information with which to build a computer model, or to incorporate such data into an existing global climate model (GCM). Though GCMs can only serve as rough approximations of the real atmospheric–climate system, their utility can be judged by whether or not they do truly approximate the real world under normal conditions.

In 1989 two researchers, Lean and Warrilow from the Hadley Centre in the UK, adapted a GCM so that it represented the World as if the Amazon rainforest were replaced by pasture[248]. When run with such parameters, it showed Amazonia with greatly reduced evapotranspiration accompanied by a reduction in rainfall and an increase in temperature. The annual mean value for evaporation was down by around 27%, rainfall lowered by 20%, and the temperature was up by 2.4°C. This last, if the model is correct (and it does contribute to a picture sketched out by earlier research)[249] is cause for some concern. The GCM was modelling the World as it is *today* without an Amazon forest, but in reality deforestation, should present trends persist, will result in the removal by the end of the first half of the next century of a greater proportion of current forest cover. By then, under the IPCC's business-as-usual scenario, the World will have already warmed by about 1.5°C. Parts of Brazil could have a climate close to 4°C above today's values. (Note that not only does the IPCC 1990 estimate for global mean temperature rise some 1.5°C over the 1990 temperature by the year 2050, but of the IPCC's five regional forecasts, the two nearest the tropics forecast a rise of at least 1°C *throughout* the year[20]. Models though, as mentioned before, are just models; what of reality?

The regional climatic predictions resulting from massive deforestation are serious for regions of the World currently dominated by tropical rain forest. So much so that if they did manifest themselves in reality then not only would there arguably be in all likelihood the biggest ecological disaster of modern times, but serious socio-economic consequences for the nations concerned. Predicting the accuracy of these models therefore becomes as urgent as the significance of the models' predictions. Clues to model accuracy can be found if it was known exactly how tropical rain forests channel energy and water flows in the real World. Such work is central to validating regional predictions and, indeed, to improving global predictions through refining GCMs. In the early 1990s energy and hydrological research within tropical countries is increasing. One such effort is the Anglo-Brazilian Climate Observations Study, or the ABRACOS project. The project looked at each of the components of the surface energy and water balance for both tropical rain forest and pasture ecosystems. Comparing energy flows in forests and those in pastures showed that, during the day, the forest reflected less solar radiation and emitted less thermal radiation than cleared forests which had been turned into pastures: i.e. during the day pastures were far warmer than forests. At night the situation was reversed with pastures emitting more thermal radiation, so cooling down faster than the forest. Also, during the day the variation in temperature in the cleared pastures were twice those of the changes in temperature found in the forests. At night the ABRACOS team found the climatic contrast between forests and pastures visual and striking; in forest-cleared pastures the fall in temperature was so great that it approached saturation level, producing a fog[250].

Not only does the clearing of rain forests result in local climatic alterations (indicating that extensive clearance might have, as computer models suggest, regional climatic impacts), but the act of burning forest and savanna (again for slash-and-burn), might also affect the weather regionally. French researchers Hélène Cachier and Joëlle Ducret, have found the air in the Northern Congo rain forest affected by biomass burning in a way that might well be altering the region's weather[25]. In 1989 they collected 46 rain samples over one year (June 1988/9) at Enyele (2°5N, 18°E) in the primary forest of Northern Congo. This is an area almost unaffected by local human activities yet in every rain sample they found black carbon which ultimately must have originated from combustion sources, most likely biomass burning. Biomass burning would appear to affect equatorial rains all year round. The researchers also suggested that the particles they found, of black carbon, might act as cloud condensation nuclei. If they are right then the act of biomass burning (in addition to changing the local environment from forest to pasture) could be altering the energy radiative balance and the precipitation regimes in the tropics.

In terms of the global atmospheric perspective, over the past 135 years, carbon dioxide from tropical regions accounts for two to three times the input from all other regions due to land-use management (or mismanagement). As was discussed earlier, the IPCC[20] concludes that between 0.6 and 1.6 Gt of carbon a year are emitted as a result of deforestation and human land use change. This puts the carbon contribution from forests and land use on a par (or in the same order of magnitude at least) with the increase in atmospheric carbon dioxide. However, such a comparison can easily be misleading. If somehow we halved the carbon emissions from deforestation and land use, this would not mean that the atmospheric build-up of carbon dioxide would be halved. Carbon dioxide emissions from fossil fuel burning is three to four times that from forests, so fossil fuel use dominates the greenhouse equation, not deforestation. Having said this, improved forest management (tropical to boreal) can be a boon in the greenhouse world. Forest management can provide a partial control that would help to lower the rate of increase in global warming, not to mention reverse regional impacts.

CAUSES OF DEFORESTATION

Forestry may well prove to be an asset for life in the greenhouse, but its benefits will not be manifest while present trends continue. In order to buck these trends, and establish new ones tailor-made for the greenhouse world, the root causes of deforestation have to be nullified.

Much has been written on the environmental problems of the Third World and on World Development. As has already been mentioned, development and the environment have received the attention of international studies[2, 10] and World leaders[3, 17]. That the same developmental issues are in the main still with us is a testimony both to the scale of the difficulties and that traditional solutions are not appropriate. The problem, in essence, is one of human ecology and human economics. Human ecology because of the demographic failure of environmental resources to sustain a large growing planetary population. Human economics because of the value we place on certain, regional, human populations and environmental resources.

The growing World population needs to be fed, kept warm, and clothed. Crops need to be grown and fuel found. More people means more fuels need finding and more crops grown. The result, worldwide, is more pressure on the environment and on maintaining the sustainability of natural resources. The tendency is not just one of where more people live, greater destruction of the forest follows – some areas of deforestation have low population densities – but there are broad correlations. The South American country of Colombia originally had some 700 000 km^2 of tropical forest cover (equal to roughly a quarter of neighbouring Brazil's original forest). By the late 1980s Colombia had lost about 74% of its primary forest. Colombia's population density, which averages 25.5 inhabitants per square kilometre, is half as great again as Brazil's 16.2 inhabitants per square kilometre. Brazil has lost only 37% of its primary forest – half that of Columbia, indicating that there is a link between population and deforestation. Similarly, in the Caribbean population density and deforestation go hand in hand (see Table 9.2).

Table 9.2 Early 1990s estimates of population density and forest cover in the Greater Antilles, with mainland comparison[252]

Location	*Approximate population density per km^2*	*Estimated forest cover (%)*
Tropical S. America	20	45
Tropical Central America	45	28
Cuba	80	18
Haiti	230	3
Dominican Republic	140	9
Jamaica	210	5
Puerto Rico	350	8

Although playing with numbers like this helps to identify trends, it does little to illuminate the root cause of the problem. People have been living in forests for millennia without harming the integrity or eroding the species diversity of such ecosystems. So what is causing the twentieth century spurt in deforestation? Part of it does boil down to the numbers game in terms of native population densities, but it would be very wrong to think of a nation's population as the sole culprit. In a real sense the population the forests support is far greater than those who happen to live in it. Forests help to support the lifestyles of those living outside of the forests in the cities. After all, it is the policy makers in the towns who decide on how the forests are used and who manage the economy. Furthermore, the populations of the rich developed nations also materially benefit from the forests. Those exquisite items of mahogany furniture that adorn countless living rooms in the developed nations have their origins in tropical forests. If the owners of the land cleared for ranching can sell their beef for export, then again an overseas population is in part being supported by land that was forest. Looking at the forests this way it is easy to see how the carrying capacity of such regions is being exceeded. Indeed, such is the scale of the profits to be made, with everyone at the top wanting a proverbial piece of the action, there has been a tendency in some areas to cut out the middle man and to directly displace the local population, which then goes on to over-exploit the remaining marginal lands.

The problem of deforestation is therefore a morass of ecological, developmental, demographic and economic difficulties, with each country having a different slant, or variation. Nonetheless, there is a theme running through all the various deforestation scenarios: one of value. When deforestation takes place it is invariably because the benefits of deforestation are perceived to outweigh those of the costs. To the farmer, granted at a stroke a low cost, or even free, area of forest a quick return regardless of the long-term consequence is often more attractive than the prospect of a low rate of return, albeit a sustainable one over a longer period of time. Then again, remote markets may determine the value of the forest both pristine, managed or felled, in its own market terms rather than in any real sense. The developed nations' hunger for cheap timber, beef or soya may seal forest fates at the behest of their own market movements.

In 1991, Christmas (the developed western nations' traditional season of charity an(goodwill) exemplified how markets far from tropical forests can directly affect their management. That year, Christmas saw the trade price of Brazil nuts fall from $1.20 a pound to $0.80. The harvesting of Brazil nuts is considered a sustainable use of fairly undisturbed forest. Part of the reason for the price slump was aid for the disastrous crop the previous season, but a compounding factor was the early-1990s World recession. The previous season's poor crop meant that the nut gatherers, the *castanheiros*, were unable to pay off their previous season's debts and so in turn obtain, on credit, goods and supplies to see them through to the next harvest. The previous season's poor harvest also meant that exports were made up from nuts of lower quality, and this coupled with the World recession, depressed demand. Without their credit the nut gatherers had to expand their smallholdings' slash-and-burn farming and cattle ranching, neither of which was good for the forest[253].

Stable pricing at a level that provides those who manage forests sustainably is one key factor that will secure such practices. Another would be to ensure that those who use the land have security of tenure so that its management will be carried out with an eye to the future and not just the present. While widespread conversion of intact forest to agriculture is neither ecologically nor, in the long term, socially desirable, rural populations can only be expected to continue in quick return, nonsustainable practices unless alternative land uses become more attractive[254]. Fortunately there are sustainable alternatives.

At the 1991 annual meeting of the American Association for the Advancement of Science, Robert Mendelssohn, of Yale University School of Forestry and Environmental Studies, took a positive view of tropical rain forests. Aside from providing genetic resources for agriculture and biochemicals for the pharmaceutical industry, tropical forests can be used to harvest a range of products. Mendelssohn, together with a team from the New York Botanical Gardens, had been studying the various profitable products obtainable from tropical forests in five countries: Belize, Brazil, Guatemala, Indonesia, and Peru. Profitable products from living forests traditionally include latex, rattan and the kiwi fruit, but there are far more. In a study conducted near Iquitos, Peru, it was found that edible fruit, nuts, oils and rubber from five acres of tropical forest generated a financial return twice as great as that from cutting five acres of trees: what is more, such income is sustainable. In Guatemala the extraction of the raw material of the *xate* palm yields a higher income per unit area than agriculture.

Ecologists, such as Mendelssohn, recognise that such harvesting does not preserve forests in their pristine state (and that there will always be the need for wildlife reserves), but that there are viable land-use alternatives that will maintain tree cover, and, while cover is being established, such areas provide a sink for carbon dioxide[255].

REFORESTATION: GREENHOUSE IMPLICATIONS

Given that deforestation is actively contributing to anthropogenic climate change, it follows that reforestation (turning land that was once forest back to forest), and afforestation (turning land that historically never was forest into forest), could perhaps ameliorate matters. The question is one of whether deliberate human policies can drive a net increase in the biological storage of carbon such that the rate of increase in atmospheric carbon dioxide is significantly reduced? There is also the question of using forests as a greenhouse neutral, source of biofuel; which indirectly has a beneficial greenhouse effect through offsetting fossil fuel consumption.

Natural (steady-state) climax forest is greenhouse neutral. As trees die naturally, decomposing to release carbon dioxide, or as forests experience natural forest fires, so room for new growth is provided which will absorb the carbon dioxide released. So the whole question of forests and the anthropogenic greenhouse effect turn on four principal factors: forestry management, deforestation, reforestation and afforestation. Each of these – as they relate to the World's forests and atmospheric carbon dioxide – depend on three other factors:

(1) land use (primarily the amount of deforestation and land available to become forest);
(2) the rate at which carbon can be stored within growing forests (and their soils); and
(3) the fate of carbon in growing forests.

In addition there is the old question of cost (and benefit). For example, the cost of turning agricultural land into forest (while attempting to feed a growing World population).

One recent estimate[256] of the area of once-forested land worldwide (both tropical and boreal) is of the order of 800 million ha (8 million km^2). However, this estimate is not one of the land available for reforestation as it ignores the cost of giving up agricultural and pastoral land that such a transition would imply. A figure one quarter of the size, 200 million ha, might be a more realistic estimate of the area available for forestry.

Given 200 million ha, there are a number of optional forestry regimes available, each with its own implications for the greenhouse climate. Straightforward replanting will result in net carbon dioxide absorption from the atmosphere, but this will continue only until the forest has reached maturity and its climax state. On the other hand, if wood is taken out of the forest and used in a product that is then stored[257], e.g. as paper in personal or municipal library books, carbon is removed from the forest without becoming atmospheric carbon dioxide. This then allows new forest plantings to take place, drawing down more carbon dioxide.

There is tremendous potential for storing carbon in growing forests. Forests convert atmospheric carbon dioxide into a solid form of carbon store – wood and wood products can be cut and stored, ideally for an indefinite period. In a perfect world, the wood or its products could be buried so as to replace the fossil fuels[258]. Carbon from forests is already being bought and cost-effectively stored for many years, decades even. Furthermore, there is considerable scope for - additional storage. Paper and wood are both about 40% carbon by weight. Books are just one way of storing carbon dioxide photosynthetically drawn back out of the atmosphere, but they can have other advantages. Libraries are all very well as large single stores, but book collections in homes can help save energy. A wall of books alters the thermal properties of a house: it restricts heat flow and can help insulate a room. Encouraging book reading is not only clearly of benefit to society, but in providing insulation can help keep dwellings warm, so offsetting energy consumption and fossil fuel combustion. Other ways of storing forest carbon primarily involve the use of timber for furniture and construction. The use of carbon stores to combat the greenhouse effect may seem frivolous, but the principle is sound. They are but one tool that can frequently be self-financed.

A number of efforts have sought to estimate the actual and potential rates of carbon accumulation in natural woodlands and in tree plantations: the latter being more relevant to the above storage scenario as production is significantly higher, and because any deliberate attempt to sequester carbon will require management. The Oak Ridge National Laboratory (USA) has conducted a 'Short Rotation Wood Crops Program', under the sponsorship of the US Department of Energy, in an attempt to determine the maximum rate of carbon accumulation[256]. The range of carbon storage (including in soils) varies from 7 tonnes ha^{-1} to 17 tons ha^{-1}. Higher rates of carbon accumulation occur in plantations of between 5–10 years of age; this cannot be sustained in older plantations because the accumulated biomass gives off carbon dioxide due to increasing (per unit mass) respiration. Assuming for a minute an estimate of 7.5 tonnes per ha^{-1} yr^{-1} for managed forests, and that the land available for forestry worldwide as 200 million ha, then some 1.5 Gt of carbon a year could be taken out of the atmosphere. This would roughly be enough to counter the carbon contribution to the atmosphere from current deforestation. Halt deforestation itself and the accumulation of carbon would offset about a quarter of the carbon released by fossil fuel burning.

Given the limitation of land availability, the question of managing land to maximise carbon accumulation becomes important. As part of this management the ultimate fate of the carbon is critical. Broadly, we have two options. At one extreme, plantations could be grown rapidly for 5 years, then harvested and used entirely for fuelwood. Here, there will have been a net removal of carbon dioxide from the atmosphere, but all of it will be returned to the atmosphere at the end of the rotation. Unless the fuelwood replaces an equivalent amount of fossil fuel that would have otherwise been consumed, such plantations can only cause a brief delay in the atmospheric accumulation of carbon dioxide. At the other extreme, consider a forestry system in which carbon accumulation continues for 20 years, at the end of which most of the carbon is harvested and utilized as wood in construction. In this case the carbon fixed remains in storage for additional decades to centuries, and the next increment of wood twenty years later is added to that storage. Wood could also be buried in bogs, ocean sediments or anoxic (without oxygen) landfills. Such plantations could result in a longer-term reversal of atmospheric carbon dioxide accumulation. This may seem extreme but it might well be unlikely that (even with an unprecedented increase in demand) long-term demand for forest products could be sustained[256].

Currently, we are a long way away from creating such significant increases in forest area and products. The present form of carbon forestry concerns slowing the rate of deforestation and increasing productivity. The Malaysian timber company, Innoprise Corporation, is engaged in two such initiatives. The first was established in 1991 with the support of the FACE (Forest Absorbing Carbon Dioxide Emissions) Foundation in The Netherlands, which was itself established in 1991 by the Dutch Electricity Generating Board with the aim of absorbing the equivalent carbon dioxide as would be released by a large power station over its 25-year life span. The initiative consists of the reforestation ultimately of some 30 000 ha. The second initiative is jointly run with the New England Power Company and attempts to reduce the impact of logging by saving vegetation (hence carbon) from accidental damage during logging[259].

Finally, one pioneering scheme in Guatemala is attempting to use reforestation to absorb the 387 000 tonnes of carbon emitted by a 183 MW coal fired station in Connecticut, USA. The 10-year project will cost about $16.3 million and involves the planting of 52 million trees[260].

REFORESTATION: THE HUMAN DIMENSION

The human dimension to a major worldwide forestry programme would be considerable and will only briefly be touched upon here. The international - political effects might include additional stress between the developed countries and the less-developed nations, especially if the developed countries continued to use fossil fuels in a profligate way, while the less-developed countries grew most of the plantations. Forests will, however, no doubt be welcomed in those regions already suffering a shortage of fuelwood.

Perhaps the most significant question is whether there is mileage in the pilot forest plantation schemes mentioned at the end of the last section, so that forests can realistically be used by humans as a sink for carbon dioxide? In a report[261] produced for the Electric Power Research Institute (EPRI) (California), it was concluded that up to 2.45 Gt a year of carbon might be accumulated by a worldwide forest programme, but that only 0.59 Gt a year might be stored if all new forest lands were merely left to regenerate, or develop 'naturally', before harvesting – the low capital cost option. Table 9.3 shows how this potential might be broken down.

Table 9.3 Potential for increased storage of carbon by increased forest area[261]

	Natural regeneration	*GtC/yr*	*Plantation*
US 14% agricultural land	0.03		0.16
European and Soviet agricultural land	0.06		0.32
USA & Europe urban land	–		0.10
Chile, Mexico, Argentina, Australia			0.02
China	–		0.21
Tropical Savanna	–		0.14
Conversion of 90% fallow agricultural land	0.50		1.50
Low	0.59	High	2.45

From Table 9.3, it would appear that the largest single part of this managed biotic carbon sink consists of the conversion of 90% of the land currently used for long-term fallow agriculture. A mid-1980s FAO estimate puts the tropical land area of Asia, Africa, and the Americas devoted to long-term fallow agriculture at over 200 million ha. If just 10% of this land used state-of-the-art agricultural techniques then food production would exceed that currently obtained by more primitive methods, or so the theory goes (the Chinese have arguably gone a long way to demonstrate this). With concentration of intensive agriculture on just 10% of the land area, the remaining 90% could be used as a carbon sink. However, the human, and indeed ecological, costs of this exercise are uncertain and at the very least it would require a massive re-direction of current investment. It was further suggested that if most of the forest product was used locally for energy, then the plantation establishment would not require subsidy. The principal uncertainty in the development of such carbon sinks is associated with the rate at which state-of-the-art agricultural technology can be introduced.

The next largest components of the postulated carbon sink, are the agricultural lands of Europe and the USA. In the USA farmers are paid to withhold, on average, 10–15% of the land from planting. In 1988 this comprised of 22 million ha, not to mention an additional 10 million ha set aside as erosion-prone cropland. Consequently, over 30 million ha of US agricultural land might be converted to forest without any lowering of agricultural production. In the EEC, with its agricultural surpluses and policies of set-aside, it is not difficult to see a similar proportion of land being freed for forestry. On the other hand, in areas of Eastern Europe and Russia it is not the extent of agriculture that will ensure that land is available to become a carbon sink, rather that the agriculture system is inefficient and even a few percent improvement on 86% of their farmland is achievable. Finally, though the worldwide growth in towns and cities has taken much land away from farming, there is no reason why some of that land could not be used to grow trees. Suburban areas make up 2% of the land area in the USA. If over the next two decades the total number of trees in suburban-urban areas doubles, then it could provide a carbon sink as great as 0.1 Gt of carbon a year. A similar order of value would be achievable in Europe and could be subsidised in terms of the aesthetic value and improvements gleaned by urban dwellers themselves.

Finally, the EPRI report[261] concludes that, in addition to increased land area devoted to plantations and forests, more carbon could be sequestered from the atmosphere through improved existing forest management and fertilization. The EPRI report estimates that an additional 1.2 Gt of carbon a year could be sequestered in this way making the size of the total EPRI carbon sink upto 3.65 Gt of carbon a year. Of course the EPRI estimate is largely what could be obtainable in an ideal world that decided to heavily invest in tree planting. However, we are not in an ideal world, but the report indicates that 1.5 Gt of carbon a year would appear to be a reasonable goal in a world geared to greenhouse realities. Of course it has to be remembered that 1.5 Gt would have to be harvested each year to allow for replacement forest growth (hence CO_2 absorption). The harvested wood either has to be used as wood products (paper, construction etc.) or as a greenhouse-neutral fuel displacing fossil fuel consumption.

RECYCLING PAPER

Before leaving the role of trees in the greenhouse world, it is perhaps worth emphasising that it is necessary to decide what needs to be recycled and what taken from the forests. Forests in the greenhouse world need to be managed in a sustainable way and – if used as a 'tool' to sequester atmospheric carbon – can enable carbon to be taken out of the forest system and stored[257]. Forests can therefore become a conduit for the transfer of carbon out of the atmosphere and to useful end-products. The big 'but', as discussed earlier, centres around keeping the carbon in solid 'forest products' (paper and timber products), for it is all too easy for the carbon to return to the atmosphere be it through burning or biological decay. Here recycling can help but, contrary to popular belief, recycling should not become an end in itself. Recycling too has its costs in terms of labour, chemical processing, and energy which have to be set against the benefits it provides.

Contrary to many lay-environmental movement campaigns, 100% paper recycling is not feasible, since recycling ultimately shortens paper-fibre length to a degree which renders it unsuitable. The top priority in the greenhouse has to be to ensure that forests are used in a sustainable way when exploited for their resources. But the lay-environmentalists are arguably right to campaign for more recycling as clearly the present mix of virgin to recycled wood pulp in the developed countries' paper markets is far from ideal. In the greenhouse world we must keep our goals (to minimise the degree and rate of climate change) to the fore. Whilst unrestricted recycling is as undesirable as no recycling at all; recycling paper does keep carbon locked in its solid form, and throwing paper into landfill sites does result in it breaking down into greenhouse gases: carbon dioxide and methane. The other most common fate for paper products is burning; that too, obviously results in carbon dioxide generation and nullifies the atmospheric carbon sequestering benefit of the exercise.

So why is the paper recycling business, at best, marginal in most developed nations? Even in the early 1990s there are myths pervading the print industry: for example, that good quality paper cannot be made from recycled material. By 1990 there were several brands of recycled coated papers available that were suitable for high quality colour printing[262]. The difficulty has not been one of production technology, but investment and the market. Back in 1992, the UK paper market still could not provide quality recycled paper at a competitive price. While at the end of 1996 there was a year-long shortage of wood pulp and so paper prices rose by over 20% there was not enough recycled paper available and so the prices were high. These are market-based problems: while the demand for recycled paper is low, the costs will be high; while the costs are high, the market for recycled paper will remain depressed compared to virgin stock. (A situation which reflects that of the new low-energy light bulbs.) There are several ways to increase the proportion of recycled paper used including:

(1) *Taxes*. Taxing virgin paper use would make recycled paper use more attractive. The important thing is to determine what exactly is greenhouse friendly paper use – we want forest products to be used, provided that the forest is managed sustainably. The wood pulp's end-use must be taken into account. Short-lived paper products might be taxed as these would soon be thrown away or recycled (the former returning carbon to the atmosphere on decomposition or burning, while the latter nearly always generate greenhouse gases since recycling itself requires energy!).

(2) *Voluntary agreements*. Firms that consume large quantities of paper can

come together to voluntarily increase the proportion of recycled material in their products. By themselves, voluntary agreements are difficult to arrive at when the firms involved are making effectively identical products. The temptation for a firm struggling to survive to lower its costs by abandoning recycling becomes unavoidable. Office stationary, for instance, would be a prime candidate for a high-recycled content, but such arrangements are unlikely. On the other hand newspapers are not identical, and so competition is in theory not solely price-based.

In 1991 the UK newspaper industry managed to come to a voluntary agreement with the British Government to increase the proportion of wastepaper in newsprint to 40% by the year 2000 – with the backing of the British Newsprint Manufacturers' Association. But in 1991 newspapers were already using a 30% recycled content, so the 40% target was easy to attain. UK newsprint paper consumption, peaked at 2.1 million tonnes in 1989 but declined slightly in the early 1990s. Whereas the early 1990s annual UK production of newsprint containing recycled fibre was about 700 000 tonnes per year, only 550 000 tonnes of this was bought by UK newspapers with the remainder going to export. Sources within the industry said that the publishers were reluctant to become too dependent on UK mills as this might lead to a lack of competition[263]. It would therefore appear that international trade agreements and policies are required. (This will be returned to in the next chapter.)

(3) *Capital support.* Since part of the problem is to make recycled paper more competitive, one obvious solution is to kick-start the commercial supply of recycled pulp by providing interest-free government (or international) loans and/or grants. UK 100%-recycled fibre paper mills (capable of processing 250 000 tonnes per annum) would require a construction investment of the order of £250 million (1991 prices)[263].

Whatever assistance we, or the paper industry itself, gives to recycling paper, it has above all to be done in an environmentally sustainable way. The early 1990s saw a rapid growth in products claiming to be 'environmentally friendly' as producers found that a significant proportion of US and European consumers would switch product brands in the belief that this would reduce environmental degradation but not all 'environmentally friendly' (or even 'environmentally unfriendly') claims are true. The 'environmental' labelling of products and the greening of business is slowly becoming one of the key environmental concerns in the 1990s as national economies begin to shift (albeit sluggishly) into more sustainable modes. Some businesses will accept that more of the cost of their environmental impacts need to be paid for than purely commercial concerns warrant. (In economic parlance, some firms will be prepared to internalise more of their environmental externalities than others.) Here, for the paper industry, the signs are hopeful as the environmental impact of an increasing number of its production stages are being considered.

In April 1990 ten leading UK paper merchants introduced a four-point classification scheme for their recycled products. The scheme itself pointed the way for others, though it did not recognise the benefits which can accrue through using virgin fibre. In 1991 Brands Papers (again one of the UK top paper merchants) branched out with a modified five-star scheme. The criteria included: the 'recyclability' of the paper's fibre, consumption of fossil fuels, and releases of bleaching agents and other effluents – both gaseous and liquid. To be rated on the scheme a paper mill has to answer a questionnaire which (in 1991) was only designed to 'discriminate against the worst offenders'. One flaw, of a few, in the 1991 Brand Scheme was that Scandinavian mills, using hydroelectric power, easily met the schemes 75% non-fossil production energy requirement. Furthermore, the scheme failed to account for the total amount of energy the mill used to produce a unit weight of paper[264].

Even so, that such environmental auditing takes place at all without the stick of legislation provides an encouraging prognosis for accounting in the greenhouse world. Legislation, however, is a necessary step if business in a free market is to conduct itself on the proverbial 'level playing field'.

CHAPTER 10

ENERGY POLICY

Anthropogenic climate change, virtually by definition, is a result of the way we live our lives: the inescapable logic of which is that by changing the way we live our lives, we could change the nature and degree of climate change and, of course, its environmental and human consequences. However, unlike many other human environmental impacts, anthropogenic climate change is the cumulative result of the way all members of our species live. Anthropogenic climate change is not the result of one single enterprise, business concern, government, or group of individuals. It is caused by each of us. Yet, just as we each live our lives in different ways, and potentially could live in countless others, so the options open to us are equally numerous. In short, the way we govern and police ourselves is crucial in determining how great, or how little we will perturb natural systems. Policy, therefore, is the key, and greenhouse policies relate primarily to energy.

WHY ENERGY POLICY?

Carbon dioxide alone accounts for over half of anthropogenic global warming: the IPPC calculated[20] that, of the greenhouse gas releases in 1990 alone, 61% of the resulting climate change over the subsequent 100 years would be due to carbon dioxide. Here the largest anthropogenic contribution, between two thirds and three quarters of this carbon dioxide, is a direct result of the way we generate our energy. In addition, a proportion of the anthropogenic nitrous oxide and methane emissions are also a consequence of human energy supply activities. The way we handle energy is therefore central to, indeed inseparable from the greenhouse issue. Furthermore, given past trends worldwide over the last century, it is obvious that our species' production and consumption of energy has been steadily increasing.

In the greenhouse, our dependence on energy has grown in two ways. First, on a per capita basis worldwide energy consumption has grown, largely a result of the Fundamental Economic Problem of infinite wants outstripping finite resources (see Chapter 1). No matter how much resources (considered here in the fundamental form of energy) are deployed to meet human wants, the wants are never satiated and so there is pressure to consume even more. This global per capita rise has been considerable. Looking at the type of energy resources traded on the World market (i.e. excluding such activities as dung burning in the Third World), global per capita primary energy consumption has increased from about 1.2 tonnes of oil equivalent a year in 1966 to around 1.45 tonnes of oil equivalent in 1991: an increase of over 20% in a quarter of a century[134]. Second, compounded by a simultaneously growing population the gross World consumption of primary energy has grown even faster – by a little over 30% in just 15 years. Clearly, when dealing with finite resources, this is not sustainable indefinitely. The need to satisfy human hunger for energy in a controlled way is, therefore, becoming more and more acute.

ENERGY POLICIES FOR THE GREENHOUSE

There is no one single preferred energy policy option that can be readily identified as being the optimum policy for the greenhouse world. We could for instance, take a supply and demand approach. One might divide the World into socio-economic groupings to discern individual trends within the broad, global picture. Western Europe and the former Soviet block have seen a steady rise in both gross and per capita energy consumption: the latter from about 2.5 tonnes per capita in 1965 to roughly about 4 tonnes per capita in 1990. Apart from North America, the rest of the World has also seen a steady growth in both gross and per capita consumption: per capita consumption increasing from roughly 0.5 tonnes of oil equivalent to around 0.75 tonnes of oil equivalent. Only North America showed any signs of stabilising its energy consumption over this time, but here the sting in the tail is that annually the North Americans each consume 7 to 8 tonnes of oil equivalent: roughly twice that of European, most other OECD nations, or the former Soviet block, and eight times that of anyone else.

These trends broadly suggest that all else being equal, and with unlimited energy resources, the rest of the World would continue its overall and per capita growth in energy consumption throughout the next century and the overall world energy consumption would have become several times (over 1000%) greater than today. Of course all else is not equal, and energy resources are not unlimited.

If, therefore, we are to police our energy resources and have a coherent energy policy, the developed nations should curb (through rationing, taxation or whatever) their profligate energy consumption before they run out of indigenous resources, and accept that the non-industrialised countries will, for many decades to come, increase their overall and per capita energy consumption. The alternative is to either have an energy policy that encourages the industrialised countries to consume more and more energy, or to have no energy policy and continue with business-as-usual. These both exacerbate the pressures on dwindling energy reserves and are clearly non-sustainable. In 1994 the industrial OECD nations had less than 25 years of indigenous oil and natural gas left in the ground, and were already importing considerable amounts of energy resources from the rest of the World, which only has about half a century's worth of gas and oil; while coal has about 200 years supply left[135].

Yet it is perfectly possible for our species to continue to burn fossil fuel, yet at the same time lower carbon dioxide emissions. It is not the burning of the fuel that is of concern in the greenhouse world, merely the release of greenhouse gases into the atmosphere. Technically, there is nothing to prevent, say, electricity generators from scrubbing exhaust fumes. Though, economically, the cost of removing 90% of carbon dioxide from exhausts would more than triple the cost of electricity generation from fossil fuels[265]. Conversely, regulated energy policies based on low energy strategies have a lower net cost and it is perfectly possible to lower energy consumption without cutting back on other areas of economic activity. Indeed, we are so wasteful in the way we use energy that the act of curbing emissions through increased energy efficiency can frequently prove to be economically profitable.

The argument that such low energy strategies *are* cost effective (see Chapter 8) are not new, even though the short-term cost of employing really high efficiency measures is prohibitive. (The cost, per joule of energy saved, from quintuple-glazing your home is far higher than that from just double glazing.) To be cost effective, the cost of adopting energy saving and energy efficient measures must be less than the value of the energy saved. Indeed, the way we have been using energy (and other) resources is so wasteful that there are many measures that are cost effective.

Using energy resources efficiently provides a central theme to greenhouse energy policies, but in itself is not enough: energy conservation and low energy strategies alone cannot lower energy consumption and the generation of greenhouse gases as we enter the twenty-first century. There are two reasons for this. First, not all forms of energy production result in greenhouse gas emissions. For any energy policy to succeed in greenhouse terms, it has to acknowledge the benefit of the so-called 'greenhouse friendly' technologies which can lower emissions. Second, people do not want to consume energy for its own sake, they want to go from 'a' to 'b', keep warm, listen to music, cook meals, etc. People only want the service that energy consumption brings. Energy conservation does not affect the nature of these services but does lower their cost. Here enters the law of supply and demand: lower the cost of the goods and services and they become more affordable so that consumption increases. History demonstrates this well. Though energy efficiency has improved throughout the 1970s and 1980s, most nations' energy consumption was only temporarily lowered by the 1973 oil crisis, consumption resumed its upward trend a couple of years later. Looking at a developed nation's energy consumption in terms of energy per dollar GDP, then most countries saw a decline in their energy/GDP ratio immediately after the 1973 oil crisis before rising again until the next oil crisis at the end of the 1970s. Even countries with a significant proportion of indigenous energy resources (such as Sweden with its HEP, and the UK with its coal) were not immune, despite the background of improving energy efficiency and conservation technology.

The post-1973 lowering of energy consumption owes at least as much to the energy price as to efficient use. Greenhouse energy policies, therefore, must include financial considerations and economic policy which reflect the cost of greenhouse impacts in the price at which energy is sold[266]. Such pricing features have manifested themselves most commonly in the form of the so-called 'carbon tax'. Given policy packages featuring the above elements, the key question is whether they are cost effective?

Shortly after the 1988 Toronto Conference on the Changing Atmosphere, the Canadian Federal and Provincial Energy Ministers received a consultancy report on the feasibility of meeting a target of a 20% reduction in carbon dioxide production by the year 2005. It concluded that Can$100 billion could be saved in the process through reduced energy bills from improved energy efficiency; monies that could be used to implement the energy efficiency strategy in the first place.

Whatever the agreed energy policy for the greenhouse world, it is clear from the gross and per capita emissions of greenhouse gases that the developed nations must lead in formulating policy if anthropogenic climate change is to be lessened. With the worst greenhouse offenders being the USA and Europe, the priority is for the formulation of their greenhouse energy policies. Yet the USA and (specifically early in the 1990s within Europe) the UK are finding it difficult to instigate such policies. Indeed, as we shall see, both the USA and the UK have for over two decades failed to successfully implement a national energy policy, let alone one with greenhouse considerations.

USA ENERGY POLICY

If any single thing in the latter half of the twentieth century could shake the USA out of its energy-guzzling infatuation and demonstrate the need for a successful energy policy, then it should have been the 1973 oil crisis. OPEC oil then increased in price from $1.77 to $7.00 a barrel, and nearly half of the annual USA oil demand (818 mtoe) was met by imports, only then did they begin to realise how exposed they were to the vagaries of the World's oil market. They were paying real energy costs and experiencing immediate economic and political pressures: unlike those imposed by greenhouse impacts where the cost and impacts are obscured from the consumer. As a result, the 1973 crisis prompted President Ford to propose what he called a 'series of plans and goals set to insure that by the end of [the then] decade Americans [would] not have to rely on any source of energy beyond [their] own', these proposals were packaged in what the President named 'Project Independence 1990'. A laudable goal, but how realistic was it? By 1990 40% of the USA primary energy demand was still being met by imported oil. Of course it is easy to criticise this failure with hindsight, but even at the time energy policy experts had their doubts. David Rose was one such, and in January 1974 he condemned USA energy policy as unrealistic and unworkable without the 'application of Draconian measures'[206]. The thinking behind the 'Independence 1990' initiative was, he opined, sound but in itself the deadline was far form pragmatic and the policies were decades overdue. The question being asked by those concerned with energy in the USA in 1974 was whether the then President's policy ideals would be formally adopted and implemented by Congress and would they succeed?

The finite nature of energy resources, and their security as far as the USA was concerned, was predicted by energy experts as long ago as the late 1940s. The oil crisis served to highlight this finite nature, though was not in itself a result of it. At the time of the oil crisis Rose noted estimates of the Western Hemispheric energy gap (between nation's production and consumption) of nine million barrels per day. To put this into a global perspective, this shortfall was equal to 13% of the 1974 world's refinery throughput[134]. The scale of the problem has been too vast for the institutions charged (or left) to deal with energy policy. For example, industrial rates of economic return typically result in time horizons of a decade or less, yet the prime length of economic activity of fossil energy resources is longer than this (cf. 1995 Reserve/Production ratios for oil and gas of 42 and 65 years respectively[135]). Energy policy has to operate within these longer timescales if the cost of future scarcity is to be recognised in the (plentiful) present (see Figure 10.1). David Rose, writing in 1974, did not have the benefit of hindsight. Which brings us to the question of how USA energy policy fared between the 1973 oil crisis and today; did it succeed in its goals of 'Independence 1990'?

In 1990 the USA produced some 417 million tonnes of oil and 561 mtoe of coal, yet though it only(!) consumed 481 mtoe coal so providing a coal surplus of about 80 mtoe (17%), oil consumption was 782 million tonnes incurring an oil deficit of 365 million tonnes (or 47% of production). Including natural gas figures does not help since the 1990 USA consumption of 486 mtoe exceeded USA natural gas production by 23 mtoe (or 5%). In short, in 1990 overall USA fossil fuel consumption exceeded production by 21% or 308 mtoe. Non-fossil fuel domestic primary energy resources did contribute to the USA energy budget. There was a 156 mtoe contribution from nuclear energy and a 72 mtoe contribution from HEP: in other words without the HEP and nuclear contribution the USA would have required an additional 228 mtoe from another source and this would almost certainly have been from fossil fuels[267]. (Note: subsequent refinement of energy data affects the above figures – typically within 1%.)

By 1990 the USA had seen its peak production of oil. Domestic oil

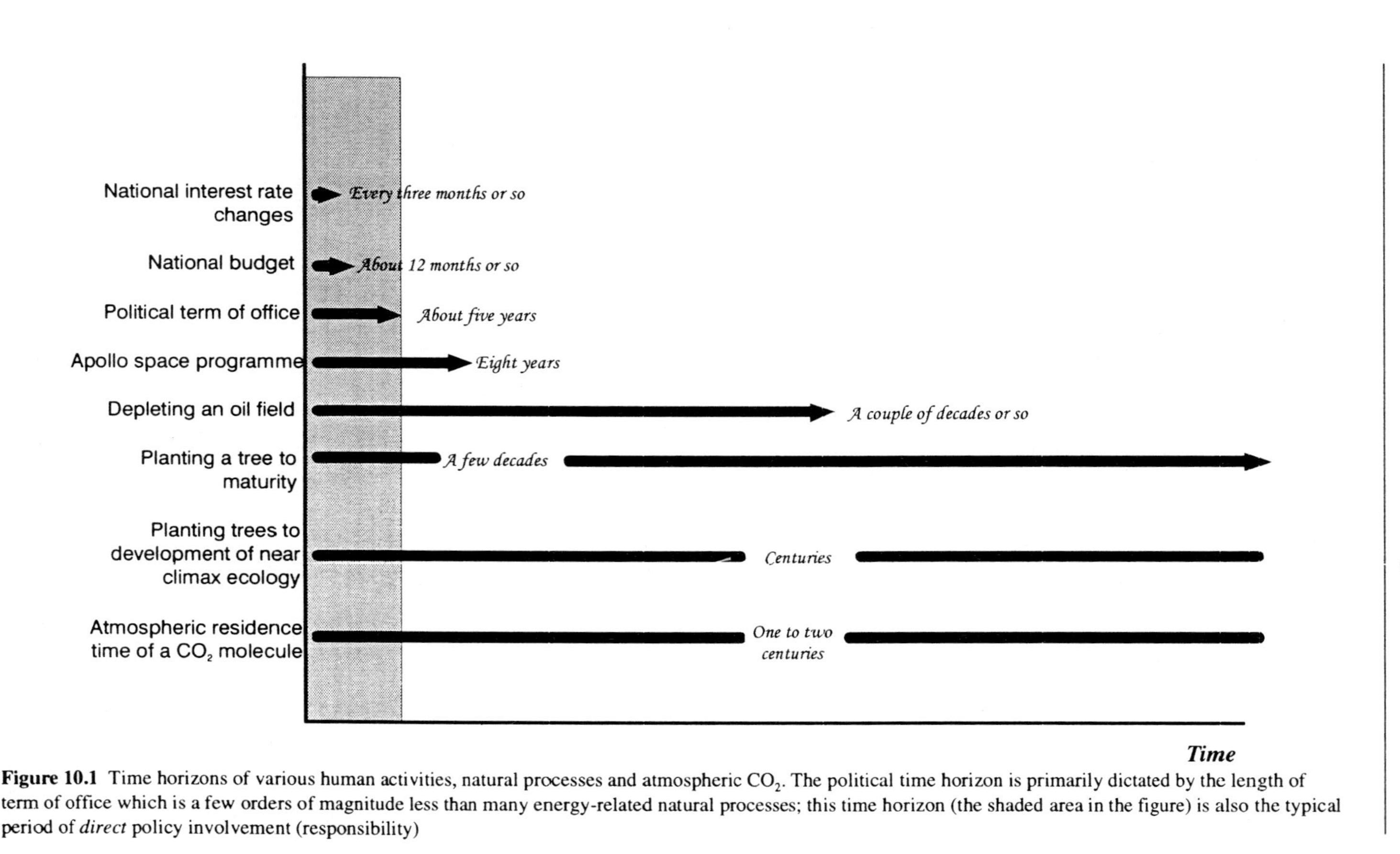

Figure 10.1 Time horizons of various human activities, natural processes and atmospheric CO_2. The political time horizon is primarily dictated by the length of term of office which is a few orders of magnitude less than many energy-related natural processes; this time horizon (the shaded area in the figure) is also the typical period of *direct* policy involvement (responsibility)

production had declined by about 14% during the 1980s and domestic natural gas production was down by around 11%, with only coal production growing. But, surprisingly, US natural gas production did increase from 1992 through the mid-1990s to the time of writing (1997); though the US continued to consume more than it produced. Again aside from coal (with a 1990 North American Reserve/Production ratio of over 260 years), its domestic oil and gas reserves are rapidly dwindling – the USA 1996 oil and natural gas R/P ratios were both close to 10 years. Even if present proven reserves were roughly doubled (due to new finds and extraction techniques) domestically the USA will be physically unable to maintain constant current production levels much beyond 2010. Finally, the USA energy deficit is predominantly in the oil sector, and this has profound implications for the nature of its car-based society.

Such is the benefit of two decades worth of hindsight, but even just seven years after David Rose's 1974 article, the signs were that the President's USA energy policy lacked the substance necessary to deliver the goods. In 1980, Raymond Siever in his introduction to a reprint of the Rose paper, noted that: 'when Rose's article was written, in late 1973, there was no USA energy policy. In mid-1979, there was still none. How long will we wait?'[268]

Regrettably, a decade later, in the early 1990s, the answer remained unchanged. The USA still has no coherent and workable national energy policy to tackle the twin concerns of profligate energy use – the finite nature of the resources and the environmental impact of such profligate use. In 1992 the Energy Policy Act was passed by Congress. It did go some way to tackling the USA strategic energy position and climate change concerns, but was still far from an effective tool in reducing USA per capita carbon dioxide levels or decreasing the nation's dependence on imported energy. The Act constrained oil companies restricting federal areas available for oil exploration (primarily in Alaska's Arctic Wildlife Refuge) though small independent oil producers gained some tax relief. The Act encouraged renewables and natural gas (which provides slightly more energy per unit of carbon dioxide emitted), through provisions for more R&D spending and the extension of tax credits. The USA Department of Energy estimates that by the year 2000 annual savings would be of the order of the equivalent of 2.5% of the 1994 total USA primary energy consumption, and could be as much as 6% by the year 2010. Such a saving, though significant, is small compared to the theoretically practicable savings potentially available: the Energy Policy Act ignores efficiency in cars, and lacks the economic incentives proposed by energy efficiency groups[269].

So what of greenhouse implications? Since those who formulate USA national policy find it so difficult to come up with a rational low energy policy, it is unreasonable to expect them to formulate and adopt an energy policy incorporating greenhouse considerations. In the first few years after the publication of the 1990 IPCC report on climate change, USA political leaders perceived the greenhouse issue as a problem that did not warrant much attention; at least as far as energy policy was concerned. Indeed, while the subsequent IPCC reports were being prepared there was considerable 'lobbying' from the USA to protect the fossil fuel industry. In 1994, energy lobbyists encouraged the USA State Department to raise concerns over procedural matters. It appeared to many of those who accepted the 1990 IPCC report that, having failed to win the scientific debate, critics were using procedural error to discredit the IPCC process[270] as it worked to produce its 1995 assessment. Fortunately, since then, things are beginning to change and in 1997 the signs were that the White House was beginning to accept greenhouse concerns.

EUROPEAN ECONOMIC COMMUNITY ENERGY POLICY

The fifteen countries that make up the European Union (EU), make up about 14% of the World's total energy demand and emit a similar proportion of the World's anthropogenic emissions of carbon dioxide. In 1990, the EU's energy consumption was about 1148 mtoe, of which: oil contributed 45%; coal 20.5%; natural gas 18.5%; nuclear 14%; and others (primarily HEP) 2%. Though each member country has developed its own way of consuming and using the various energy resources, the EU Parliament has a broad framework that has increased in complexity and authority over the years.

The EU energy policy framework, as it relates to the greenhouse, had its origins in the first EU 'Action Programme on the Environment', which was adopted on the 22nd November 1973. Though EU environment policy has evolved with time, its basic tenet has remained the same: to help bring economic expansion 'into the service of Man' by preserving the environment in which he lives and managing the resources upon which economic expansion depends. For unless there is a reasonable and sustainable use of resources there will, in the end, be nothing to sustain the economy. This was reaffirmed in the 'Second Action programme' (1977–81). Importantly, as echoed in virtually all EU policies that relate to the environment, is the *polluter pays principle*[271]. As mentioned before, this principle encourages the 'less invisible' environmental costs (or externalities) associated with economic activity to be paid for by those who incur them: or as economists say, to internalise externalities.

As Europe evolves (with the unification of Germany in 1991, and the forming of the Single European Market in 1992), so European policy has to allow for the vagaries of member states, their economies, indigenous resources and historical development. Looking at the energy policy formulation, as it relates to greenhouse impacts, one of the greatest hurdles was the creation of the Single European Market in 1992. This move was laid down in the 'De Lors Plan' of 1987 and aims to remove all economic barriers to economic competition between member nations. Here the European goal was to create a Community-wide free market, but in terms of European energy management this conflicted with most nations' energy policies. It also, on one level, conflicts with the principal tools available to those formulating energy policies in the greenhouse – the polluter paying for greenhouse impacts. Completely free markets are by definition unregulated, and as previously noted, encourage externalities.

Meanwhile, the 1973 oil crisis not only had a profound economic effect on the USA economy. The crisis' reverberations were felt by all developed nations, including Europe. The drift was decidedly away from a free market for energy.

In 1974 France adopted its *toute nucleaire* strategy; a heavy investment to replace oil-based electricity production with nuclear. While worldwide oil consumption held its own throughout the decade straddling the crisis (up 32% between 1970 and 1980), France was able to make significant cuts a few years later once its nuclear stations had come on line (France's oil consumption lowered by over 19% from 1980 down to a 1990 annual consumption of 89 million tonnes). French nuclear generation increased dramatically by 377% in the decade to 1990 to 61.1 mtoe[134].* The Germans too accelerated their nuclear programme and, in addition to non-free market investments, also introduced the *Kohlepfennig* tax on electricity to subsidize domestic coal production. The UK response to the 1973 oil crisis was to quickly develop North Sea oil reserves, and to promote energy security by investing heavily in coal mines in the late 1970s ensuring that the UK electricity supply industry bought UK coal (even though cheaper imports were available). Denmark, which in 1972 was over 93% dependent on imported oil for its primary energy requirements, was one of the EU countries most sensitive to the 1973 oil crisis[272]. It formulated a coherent energy strategy implemented primarily through two major parliamentary acts that: i) encouraged the development of the North Sea fossil fuel reserves and, ii) provided rational municipal heat planning. By 1990 Denmark was producing 5.9 million tonnes of oil p.a. which met 67% of its total oil requirements. Furthermore, its improvements in energy efficiency meant that national consumption had remained almost unchanged despite an increase in GNP of over 30% between 1973 and 1990. As with the other EU nations, from the mid-1970s Denmark was able to reduce her sensitivity to external oil crises through centrally planned measures.

In the early 1990s such central controls ran against the basic free market philosophy of the Single European Market and, as a goal for European energy sectors, the Single Energy Market. Up to then, because of its special significance, the member states' energy sectors were largely left out of the implementation of the Treaty of Rome (which created the European Common Market).

* More recent estimates of France's nuclear production for 1990 give a figure of 81 mtoe to allow for fossil fuel station's energy efficiency.

Three Single Energy Market proposals in particular, highlighted the conflict. These were:

(1) **Common carriage**. The abolition of the monopolies by which the gas and electricity utilities distribute their products, in order that independent energy utilities might also market their products.
(2) **Price transparency**. The freedom of energy pricing to be open to inspection (as to how costs were arrived at), to prevent tactical undercutting of price to major (primarily large industrial) customers.
(3) **Open procurement**. To allow competition between nations for the purchase of energy supply technology.

The problem with all three is that they require central inspection and control; in other words regulation[273]. By the early 1990s only the UK within the European Union had the regulatory mechanisms to ensure common carriage – the UK had allowed for this with the privatization of the gas and electricity supply industries. Even if the other nations were to create their own regulatory mechanisms, a Community-wide monitoring and enforcement system would need to be established to ensure common rules and standards between the member states. Price transparency, virtually by definition, requires monitoring and regulation if it is to be effective. Similarly it follows that effective open procurement requires a mechanism for monitoring and control, but here there is the additional problem in that the number of players has been decreasing in recent years. With mergers between the energy supply technology industries (such as between UK's GEC and France's Alsthom) becoming more common, competition is reduced. Furthermore those mergers between EU and non-EU companies (such as between Italy's Ansaldo and the Swiss-Swede ASEA-Brown Boveri) serves to erode the trade tariff barrier surrounding the European Union.

It seems that the EU is split in its mind over energy policy: on the one hand it is striving for a free market and to ensure fair competition (which is difficult given the near monopolistic relationship many energy suppliers have with the end-user), while on the other the EU is imposing regulations and trappings of a centrally planned economy. Yet, ironically, this state of affairs helped to facilitate early 1990s EU policy proposals to allow for the environmental impacts of energy consumption to be paid for by those using the energy – the internalisation of environmental externalities through the 'polluter pays principle'.

By 1992, prior to the United Nations Conference on the Environment and Development (UNCED), the EU was accustomed to the notion that it could not have private (non-state) run energy sectors to allow for competition without regulation; and that fair competition and regulation go hand in hand. If the EU does not establish the regulatory framework to support a Single Energy Market then the alternative will be for more centrally controlled energy sectors; in other words regulation and more centralised planning again. From a greenhouse perspective, regulation is a valuable policy tool for the lowering of greenhouse emissions. It need not replace a free market, but provide controls within which a bounded free market can operate.

Regulation implies standards to which whatever is being regulated must comply. Here the EU has managed to set policy standards, or goals, which it is urging energy utilities and users to work towards. In October 1990 EU Energy Ministers set a target of stabilising European CO_2 emissions at the 1990 level by the year 2000. To assist energy users in meeting this target, the EU created an energy efficiency programme, SAVE, to run for 5 years and with a budget of 35 million ECUs. A second programme, called ALTNER, supported research into renewables. However, these programmes alone are unlikely to meet Europe's self-imposed targets. SAVE is expected to cut CO_2 emissions by 500 million tonnes by the year 2010. Complementary measures will be required and one of the most favoured is that of an energy or 'carbon' tax.

During the run-up to the 1992 United Nations Conference on Environment and Development (UNCED) the UK had already accepted EU proposals for some centrally imposed goals for reducing carbon dioxide releases: in part realised through polluter pays policies of financial discouragements – energy taxes. At the time the UK and Europe's promotion of energy tax proposals contrasted sharply with that of the USA. UNCED was to see fierce discussion and negotiation over international regulatory energy policy and the role of energy taxes. In a way this was surprising, for the UK's approach to energy policy had been similar to that of the USA. For instance in 1993 at a meeting of the EU member states, the UK was the only one to refuse to commit itself to a carbon tax[274]. Subsequently, the EU was to continue to move towards implementing a carbon tax, though in 1997 it is still only being discussed and remains just a policy goal.

Meanwhile, as we move towards the year 2000, the EU seems to be having difficulty in staying on track for its agreed policy goal of stabilising emissions at the 1990 level. By 1993 the EU's own analysis indicated that by 2000 it would have exceeded its target by 4%[275]. In 1997 these fears remain justified.

UK ENERGY POLICY

In absolute terms, after Germany, the UK is the second largest European consumer of fossil fuels. It would not seem unreasonable if it had a thoroughly though-out energy policy. Indeed, after the 1973 oil crisis the UK Department of Energy produced a number of discussion documents (or 'Energy Papers') on various policy aspects. In 1977 Energy Paper #22[276] specifically dealt with energy policy. It defined the objective as to secure the nations energy needs at the lowest cost in real resources, but consistent with national security, environmental, social and 'other' objectives. It acknowledged that energy planning requires a broad time horizon, albeit flexible. It also underlined the desire to meet UK international obligations in the European Economic Community, and the International Energy Agency. Finally, it outlined the key criteria for a long-term energy strategy:

- developing guidelines for the future expansion of coal;
- ensuring UK reactor technology was sound for a rapid, post-1990, expansion if required;
- having regard for the long-term role of North Sea oil and gas;
- increasing efforts to secure all cost-effective savings of energy;
- ascertaining the viability and potential contribution from renewable energy sources.

It also noted that, 'there [was] an argument for devoting some of the benefits to come from North Sea oil to investment in energy production and conservation'[276]. This energy paper was of particular importance as it preceded a governmental energy policy Green Paper[277] the following year. (Green Papers are discussion documents, whereas White Papers are policy documents that outline legislative intent.) The 1978 Green Paper was the true landmark in the UK energy policy for it represented the next most significant official statement on UK energy policy since the previous White Paper on fuel policy in 1967: the UK, like the USA, has no real energy policy.

The 1978 Green Paper[277] endorsed the criteria for the formulation of a national long-term energy strategy presented in the previous years's Department of Energy document. It also addressed each of the primary fuels, and particularly concluded that there was an urgent need to allow for the dwindling potential for the World's oil reserves to sustain economic growth and to switch to other fuels. It said that: 'This change was in no way comparable with previous changes in the predominant source of World energy – from wood to coal, from coal to oil – since these changes were not in general dictated by insufficiency of supply of the old fuel but took place as and when the new fuel appeared more advantageous. By contrast the change away from oil will be an enforced change and subject to the constraints of timing. We may have to accept that the available alternatives will be less convenient, more expensive or less environmentally attractive than the use of oil. Careful planning will be required if we are to effect the transition successfully and without damage to the World economy.'

Over one and a half decades later, the UK has failed to address the underlying issues presented in 1978. It, in common with other EU nations, was trying to reconcile a free market philosophy with the antithesis of regulatory constraints required for such a market approach. Spending on encouraging greenhouse-friendly practices were cut. For example, the UK Department of Energy's spending on energy conservation fell from £24.5 million per annum in 1985 to £15 million in 1990.

In 1982 the then UK Secretary of State for Energy, Nigel Lawson, referred to what was to be his Government's policy through to the 1990s: 'Energy is a traded good ... our task is to set a framework which will ensure that the market operates in the energy sector with a minimum of distortion'. Since then the UK Conservative governments have privatised gas and electricity (save for nuclear-generated), and in 1992 privatised coal. Commenting on this, the energy specialist Professor Ian Fells, wrote the following in an article entitled 'What price an energy policy?': 'The Government has introduced desperate measures to inject competition into these naturally monopolistic industries. Some of the measures are so Byzantine in their complexity that it is difficult to judge their effectiveness'. Of relevance to the greenhouse issue he went on to say, 'protecting the environment costs money – a lot of money. Without regulation there is little incentive to spend money on, for example, plant flue-gas desulphurisation at a coal-fired power station. The cost of the desulphurisation plant will have to be recovered. The overall efficiency will be reduced ... In all, the electricity that is generated will cost more. Altogether, it is an unattractive and expensive prospect, and no amount of improved 'green image' of the industry can compensate for the increased costs. The industry must be directed to clean up pollution by enforceable legislation'[278].

Professor Fells was not alone in his criticisms as to how the government related energy policy and the environment, specifically the greenhouse issue. Two years earlier, and just prior to the publication of the 1990 IPCC report[20], the Parliamentary Energy Committee of the House of Commons noted that 'market mechanisms alone will not produce an adequate response to global warming. It concluded that it would be inexcusable if pusillanimity, and the inability of the governments of the World to plan long-term (see Figure 10.1), allowed irreversible and disastrous global warming to occur for want of the means or the political will to take effective action to curb it .... It would be irresponsible not to adopt targets for reducing greenhouse emissions, especially when the costs are so modest'[279].

In the run-up to the 1992 United Nations Conference on the Environment and Development (UNCED), the UK was for a while leading Europe in arguing for an international carbon tax. By contrast the USA had adopted what might be thought of as the typical free-market attitude of reluctance to address the issue. Indeed right up to UNCED there was doubt as to whether President Bush would attend. Yet the so-called carbon tax is a key option to lowering greenhouse emissions, and to eke out finite fossil fuel reserves.

At the 1992 UNCED the USA and the UK were just two of the 150 countries to sign the UN Framework Convention on Climate Change. At the time the UK pledged to produce a programme of measures to implement the Convention's commitments. In January 1994 the UK Government published *Climate Change: The UK Programme*[280] which set out its policy proposals to combat anthropogenic climate change so making the UK the first country in the World to show in detail how it will reduce its emissions. At the heart of the Framework Convention is the requirement that parties must return their greenhouse gas emissions to 1990 levels by 2000. This can be interpreted two ways. Either the sum total of greenhouse emissions should return to 1990 level, or each individual greenhouse gas should be reduced to that level. Interpreting the convention in the former way means that all the greenhouse gases would be considered together in a 'basket', and would allow trade-offs between different greenhouse gases. Alternatively, if each gas is considered separately, then for some gases that had low emissions in 1990 there would be little room to manoeuvre. It was in this latter way that the UK Government's 1994 programme for climate change interpreted it.

The first casualty of the 1994 UK strategy was the chemicals giant ICI's plans for CFC replacements. It had invested in plant to produce HFC-134a as a substitute for ozone-depleting CFCs. HFCs are less harmful to ozone but are still greenhouse gases and production levels were minimal in 1990. If emissions were not to rise above the 1990 level then increased production of HFCs as an alternative to CFCs would not be possible.

The other casualty was the concept of a carbon tax. This did not feature as such in the Government strategy, instead Value Added Tax (VAT) was placed on all energy consumed from all types of fuel, not just fossil. VAT revenues (at 17.5%) were expected to be of the order of £1.25 billion a year and expected to save 1.5 million tonnes of carbon dioxide a year by the year 2000, or 15% of the strategy's estimated carbon saving[281]. But the VAT was not a direct carbon tax, non-carbon fuels were also taxed, and if it were truly part of a commitment to carbon dioxide abatement then the revenue would be ploughed back into carbon abatement measures, such as subsidies for improved energy efficiency. However, so unpopular was this energy tax that the Government did not fully implement it and at the time of writing, in 1997, VAT on energy was 8%.

The UK strategy was criticised for having no teeth, relying heavily as it did on voluntary measures and the 'dash for gas'. The environmental news journal, the *ENDS Report*; stated that the strategy 'left the [UK] government open to the charge that it will make the task of reducing carbon dioxide emissions more painful in the early years of the next century'[281].

Nonetheless the UK government of the early-to-mid 1990s did implement some measures that encouraged non-fossil fuels and the development of new technologies. The first was a levy on electricity from fossil fuel stations to subsidise nuclear power generation. This fossil fuel levy (set at 11% in the early 1990s) was designed to bolster the nuclear power industry to make it viable for privatization, and to ensure that the UK electricity supply industry did not have to rely too heavily on coal-fired stations. But this levy was at best a short-term measure as under European (Energy Charter) regulations it could not continue after 1998. Even so it did much to bolster the UK nuclear power industry and so its contribution to UK electricity supply did displace some of the need for electricity from fossil fuels.

The second measure was a series of Non-Fossil Fuel Obligations (NFFOs) which guaranteed the sales of electricity from a variety of non-fossil fuel sources. This measure was included in the 1994 UK *Climate Change* programme. The NFFOs oblige the (UK) Regional Electricity Companies to purchase electricity from producers using new non-fossil fuel energy sources at prices the individual producers negotiate with the Department of Trade & Industry. The producers, having had their projects approved by the Department of Trade and Industry, then have to obtain local planning permission for their projects. For the UK government's third Non-Fossil Fuel Obligation order (NFFO-3) in 1995, some 627 MW of capacity were approved of which it expected some 300–400 MW would obtain final approval once standard planning permissions had been obtained from local authorities[282].

Table 10.2. Summary of the UK third Non-Fossil Fuel Obligation (1995)

Technology	*Capacity (MW)*	*Number of projects*
Wind		
Over 1.6 MW	145.92	31
Under 1.6 MW	19.71	24
Hydro	14.48	15
Landfill gas	82.07	42
Municipal and industrial waste	241.87	20
Energy crops & agricultural & forest waste	122.87	9
Total	626.92	141

The NFFOs by themselves are not a major reducer of the UK carbon dioxide emissions: the generation capacity the third NFFO provided was roughly the same order of magnitude as 1% of the UK's electricity production. Nonetheless, considering that in the early 1990s just 2% of the UK's electricity was generated from renewables, the NFFO policy does provide comparatively major support for UK renewable technologies. Furthermore, and more importantly, it allowed new renewable technologies to be developed so that in the future these might contribute to the commercial electricity generation mix. While the UK Government was content to nudge the energy market, it was not happy to spend money on renewable energy R&D, although, in 1992, the Government's own Renewable Energy Advisory Group reported that as much as 20% of UK electricity could be met by renewables by 2025 (equivalent to 10 000 MW of capacity). In 1994 the Government proposed a long-term strategy of renewable energy budget cuts from £22.6 million per annum in 1993/4 to £10 million in 2005 and expected the private sector to invest a further £3 billion. Of course, policy is ever-changing. In 1996 the UK Government turned its back on individual greenhouse gas considerations and adopted the 'basket' approach. However, the 'separate' or 'basket' approach question is likely to remain a feature of greenhouse policy debate in many countries through to the end of the century.

To sum up, the UK's energy policy as regards global warming is very weak, though it does contain some innovative components. It is possible to argue that the UK's carbon dioxide-reducing energy policy does not need to be stronger as in the short term the UK (largely through increased gas use) has managed to stabilise its CO_2 emissions to near its 1990 level. The USA contributes about 24% of the World's carbon dioxide emissions whereas, with just a fifth of the US population, the UK contributes only 2.78%. Furthermore, within the EU the UK per capita carbon emissions (as expressed in tonnes of carbon in 1990 was 2.80 compared to an EU average of 2.52, with (then Western) Germany's citizens contributing 3.56 tonnes. So that despite being one of the big four EU industrialised nations, and part of the top G7 industrial nations worldwide, the UK is already comparatively carbon-efficient. There is still a considerable way the UK could go should it wish. France's citizens, who also belong to the top four EU industrialised nations, only contributed 1.85 tonnes of carbon per capita in 1990 – largely due to a nuclear-dominated electricity supply industry (which has its own environmental concerns). Even so, much could still be done by the UK and Europe in terms of improving energy efficiency, both in terms of improving the efficiency of current use as well as in the way the nation chooses to expend its energy.

ENERGY POLICY AND THE THIRD WORLD

Just as energy policy in many of the major industrial emitters of carbon dioxide lack a coherent energy policy, so do most of the less-developed nations; the only difference being that the less-developed nations have not had the opportunity to utilise energy resources in the same way as developed nations. Those living in the developed world take it for granted that electricity is available 24 hours a day, and that there is a stocked petrol station in every village. In many less-developed countries such emerging services do not exist. Fossil fuel distribution points are often far apart and not guaranteed to hold stocks. Electricity may only be available for part of the day. Not surprisingly, the carbon dioxide emissions from less-developed nations contribute only half of the World total[283], yet, in 1990, less than a quarter of the World's population lived in industrialised countries, with the remaining three quarters in the less-developed nations. Consequently the priority is for energy policy in the greenhouse world to be addressed first by the profligate energy-consuming industrial nations. This, though, does not mean that the less-developed nations have no need for energy policy or, for that matter, that the rest of the World has no need for less-developed countries to adopt greenhouse energy policies.

A cursory examination of oil reserves would appear to place the less-developed nations in a favourable position. In terms of proven reserves at the end of 1995, Africa had more oil in the ground than either Western Europe or North America; as did the former Soviet Union combined with China. Central and South America was better off, with twice as much oil in the ground as Western Europe and North America combined[135].

With regard to production/reserve (P/R) ratios, in 1994 North America and Western Europe had about a decade's worth of oil (at that year's consumption rates) left in the ground, Africa had over two decades worth, and Latin America over four and a half decades. (Remember that this does not mean that the countries of these continents have this number of years of oil left, just that the then P/R ratio was a snapshot of what would happen if the then trends continued.) However none of these regions comes close to the proven oil reserves of 1010 billion barrels (or 43 years worth of consumption at 1994 rates), the largest geopolitical proportion, some 65% (660 billion barrels) to be found in the Middle East. Indeed, if the Middle East was to cease its exports to the rest of the World, and just produce oil for domestic consumption, then at 1994 rates it would have enough proven reserves to last itself over 93 years[135].

Altogether the distribution of the World's fossil fuel deposits gives the non-industrialised nations an advantage over Brandt's rich industrialised nations, at least for the short-term. In the medium- to long-term it is quite a different story. The less-developed nations are seeing population growth at rates far faster than in the developed nations (which have almost stabilised). Moreover, as the average standard of living within the less-developed countries is so low, their aspiration for a more Western lifestyle means that their energy demands are likely to soar. In fact this has already begun, and most developing nations that do have a coherent energy policy, have one centred around the provision for a rapid expansion of energy supply.

Three years after the 1980 Brandt Report the Brandt Commission wrote a follow-up report[14] pointing out that the reasons for the rich Northern nations to help the poor South are not borne of altruism, but need – North and South are mutually dependent. In greenhouse terms if the rich Northern nations can learn to use energy with maximum efficiency and minimal environmental impact then not only will they become more self sufficient, but they can trade the technology (which only the wealthy nations can afford to develop) for some of the South's resources. The World Commission of Environment and Development in 1987 took this further in saying that while this short-term exchange was going on, both the North and South should ideally attempt to move to a more sustainable, resource-efficient mode.

PRIVATE TRANSPORT

If a single greenhouse symbol of our twentieth century global culture's failure to see beyond the present were required, then one need look no further than the nearest road and the car. The car is arguably the most aspired to form of transport throughout the World. Not only do by far the majority of the world's 500 million cars and light trucks run on finite fossil fuels, but road transportation (all forms) is responsible for between a quarter and a sixth of most countries' carbon dioxide emissions (over 20% for the UK in 1994), and just under 20% of all carbon dioxide emissions Worldwide. Yet by whatever criterion, be it cars' sales worldwide (about 600 000 new cars per week in 1990), or the proportion of passenger-miles travelled by car, or even the absolute number of passenger-miles travelled, car use is growing relentlessly and with it its contribution to global warming. Nowhere is this illustrated better than in the UK. In terms of the industrialised nations, in 1990 the UK had over 300 cars and taxis per 1000 population. Japan, Spain, and Denmark had between 200 and 300 cars and taxis, (former) West Germany and Italy had over 400, and the USA just over 450 per 1000 population. Of course, where there are cars, there have to be roads[284].

Poignantly in 1989, just three years after UK oil production had passed its 1980s peak, the British Government published a transport White Paper to announce 'a greatly expanded motorway and trunk road programme'[285]. Note that this White Paper was not announcing an expanded motorway and trunk road (the UK term for a regional main road) programme, but expansion of the existing programme. Indeed the new expansion represented more than a doubling in funding, with some £6 billion (1989 money) being added.

Given that the World oil reserves-to-production ratio in the early 1990s was a little over forty years proven reserves, or given greenhouse considerations, it might be expected that the UK would have a regard for other more efficient and less polluting forms of transport. This, broadly, has not been the case. The 1989 UK transport White Paper identified a problem of traffic congestion on principal highways, and proceeded to tackle it purely in road terms without reference to other forms of transport (such as rail), and assuming a business-as-usual forecast for the future. Under this scenario the White Paper predicted that in the 37 years from 1988 to 2025 the increase in UK road *car* miles would be between 82% and 134%: in short that car traffic would double. All other forms of traffic would also increase except for buses and coaches which the report assumed would remain at the same level up to 2025. Yet of road passenger traffic, an increase in the use of buses at the expense of cars would lower both congestion and greenhouse gas emissions. The transport White Paper relied on an overall forecast for *all* traffic that assumed that there would be an increase of between 83% and 142% over the 1988 figure of 205 billion vehicle miles[285].

For the UK the greenhouse implications of these trends, if realised, are considerable. Already UK transport contributes over a fifth (over 100 million tonnes per annum) of the nation's total carbon dioxide emissions. Transport is now the largest of the five energy end-use sectors: larger even than the iron and steel, and the 'other' industries sectors combined[168]. In 1990, within the UK transport sector some 48.7 million tonnes of oil equivalent was consumed, of which over 99% was directly in the form of petroleum fuels. On a more mundane level, and broadly speaking, in industrialised nations the average family car emits around about its own weight in carbon dioxide each year (though this neglects the fossil fuel energy that went into the vehicle's construction).

The USA has the highest level of car ownership, the rest of the World is catching up fast, so that although the USA still has the largest car fleet, its share has fallen from 77% of the World total in 1930, to 63% in 1960, and 35% in 1986[282]. OECD estimates suggest that by the end of the century Western Europe will overtake the USA in the number of automobiles on its roads. Moreover, from a 1980 base, by the end of the century the Third World nations will have tripled their car populations. In the early twenty-first century energy considerations and environmental impacts, arising from the World's car fleet will become increasingly more difficult to ignore, yet at the moment car transport and the car industry seems to be actively encouraged at national level by virtually all nations. The only meaningful exceptions being either when shortage arises (rationing cards in the UK after the 1973 oil crisis), or when the environmental externalities compromise human welfare (car exhaust regulations in Los Angeles, USA). At present the transportation sector accounts for two-thirds of the oil used in the USA, over 40% of the oil in Western Europe, half the oil in developing countries, and an estimated 40% of oil in the former USSR – hardly insignificant proportions. Car use dominates the transport sector and (in 1990) contributes to well over 15% (and increasing in the late 1990s) of the World's carbon dioxide emissions: in the USA alone motor vehicles account for almost one quarter of the CO_2 emitted[287].

Despite the obvious scale of the growing problem that cars pose, very little has been done other than, importantly, the improvement of engine efficiency. There has been an approximately 30% improvement in Western European car fleet efficiency between 1977 and 1988 to 7–8 litres per 100 km, and a staggering 47% improvement in the USA car fleet's efficiency to 9 litres per 100 km[288]. In greenhouse gas emission terms the gains from improved efficiencies have been swallowed up by an increased dependence per person on the car, and more significantly because of a 730% rise in the World automobile fleet between 1950 and 1993. Although other forms of transport release less carbon dioxide than cars (underground and regional rail 90% compared to cars (100%) per passenger mile, and urban buses and inter-city rail less than 60%) little is done to encourage non-car transport. Not only is fossil fuel car use an integral part of the greenhouse issue but so are the policy controls (rather the lack of them). Such views governments and societies have had towards cars, have mirrored those of most nation's energy policies.

ENERGY POLICY, GOALS, AND CLIMATE CHANGE

The overall purpose of any greenhouse energy policy is to 'account' for the cost of generating greenhouse gases so as to pay for their mitigation. For our purposes these costs can be considered in two (overlapping but inter-related) ways: the quantifiable economic costs and non-tangible costs.

Everyone uses energy and everyone is involved in the domestic consumption of energy, so taking a domestic perspective to illustrate the greenhouse time dimension – though an industrial, or transport, view would suit just as well – two things happen over time when a fossil fuel cooker is used. First the cumulative greenhouse gas burden on the planet increases. This leads to marginally increased costs from the slight warming that has taken place. Second, there are less fossil fuels left in the World to consume. In perfect free market theory the law of supply and demand should apply, so that at the end of the year with marginally less fossil resources available, the cost of subsequent fuel should be correspondingly increased. But the free market is not quite that free, and there is a role in greenhouse policy to ensure that this 'increased scarcity' cost is recognised.

The use (by an average Briton) of a third of a tonne of oil equivalent of gas in a year may seem insignificant compared to the 90 or so billion tonnes of oil equivalent of gas known to be accessible in the ground. Yet the trend of gas consumers worldwide is to participate in the complete depletion of the resource over a period of about 60–70 years. Markets do not quite work (nor are resources consumed) in this way but there have been analogous circumstances (simulating oil shortages resulting in price rises) in the past, so the overall principle of prices being influenced by supply and demand has some merit. Even so supply-demand price calculations are often arbitrary. For instance, over what time period does one make such a calculation? On a century level then the World's gas resources would be severely eroded. On a daily basis the proportion of the World's stocks of gas would scarcely be touched (on the order of a trillionth of 1%), so its scarcity would be virtually unchanged. Fortunately, there are other ways of accounting for such environmental resource costs, especially as the real market has difficulty in recognising scarcity.

Despite the obvious logic of the scarcity argument, in the world of the oil market, and in real 1990 monetary terms, the value of a barrel of crude oil has remained broadly constant, wavering about a mean 1995 dollar value of some US$12 for over a century[135]. If anything over this time the price has tended to decrease – the exact logical opposite if scarcity was continually increasing!

The accounting of 'non-tangible' costs of greenhouse emissions are harder to incorporate into a greenhouse policy, simply because the values are difficult to quantify. However, one other way of looking at these is to say that in absolute terms any value attached to a natural ecosystem, even a value as low as a cent a year, would in the fullness of time become considerable and tend (assuming humans consumed resources indefinitely) to infinity*. Infinite costs are clearly ludicrous. Economists find infinite values in the market place as difficult to deal with as physicists do with the infinite values associated with singularities in the cosmos – laws, rules, guidelines ... whatever, including infinity breaks them down. Fortunately, we can sidestep this explorative approach into a future of unknown benefit, by taking what is known as a normative approach by bringing the costs of preservation for the future into the present. This task is easier since the shopping lists (as to what is needed to be conserved) have already been compiled. We have, for example, already identified through the 1980 World Conservation Strategy[2] priority biogeographical provinces that require protection, and *Caring for the Earth* (the World Conservation Strategy 1991[16]) has provided a timetabled action list of targets and goals for individual nations as the World moves towards a sustainable mode. Meeting these targets has been costed.

At the end of a decade after the implementation of their proposed programme, the World Conservation organisations costed the major aspects of the World 'strategy for sustainable living' at US$161 billion a year (1991 money). Of this, the costs of providing immediate greenhouse benefits include: US$8 billion (5%) on reducing deforestation (and in the process conserving biodiversity); US$9 billion (5.6%) on increasing energy efficiency; US$27 billion (16.7%) on developing renewable (greenhouse friendly) energy resources; and US$50 billion (31%) on improving energy efficiency. The remaining 41.7% of the US$161 billion on the less immediate greenhouse determining factor of stabilising population, and encouraging the circumstances that tend to reduce population growth (education, health provision, soil protection, and reducing Third World debt) and in turn the human-induced greenhouse effect.

US$161 billion a year seems a considerable sum, until, that is, it is compared against other global bills. Willy Brandt estimated that in 1986 alone over US$1000 billion was spent on military purposes[15]. The World Conservation Strategy bill of US$161 billion is realistically probably no more than a tenth of the early 1990s global annual arms spend.

In fossil fuel terms US$161 billion represents about 28% of what was spent in 1990 on crude oil worldwide. If oil sales alone had to bear the burden of the

* The counter to this is that the value to us today of engaging in a cent's worth of conservation for someone living in the next century is actually less than a cent – it is depreciated. This is so unreal that one can say that if a new life-saving drug is found in the future, having conserved an ecosystem in the present, then (as the value of human life is immeasurable) the benefits of conservation are again immeasurably great and so worth paying for.

principal World Conservation costs then a barrel of crude oil, priced at about US$25 (1990), would increase by US$7.1. This works out at about 17 cents per US gallon. Oil itself represents less than half (44.1% in 1990) of all fossil fuel consumption[267], and any greenhouse fossil fuel tax would in actuality probably not be restricted just to oil.

The proper calculation as to how much extra we would have to pay for oil to finance the 1991 World Conservation Strategy's estimated annual bill is more complicated, the above estimate gives us a broad feel for the approximate scale of such levys, or carbon taxes. One of the complications is that we use crude oil not just as automobile fuel, but for agrochemicals and plastic feedstocks too. However the uses are far from the public's mind: members of the public, if not the media, are more concerned as to the cost required to run their cars. Our broad approximations are easy for people to relate to. Furthermore the EU 1992 carbon tax proposal was to introduce a $1 a barrel oil levy in 1993 and slowly increase this to $10 by the year 2000. The EU proposal neatly straddled the $7.1 a barrel (17 cents a US gallon) WCS-2 estimate arrived at above – which if nothing else makes this carbon tax estimate realistic if applied worldwide. (However, in 1997 these EU carbon tax proposals still remain just proposals.)

Take a greenhouse levy on fossil fuel of $7.1 a barrel of crude and apply this levy in the real world and you have to start asking some serious questions. First, does this fixed levy apply to developing nations as well as the developed nations? Clearly, less developed countries will find it harder to pay the same as developed nations. Politically one may avert this by skewing the levy, so that it is no longer fixed and developed nations pay a slightly higher levy than those in the Third World. Such are the vagaries of twentieth century free market economies that even internationally traded goods on World markets, such as oil, are sold to consumers in countries of similar economic circumstances (the same World Bank grouping) at markedly different prices. Let us look at such differences in the light of our greenhouse levy, starting with the consumer price of automobile gas (petrol) in the USA.

By 1992 the World Conservation 1991 Strategy (Caring for the Earth)cost calculation of 17 cent per US gallon would have, with inflation, broadly equalled 19 cent. In the summer of 1992 garages in Washington DC were charging $1.30 a US gallon of unleaded petrol. Add on the 19 cent World Conservation levy (and inflation) and the gallon price of unleaded fuel increases to $1.49, an increase of nearly 15% (which of course is making the assumption that a gallon of crude entering the refinery equals a gallon of crude leaving the refinery). One might think that we have theoretically solved the problem of paying for the environmental degradation cited by the World Conservation Strategy (WCS) team but only in the USA did a 19 cent rise in petrol price represent an increase of 15%. Elsewhere this levy would have represented a different percentage burden, at the same time the same amount of the same petrol from a garage in

Table 10.3 Automobile gas (petrol) is 40% cheaper in the USA compared to the UK, as on June 1992 prices and exchange rate. Prices from retail outlets in Washington DC (USA), and London (UK). Such market discrepancies distort demand, and the impact of possible internationally determined fossil fuels levies (carbon tax).

Automobile fuel	*Price in UK*	*Price in US*
1 US gallon	£1.75 ($3.24)	$1.30 (£0.70p)
1 UK gallon	£2.10 ($3.88)	$1.56 (£0.84p)

England was US$3.24 (see Table 10.3). The reasons for this divergence arise from differences in taxation between nations, and in the cost of domestic oil production (oil from the North Sea is more expensive than that from Texas). Consequently a 19 cent rise in price per gallon in the UK would only represent an increase of 5.9%. Clearly, an internationally fixed levy per gallon would impart a smaller shock to the British economy than to that of the USA. Such differences serve to illustrate (in part) why the USA attitude to a so-called carbon tax in the early 1990s was markedly different to that of Western Europe, and why at the 1992 United Nations Conference on the Environment and Development at Rio the USA was reluctant to make financial commitments. (There were other factors contributing to USA reluctance: an unprecedented budget deficit, rising domestic unemployment, and the forthcoming Presidential election.)

Accounting for (internalising) environmental costs of burning fossil fuels, and of shrugging off fossil fuel dependence, is not just difficult to undertake because it is hard to determine and quantify full environmental costs, or because of market differences between nations (or politics), but because the exercise itself has an element of feedback. Increasing the cost of fossil fuel (and that of other energy consumption to allow for the associated environmental impact) not only provides funds for environmental protection and the extension of finite resource lifetime, but directly affect the rates at which energy is consumed – which after all is in part what is desired. Pricing at a higher level lowers the quantity bought. Unfortunately the percentage lowering of demand due to a 1% increase in energy price is difficult to ascertain. Part of the energy saving would arise from the consumers' own increased investment in energy conservation – at higher prices it pays to save energy. Part of the lower demand arises out of consumer choice to use less energy by turning off lights, living in cooler houses, and cooking more efficiently. The way consumption lowers in response to price increases is known as the elasticity of demand. In the unlikely event of energy prices doubling but leaving demand hardly changed, then economists would say that the energy market was 'highly elastic'. At the other extreme, if energy prices increased only very slightly, but demand fell dramatically, then the energy market would be considered to be 'highly inelastic'.

One of the problems that energy policymakers face is that the 'elasticity' of energy markets not only varies between fuels and regions for small changes in price, but the effects of greater changes is unclear. Many factors come into play, for instance, building owners might be motivated to invest in insulation to conserve energy to lower their energy costs, but those renting buildings will be less likely to make such a commitment unless they can be assured that they will be in occupancy over the necessary time required to realise the energy savings.

These and the other factors place many obstacles in the path of creating and implementing a greenhouse policy that accounts for greenhouse impacts, or encourages energy conservation, and – as an indirect consequence – promotes the longevity of finite fossil fuel resources and the transition to a more sustainable World economy. This is not to say that such obstacles cannot be overcome: a greenhouse policy can, arguably, be made to work. Even in the USA – which has been so against creating a greenhouse policy – national academic bodies call for a greenhouse policy and claim that it can even be profitable. In 1991 the Committee on Science and Engineering and Public Policy published a report[232] requested by Congress showing that the USA could cut emissions of greenhouse gases by between 10% to 40% at little or no cost – 'Some reductions may even be at a net saving if the proper policies are implemented.' The argument being that even if energy taxes do result in some job losses, other jobs will be created in new energy conservation industries. Other benefits include decreasing national dependence on fossil fuel imports, and the development of 'greenhouse-friendly' technologies that can then be exported.

In the early 1990s, the phrase 'paying for greenhouse costs' deterred many developed world politicians. Those that most ardently refused to acknowledge the need for such payment had an electorate whose consumption of energy was the largest in the World (both in absolute and per capita terms): the people of the United States of America. At that time, in the run-up to the Rio United Nations Conference on Environment and Development (UNCED), the developed nations had a range of individual responses to the greenhouse effect; additionally, groups of nations (such as the European Union) were beginning to co-ordinate. This policy spectrum manifested itself in greenhouse targets (see Table 10.4) which reflected upon individual national political thought as to how costly a greenhouse policy might be 'to their economies'. Some developed countries recognised that there were many ways to pay for greenhouse impacts, and that there were socio-economic benefits to be gained (such as employment in a growing energy efficiency sector) as well as costs (for example, a decline in fossil fuel sales (in product terms, not necessarily currency)). In recognising that benefits existed in lowering greenhouse emissions (without exact quantification) these nations set themselves more ambitious greenhouse timetables than the others who were more doubtful as to the potential economic benefits of such policies. Of the latter group the USA remained the most sceptical: its political leaders did not lay down any targets at all.

Table 10.4 Pre-Rio UNCED (1992) developed nation greenhouse policies summarised as carbon dioxide emission targets

Nation	*Pre-UNCED carbon emission policy targets*
Australia	No higher than 1988 by the year 2000 20% reduction by 2005
Austria	20% reduction by 2005, compared with 1988
Belgium	EU target; no higher than 1990 by year 2000
Canada	No higher than 1990 by the year 2000 (but in 1995 a rise expected by 2000)
Denmark	20% reduction by 2005, compared to 1988
Finland	No higher than 1990 by the year 2000
France	Reduction from 2.3 to 2.0 tonnes CO_2 per capita in 25 years (but in 1995 it looked like this target would be broken)
Germany	25–30% reduction by 2005 compared with 1990 (but target may not be met)
Greece	EU target: no higher than 1990 by year 2000
Ireland	EU target: no higher than 1990 by year 2000
Italy	No higher than 1988 by year 2000 20% reduction by 2005 (but a rise was forecast in 1995)
Japan	No higher than 1990 by 2000 (but small rise expected in 1995)
Netherlands	No higher than 1990 by 1995
New Zealand	20% reduction by 2005
Norway	No higher than 1989 by year 2000
Portugal	EU target: no higher than 1990 by year 2000
Spain	EU target: no higher than 1990 by year 2000
Sweden	EU target: no higher than 1990 by year 2000
Switzerland	No higher than 1990 by year 2000
Turkey	No strategies
UK	No higher than 1990 by 2005, and 20% reduction in all greenhouse gases. (In 1995 it looked like this target would be met, but in 1997 there are doubts)
United States of America	The only industrialised country without a target or timetable (a target was set by 1995 but emissions expected to rise)

Despite USA recalcitrance, by the summer of 1992 and the UNCED Rio Conference, the overwhelming body of international opinion was in favour of adopting policies to counter global warming and the Framework Convention on Climate Change was opened for signature. The Convention lacked clearly binding short-term commitments, but it did set in train policies to attempt the stabilisation of developed nation greenhouse gas emissions by the end of the century. It established a process for negotiating further measures and introduced mechanisms for transferring finance and technology to the less-developed nations. One of the UNCED successes was that in addition to the nations that already had adopted some form of greenhouse policy, the USA also signed the Framework Convention on Climate Change.

UNCED had its failures too. These in part removed much of the potential from the originally proposed Climate Convention. For instance, the UNCED

Convention on Biodiversity was not signed by the USA. This would have increased the financial value associated with standing tropical rainforests in that it called for the equitable sharing (between the developed and less-developed nations) of the benefits that flow from the utilisation of genetic resources (of which tropical forest abound). President George Bush, perceiving that signing would have a cost for the USA in unemployment terms and with an eye on the forthcoming Presidential election, did not agree to his country abiding by its terms. Another failure with greenhouse implications was a third convention on Sustainable Forestry. Here the international gulf between the developed and less-developed nations was so great that before the nations gathered for UNCED itself, the convention had been watered down to a set of 'Forest Principles'.

UNCED was a summit at which the nations of the World took two steps forward and one back: progress was made but in the main only in that the World's nations agreed that progress needed to be made. The only principal commitments were for the developed nations to limit their greenhouse gas releases, and not in the other inherently related areas. Which brings us back to the need to formulate a successful greenhouse policy.

Five years later the World's leaders were still trying to identify the core components of an international climate change policy. The non-binding goals identified at UNCED of trying to lower greenhouse emissions down to their 1990 level had only been achieved by two countries: the UK (largely through a non-sustainable dash for gas) and Germany (through have gone through unification and closing down inefficient East German industry). Then in December 1997 the World's leaders met in Kyoto, Japan, for a World Climate Conference.

Throughout the Conference the US dragged its feet. The problem was not so much that the Clinton administration did not accept the urgency of climate change issues but that it was gridlocked between the Senate and Congress. There was little point in the US agreeing policies that would not be subsequently ratified back home, so the US negotiators had to come away with a a clear win for America (for which read a result that would be considered a clear win by US politicians and voters irrespective of environment or World development concerns).

As of the time of going to press, it appears that while the EU has committed itself to an 8% cut in its emissions and Japan 6%, the US has agreed to a 7% cut by the year 2012. However these cuts, while impressive, are only for accounting purposes. The US insisted on JI (Joint Implementation) which enabled it to buy permits from other industrialised nations as well as transition countries (such as the former Soviet nations). Under JI the US could replace an inefficient power station overseas with an efficient one, and so use the saving to offset its own domestic emissions. The Kyoto treaty also seems to have allowed for carbon trading between low and high emission nations. In short, the treaty was created with built-in loopholes. Nonetheless, Kyoto was just one stepping stone of many that would no doubt follow. Its one undeniable success was that it did reaffirm greenhouse issues' position on the international political agenda, hence the prospect of being able to formulate a successful greenhouse policy sometime in the future.

SUMMARY

Most developed nations have paid little attention to energy policy; their policy concerns primarily have been to provide a regular supply of cheap energy to encourage economic growth, and to favour their countries' competitive ability on the World market. The developed nations have not made the longevity of finite energy resources, nor greenhouse issues the principal subject (if addressed at all) of their nations' energy policies, and some nations (such as the UK) have even neglected to construct a national energy strategy. Equally, the developing nations have neglected greenhouse considerations in favour of development.

Forming a greenhouse policy that internalises environmental costs and encourages the transition to an environmentally sustainable greenhouse policy is undoubtedly difficult. Identifying and incorporating such costs into fossil fuel prices is theoretically less difficult but is to some extent arbitrary – though some international agencies have attempted to list such costs and provide a programme of international goals and action.

There are many obstacles, but equally benefits to be derived from implementing a greenhouse energy policy – including in the long term, economic ones in terms of economic and strategic security. Providing the benefits of greenhouse policies outweigh the costs, then overcoming the obstacles, to their formulation and implementation, becomes a priority.

PART 3

ECONOMIC AND GREENHOUSE PERCEPTIONS

CHAPTER 11

FREEING THE GREEN MARKET

Let us wave a magic wand and, ignoring all other considerations, pretend that politicians, when it comes to energy policy, have two sincere goals: to ensure that environmental impacts through energy use are paid for, and that there is regard for future energy users in decades to come. Again, politicians are more concerned with the present than with problems that will occur some time after they have left office. This is why we need the magic wand. Yet if politicians did have the two issues of energy use impact and future resource depletion to the fore then their next problem would be to integrate these with their country's economic strategy. This integration is a very real problem of direct concern to all those involved with the management, or utilisation, of natural resources, and not just policy makers. Even on the purely academic level, it is no use environmental scientists identifying environmental impacts and various ways of reducing, or combating, these if the solutions cannot be manifested in the real world. Equally, from the policy-makers' perspective, it is no use receiving scientific advice if it cannot be acted on.

Economic strategies are defined by humans, and so are only limited in their formulation by the imagination. Only in their implementation are they limited by real world considerations, which in turn can be masked in the short term at the expense of strategy longevity. Yet by waving our magic wand we can forget problems of integration and implementation and look at what strategies there are to pay for greenhouse impacts or their avoidance. Indeed, there are so many strategies that only the briefest of introductions can be given here, however as such policies relate to costs, many of these strategies involve pricing.

PRICING THE GALLON

Economic theory is absolute in that in its fundamental form it can apply to economic strategies. If the state fixed the price of widgets at 'x' dollars, but demand kept on increasing so as to outstrip supply, then though the state's supply of 'x' dollar widgets would dry up, there would invariably arise a non-state supply of widgets costing more than 'x' dollars. This non-state supply might come from previous legitimate widget owners engaging in barter or, as has been common within the nations of the former Soviet Union, from a black economy. Either way the law of supply and demand applies despite the economic strategy governing the national market.

This distinction may seem a fine one, but it is of crucial importance. The way we can relate to natural resources as consumers is ultimately governed by economics; on the other hand the economic strategy we end up with is governed by human choice. One is a matter of fact, the other, of fashion. This last means that the human realisation of environmental economics has been coloured by little more than whim, which in turn is exactly why the price of a gallon of petrol (automobile gas) in the USA is so much cheaper than the same amount of the same product in Western Europe (see Table 10.3).

Economic strategies have varied in time as well as space. At one time it was a perfectly acceptable economic strategy to forgo much of the labour charge associated with major construction projects of national importance. Today the use of slaves to build structures like the Egyptian pyramids would, to many, be unacceptable. The present late-twentieth century view of major World economic strategies is therefore special to our time, though economic slavery still exists. These strategies are today frequently, and very broadly, considered as part of an economic policy spectrum. At one end there are the nations who have (theoretically) adopted a strategy whereby those selling resources pitch the production (availability) rate and price of natural resources (and/or transformed resources, i.e. goods). These are the so-called free market capitalist economies. At the other end of the spectrum there are the nations in which all production levels are (in theory) pre-determined as are the prices of the principal goods. Such nations are said to have 'centrally planned economies'.

The reasons that a 'free market capitalist to centrally planned economy' strategy spectrum is frequently considered are two-fold. First, most countries, even the most centrally planned, have some free market capitalist economic activity taking place (both officially, and, as with the aforementioned black market economy, unofficially). The same is also true of capitalist free market economies, for virtually all have some economic activity which is subject to: quotas (e.g. milk in the UK), subsidies (e.g. national science research in OECD countries), tax (e.g. alcohol and cigarettes in many nations), or non-financial grants (e.g. land for development in Brazil). Today, in addition to most countries adopting an economic strategy comprising of a mix of free market capitalism and central planning, politics has become increasingly dominated by economic philosophy. Whereas in the Middle Ages religious considerations dominated much of the politics of European and Asian countries, today economics dominates politics worldwide: hence the cold war split between the free market capitalist nations of the West and the centrally planned economies of the East.

Just as economic strategies are transient in time as well as space, so the economic spectrum used in commonly talking about economic strategies changes. Currently, at the end of the twentieth century, we are going through a period of great change. With the dissolution of the Soviet Union, terms like East-West have become less common, instead Brandt's term (coined in 1981) of North-South has come into vogue as a description of the two types of nation: those with developed, and those with less-developed economies. The former Soviet Union nations themselves are changing their economic strategy, with each country becoming less dependent on central planning, instead encouraging free market capitalism. In the West this is seen as a vindication of the free market system, but is it, especially if all economic strategies contain a mix of free market and central planning? From an environmentalist's perspective, the new economic spectrum to consider is one of sustainability, but it is early days yet and the accounting techniques that might be used to measure sustainability with confidence have yet to be developed.

All this may seem a far cry from the greenhouse effect, so let us consider fossil fuel, and in particular the retail price of a gallon of petrol in a free market economy. The country in which to look at our gallon arguably should be developed; as such nations produce the most greenhouse gases per capita, they will have to supply energy-efficient technology to the less-developed nations, and take the lead in any international greenhouse policy initiative. To this end let us look at how the price of a gallon of UK petrol is made up; the UK being not as energy intensive as the USA, but part of the World's second largest consumer group, Western Europe.

Of the money paid at a UK forecourt for petrol, about 6% goes to the running of the garage and to making the profits for the garage owners. A further 7% goes to the wholesalers to meet their costs – which include distribution from the refinery – and to making the wholesalers' profits. Then we come to the actual cost of the petrol itself as purchased from the refinery: this accounts for some 25% of the forecourt price. Within this price comes the cost of refining the crude oil, the crude itself, and any profit margin. The cost of crude is set internationally for NW Europe, in USA dollars (known as petrodollars), by the Rotterdam spot market – the wholesale price fixer for NW Europe. The price of crude itself is also largely determined internationally; the biggest determinants here being the OPEC cartel (Organization of Petroleum Exporting Countries). OPEC set their prices well above the actual cost of extracting crude from geological deposits. However, though this extra cost is a reality with which the World economy has to live, it is important to note that it does not go towards meeting any of the environmental costs of oil use (or oil energy use). Finally, the biggest single proportion of the cost of UK petrol, some 62%, goes to the UK government as tax. Again, although this is a real cost, none of it – let alone a significant proportion – is earmarked to mitigate the (currently unpaid) environmental costs (externalities) associated with fossil fuel consumption.

Different countries have different taxation policies regarding petrol. Lighter, or even non-existent, tax burdens result in lower petrol prices – which is predominantly why petrol is that much cheaper in the USA compared to the UK. Prices differ depending on national treasury policy and not the market.

Importantly, from a climate change perspective, the revenue from most developed nations' tax on fossil fuel does not go towards encouraging improved efficiency of energy use (though the higher price due to taxation does have a market-forcing effect towards greater energy efficiency and for a few years in the UK, nuclear power benefited from a fossil fuel levy). Overall the price of fossil fuels in developed countries, in the vast majority of instances, varies for reasons other than those of fossil fuels' finite nature, or climatic considerations. Consequently, in an international context, even in nations with so-called free market economies, fuel prices are not determined by completely free market conditions.

Yet again, the international markets themselves can be overtaken by international circumstances, and affected out of all proportion to such stimuli. Like the stock market, the Rotterdam market is sensitive to a number of outside influences, e.g. in 1990 when Saddam Hussein threatened war against Kuwait the price of Brent crude (considered to be the UK marker) went up from $16.5 to $40.70 a barrel[289]. This increase of 146% was in response to the *possibility* that both Kuwait's and Iraq's oil production would be removed from the World market. Putting this in perspective, both Kuwait and Iraq contributed nearly 7% of World production before the crisis in 1989 (some 4 865 000 barrels a day): in other words over 90% was not directly affected by the two countries ceasing to export. Fortunately this surge in price was short term, lasting three to four months, but still long enough to have a serious effect on the World economy. Overall in 1989 there was a World excess in oil production of 885 000 barrels a day which had gone into bunkers creating a reserve which helped to offset shortfalls. Furthermore, even without this reserve, the 1990 market price increase of 146% in response to a (then) hypothetical 7% shortfall in supply does seem excessive. Indeed, even with the benefit of hindsight and the knowledge that between 1990 and 1991, Iraq and Kuwait suffered a loss of production of 88.5% and 91.5% respectively, the World only lost 0.8% production, and 1991 consumption only exceeded that year's production by 2% – the shortfall being taken up by World stock supplies[134]. It is therefore difficult to consider the free market system currently controlling petrol prices as totally rational. Indeed, as OPEC is the major factor in determining the price of crude oil, the market can hardly be said to be in absolute control of setting prices.

Looking at the World's oil prices either side of the Iraq crisis an even stranger picture emerges, bearing in mind the finite nature of the fuel, and that the crisis was a reminder of the international oil market's inherent instability. The average price of Brent marker crude was $20 a barrel in 1991 compared with nearly $24 in 1990[134].

To return to our spectrum of national economic strategies ranging from free market capitalist to centrally planned economies, it can be seen that both Western Europe and the USA do not have as free a market (at least as far as petrol (automobile gas) is concerned) as might be thought. These developed countries do not have economies that are at the extreme free market end of the spectrum; just as, with their thriving black market economies the former Soviet Union nations were not totally reliant on central planning. Yet though there are non-market factors controlling the price of petrol, together with disproportional market responses to some stimuli, no national strategies have currently been created to account for (or internalise) the unpaid environmental costs associated with petrol consumption together with the cost of transferring from a fossil fuel-reliant World culture once these fuels' availability begins to dwindle.

GLOBAL WARMING: THE TRADITIONAL ECONOMIC VIEW

Central to traditional free market economics, at least from a greenhouse perspective, is the premise that having a meal today is worth more than having a meal in ten years time. To have $100 today is worth more than having $100 in ten years time because not only can it be spent immediately, it can also be invested and earn interest to provide a (real term) increase in value. Such is the exponential mathematics of interest rates that at 10% per annum, it is not worth spending more than $72.60 now to avoid a future cost – even an environmental one – of anything less than $1 000 000 in a century's time. With such arguments neither business nor governments are likely to make investments to prevent long-term environmental damage unless the prevention of damage is substantial.

This might at first seem nonsensical, but so solid is such economic logic (on a pure accounting level) that it makes perfect sense to build a major centre of economic activity on an unstable geological tectonic fault line. Despite a violent earthquake on the San Andreas fault in 1857 that devastated San Francisco, the city was rebuilt on the same spot. Providing that major earthquakes happen less frequently than once every three decades, rents need only to be surcharged at an annual rate of 2.5% of the city's buildings' construction cost for complete reconstruction to be economic, or, should it be cheaper, for earthquake protection designs to be incorporated. In short, a business (or the citizens of San Francisco) will invest in new enterprises (or construct buildings) only if the return they expect to earn will be greater than could have been earned by investing in it today. With such logic, and all else being equal, it makes sense to burn fossil fuel now rather than leave it in the ground for the future. Indeed, it is only when all else is not equal that oil is deliberately left alone: such as after the 1973 oil crisis when the OPEC countries of the Middle East deliberately cut production having realised that a keen oil market would pay prices that (in real terms) were higher than they had been since the turn of the century, i.e. a good financial return was a virtual certainty. Price is the key thing; it is all in the accounting. If one can show that at an acceptable discount rate an investment in some future gain is more economical than exploitation now, then funds would be better off not being spent on exploitation but elsewhere where they can earn interest to provide a better return, and so conserve resources for the future. How, though, does one properly account for the future?

Under a discount rate of 10% (which is not an unrealistic rate for business), long-term marginal enterprises such as sustainable forestry, are hardly viable, which is why when nations wish to tap such resources, financial support is either directly, or indirectly, given. In effect governments tinker with their discount rates. There have been arguments that social, and environmental, ventures should receive special low discount rates: indeed, governments frequently account for projects such as hospitals with low rates (again so much for Western nations running a completely free market economy). But there is a flip side. With a low discount rate environmentally damaging projects (be they roads or dams) can be constructed, whereas they would not have been if private industry had to supply the capital. Discount rates then are just as much a minefield as is assigning values to the less tangible costs and benefits associated with a project or policy. (In cash terms, how much more does one value a piece of ancient woodland over a comparatively recent one, and should conservation projects within it be accounted for with a special low discount rate?)

Post 1992 UNCED political thought is one of urging caution when it comes to global warming, the economic view to date has remained largely unchanged. The question facing economists is whether or not the costs of preventing global warming exceed the benefits of allowing it to happen. A cost of preventing global warming might be to raise the insulation standards of buildings in a city, whereas a benefit of warming might be lower heating bills in the future: we are back to cost-benefit analysis. Yet again, we can turn such questions on their head and ask, just as fairly, whether the benefits of preventing global warming exceed the costs of allowing it to take place. So a benefit of preventing global warming might be that of preventing the cost of damage caused by flooding, or alternatively preventing the need to spend money on flood control.

The question then turns from economics accounting: how does one quantify the costs and benefits of global warming? One view, that put forward by the economist Professor Nordhaus, is that the costs and benefits of human-induced climatic change largely cancel themselves out. He concurs with the view that some curbing of greenhouse emissions are cost effective, but only about a sixth. Extra curbing would, he argues, become increasingly expensive. A large reduction in greenhouse gas emissions is, he argues, unnecessary since only 13% of the USA national output come from parts of the economy that would be sensitive to climatic change. True there would be costs, such as the construction of sea walls, but these would be outweighed by the benefits, e.g. to crop productivity through enhanced photosynthesis in a carbon dioxide-rich atmosphere.

What are the true costs and benefits of preventing climate change? These unknowns separate the economic theory, that is so ideal in theory from real world accounting. One reason for the gulf between ideal theory and mundane practice is that with the economic theory everything is theoretically known; it is a theoretically perfect world. We can theoretically work out values for the future, and assign costs and benefits as we choose.

In an economic model there is perfect knowledge between consumers and producers (and within these two groups) in order to provide perfect competition. I will insulate my house to the nth degree which over my lifetime will give me a net profit. But do I really think that far ahead, or am I more concerned with a future summer holiday, or buying Christmas presents, or whatever. Furthermore, my neighbour may only insulate his (identical) house to the yth degree, still of course obtaining the same net benefit over the same time period. Because my neighbour hates do-it-yourself and so either has to pay someone to come in and insulate his house for him, or he effectively pays himself for his own time by giving himself an expensive treat once he has done the work, the cost and benefits of identical insulation are different for both of us. Knowledge, or awareness, of the costs and benefits arising from improving energy use efficiency are central to the success of free market energy efficiency programmes and, from this the perceived value of such exercises has to be recognised.

Only when we have identified the full costs and benefits of a particular strategy can we then calculate, using society's (or the free market's) chosen discount rates, the cost effectiveness, or profitability, of the exercise. If we are unaware of some of these then we will end up making an erroneous decision. A good illustration here is that of the nuclear industry: we still do not know all of the costs associated with nuclear generation. What, for example, do we do with the nuclear waste? The USA is proposing a store in Yucca mountain, and the UK has its Nirex proposals for an underground depository, but how much will these cost to build and run, bearing in mind that such constructions have to last for thousands of years? Exactly such questions caused the intended 1990 privatisation of the UK nuclear (electricity generating) industry to flounder; it had to remain governmentally controlled in the public sector while the fossil power stations were privatized. Once more, this raises another question, if some industries find it difficult to enter the private sector, and compete in the free market, could it be that there are some companies operating in the free market that should not be?

What economics does tell us is that not only do we have to identify *all* the costs and benefits associated with resource use, but that short-termism should not determine discount rates be they high or low.

MYTH OF THE ENVIRONMENTALLY FRIENDLY FREE MARKET

That pollution levels that compromise environmental integrity exist, or that some highly profitable agricultural practices also result in severe soil loss – that there are unpaid environmental costs, or environmental externalities – demonstrates that not all economic activity is 'environmentally friendly'. As frequently mentioned, environmental externalities come about when the environmental damage (by pollution or whatever) is not paid for by the economic venture that caused it.

There are several types of environmental externality[290]. Two of the main categories are 'private' and 'public' externalities. Private externalities occur when the unpaid cost is imposed on an individual or small group of individuals. For example, a mine leaking toxic water into a stream will affect only those that use the stream, and not society at large. A public externality is one where the public as a whole bears the burden of the unpaid externality. The fossil fuelled greenhouse effect, human-induced (anthropogenic) climate change, falls into this category and though the effects of climate change vary from place to place – indeed some people might even benefit – as the whole planet is (for better or worse) affected, the cost is defined as a public externality. Both private and public externalities are forms of debt, for they are real costs (pollution is a real cost) not borne by those creating them. As such, one way of looking at business economics is to say that ventures that incur externalities are uneconomic. The reason why many enterprises incur this unpaid debt is exactly because it is not paid for by the enterprise and is in effect a free non-returnable loan, or gift. Others in society pay, either in cash terms to ameliorate the impact's effects, or through accepting the lower quality of life by accepting the impact.

Some economists stress that environmental externalities are merely a failure of the market to recognise costs[291] and that this is not uncommon in the real world. Business might be tempted to be unconcerned, after all if they can get away with not having to pay for some of their production costs then that would make them more competitive in the market place. However, this is a very superficial and short-term perspective. The costs have not vanished, they are merely being transferred elsewhere (to the society that has to clean them up or live with them, paying the price of a lower quality of life). As such there is the view that environmental externalities are a failure of the market to encourage the optimum allocation of resources in the economy[292]. Without optimum resource allocation the business community as a whole suffers. An individual business will be tempted to impose externalities in others and society as a whole if it thinks that the competitive benefit it will derive from producing cheaper goods will be greater than the marginal cost it will bear as its share of the total costs (the latter being its own costs together with the externality cost society bears). This temptation can only be removed if the individual business is fully aware that all its competitors will have to pay for any environmental costs (including externalities) they impose. The implication is that they themselves along with their competitors will have to be continually monitored to ensure that the society costs are met. Businesses are wary of such arrangements for they do not always believe that their competitors will pay their fair share of costs. Furthermore, the act alone of having to meet the requirements of regulatory legislation and monitoring itself incurs an administrative cost which contributes to product price – having to increase product price is something that business fights shy of: it is almost by definition inherent in their *modus operandi*. The fundamental temptation within the free market system to get away without having to pay for externalities is central to why it is not inherently environmentally friendly.

A second reason as to why the free market in reality fails to be environmentally friendly, is that in reality there is no perfect dissemination of information between producers and consumers. The fact that information is imperfect is the whole *raison d'être* behind the advertising business. Providers of goods and services need their potential consumers to know that these goods and services exist and to encourage their consumption. Of course, if the information is provided by individual producers then they will naturally bias the information in their favour. So rife is this that we, as consumers within the free market, passively accept that many products claim to be (or at least infer that they are) 'the best' of their kind, even though there can only be, by definition, one 'best'.

That imperfect information dissemination upsets the smooth running of economies is readily apparent. There are examples where both the information and its dissemination are flawed. The London Stock Market crash of 1987 is one such. Computers were programmed to respond to a number of market indicators, and when these rose or fell to predetermined levels, the computers predicted a huge slump in the market and reacted accordingly. In this instance human misunderstanding crept in at virtually every level: choice of indicators, programmed response, subsequent dissemination to the market, and, arguably, the degree of reliance placed on this technology. Economists taking a retrospective on the 1987 crash have argued that the market had in any case overstretched itself during the mid-eighties boom, and that a slump (though perhaps not so precipitous or quite of that magnitude) would have been inevitable.

Stock markets do not just suffer computerised misinformation, they are also misdirected in the traditional way through human misunderstanding combined with the desire to turn a fast buck. In the autumn of 1991 a small biotechnology company called Anergen 'went public' with shares being sold at $8 – the closing price its first day was $7.875. All was well until a few days later the price soared to peak at $31.50 before settling at around $15. Now market fluctuations are not unknown, but there is always a reason, albeit the annual audit revealing something out of the ordinary, business or political developments. In this instance there seemed as if nothing had happened to warrant such a market response. What exactly happened is still unclear but it would appear that dealers and others had got wind of, and misread, a report in a science journal on a promising pharmaceutical that might be used to treat multiple sclerosis. The treatment was in fact being developed by a quite separate company (which perhaps regretably was not even mentioned in the journal). In another part of the same journal there was mention of an Anergen programme looking at how the body's own immune system could be cajoled into fighting off diseases – including multiple sclerosis. Clearly an analyst had put two and two together to make five. The irony is that the financial community is prepared to increase the market value of a company three fold if it *thinks* that there is a profit in it. Furthermore, as a rule the US stock market fights shy of investing in the development (as opposed to research) of new drugs; this last was the conclusion from the US Office of Technology Assessment. More convoluted a set of circumstances it is hard to imagine, but it does illustrate how a lack of information, or misinformation, undermines the free market[293].

Of greater relevance to the anthropogenic greenhouse issue, was the impact of the 1991 Gulf War on the oil markets. Iraq in 1990 accounted for just 3.1%, and Kuwait 1.7% of the total World market production for crude oil. In 1991, the year of the Gulf War itself, the two countries combined produced only 0.5% of that year's World market production. Yet due to increased production by other large producers – such as Saudi Arabia by 28.6% – the 1991 World decline in oil production over the previous year was a mere 0.8%. To put this into context the 1981 to 1982 decline in production (at the time of global economic recession) was over four times as great as this at 3.96%. In short the decline in World oil supply during the year of the Gulf War was less than an annual 'peace time' variation earlier that decade.

The Gulf War is an extreme example of how imperfect information and uncertainty affects the free market; not just price as discussed earlier, but on the World economy as a whole. Extreme though this example may be, it does illustrate the rapidity and degree of changes that can take place, and which are out of all proportion to the ideal market-influencing factors of supply and demand. It is easy to see how among these market-determining factors, that consideration for environmental costs get marginalised.

Coincident with the Gulf crisis was another free market *faux-pas*. Again, this instance concerned energy resources: specifically, the supply of electricity in England and Wales – the privatisation of the Central Electricity Generating Board (CEGB). Before 1990 the CEGB was the largest single electricity generating utility in the World. It was state-owned and met nearly all the electricity requirements of England and Wales. It had a broad base of generation. The largest single proportion of its generation came from coal: in 1988/9 about three quarters of its electricity came from coal. It also had oil power stations, but while it had the plant to generate 40% of its electricity from oil in normal circumstances, the cost of oil prohibited extensive consumption: oil usually represented about 5% of its stations' fuel consumption. Gas (though recognised as a cheap fuel) made up less than 1% of its 1984–1989 fuel consumption; largely due to the lack of an immediate medium to long-term supply. Nuclear power was making an increasing contribution throughout the 1980s, nearly up to 20%, as the first generation of magnox reactors were supplemented with the second generation of Advanced Gas cooled Reactors (AGRs). In 1988/9 the CEGB was supplying 231.9 TWh valued at £8867 million, making a trading profit of 4%, and sporting some £18 528 million in its financial reserves[295].

Unfortunately for the Government the plan did not survive the scrutiny of the financial institutions in the city (of London) to, and through, whom the CEGB was to be sold. Their major concerns focused on nuclear power and its unsteady track record: the aging magnox reactors were coming to the end of their productive life and would soon require decommissioning (an exercise dominated by cost, without much income), and the problem of nuclear waste had yet to be successfully addressed. The Government clearly understood that, without the backing of the country's leading financiers, the privatisation stood a negligible chance of success: so they altered the deal. On November 9th 1989 the then Secretary of State for Energy, John Wakeham, addressed the House of commons. The key points he made were:

- That all the CEGB's nuclear power stations were to be retained in the public sector in a new company (Nuclear Electric) which would be wholly owned by the Government. The Government also placed some emphasis on possibly extending the operating lifetime of some of the magnox plant.
- That Britain's first PWR reactor (Sizewell 'B') would be completed to establish the PWR option, but that there would be no scope for any immediate follow-on PWRs within (what is known in UK energy policy parlance as) the non-fossil obligation. (The non-fossil obligation was created to ensure diversity of electricity supply for strategic reasons. It was arbitrarily set at 20%.)
- That National Power, now minus its former nuclear component, would continue to become privatised, along with Power Generation.

The speech was a clear blow to those advocates of civil nuclear power. No more so than to the former chairman of the CEGB, Walter Marshall, who had worked in the field of nuclear power generation for virtually all his professional life. Walter Marshall believed that a privatised nuclear programme could be successful and was very concerned that, perceived as the poor boy of the generating options, that it would not receive the financial investment due to it from either a government committed to the private sector, or a future government, from the then opposition, whose energy policies did not favour nuclear power. Consequently, it was announced the same day in a letter from John Baker, Chief Executive of National Power to his staff, and containing the transcript of the Energy Secretary's speech, that Walter Marshall was 'considering his position as Chairman of the CEGB, and Chairman designate of National Power'[296]. Walter Marshall resigned a month later on December 18th. He outlined his reasons in a letter to all transition-CEGB staff:

> 'The Government asked me to become chairman of National Power because the company was to build and operate nuclear stations. That is no longer the case. The new Government policy also means that Britain will now be building a single PWR, perhaps the only one of its kind. That is not a nuclear programme I feel able to advocate or defend.'[297]

The resignation of Walter Marshall was momentous. It provided those concerned with energy policy a rare insight into the thinking taking place within the board room of the World's largest electricity generator. Marshall knew how ESIs were in the main organised and in terms of free-market discussions he said:

> 'That franchise [to generate electricity] gives the utility the severe obligation to maintain electricity supplies on demand continuously in both the short term and the long term. In exchange for shouldering that burden the utility is given a privilege, it is given a franchise which is either a monopoly or a near monopoly, in a particular geographic district. The utility then has to respond to a financial regulator, who is usually appointed by Government, and the regulator makes sure that the utility does not abuse its privilege and that it does carry out its obligations in the short-term and the long-term'[298].

The above two statements might, at first glance, appear contradictory. In the first letter Walter Marshall says that he was enthusiastic about leading a large generating company with a nuclear power component into the private sector (which implies that it will benefit from an environment of competition); in the second he says the ESIs are Worldwide monopolies or near-monopolies (that is to say that the prospects for competition are limited in that most energy consumers have little choice in who supplies them). The confusion arises through considering large suppliers of 'long-term products cum services' entering the private sector becoming more competitive: competition and the private sector frequently being associated with each other. In actuality only in one sense can ESIs compete on a day to day basis in the private sector, and that is for funding through the stock market (for which they have to compete to be seen to be good investment risks). In terms of the consumer, they provide little competition.

Walter Marshall recognised that the individual consumer has little control over the service s/he receives from ESIs; unlike consumers buying 'short-term products *cum* services'. Individual consumers can employ the free market, and its law of supply and demand, and change their supplier of breakfast cereal, or even change the nature of their breakfast to, say, croissant and coffee. On the other hand energy supply is in the main a 'long-term product *cum* service', so that if you do get dissatisfied with cooking on electricity, then to switch to gas you have to make a long-term investment in equipment: you cannot simply hop from one to the other. Of course the problem for consumers with monopolies is the lack of choice available: availability of choice enables competition and so the free market's own principal way of protecting the consumer.

Consumer protection, rather the lack of it, was behind Walter Marshall's second criticism of the privatisation of England and Wales' ESI. Marshall recognised with a monopoly that the individual consumer needs something to replace consumer choice to ensure that the buyer has access to the quality of service expected: a regulator. Should a monopolistic energy supplier fail to meet their obligations to consumers then the regulator would have the power to dismiss the Chairman and his Board. Walter Marshall has cited an exemplary instance of the effect the threat from a regulator can have. It took place in the late 1980s when the lights went out in West Tokyo. In order to mollify their regulator, and so hopefully to secure their franchise, all the senior management of the Tokyo Electric Power Company took a 10% salary cut, and every customer received a personal abject apology from a senior manager.

In the late 1980s, the (then) UK Government's favour with free market economics prompted it to free the electricity generating sector's market constraints as much as possible so as to allow the various operators to compete effectively for fuel, manpower etc. (The post-World War II UK ESIs had been tied to centrally determined fuel pricing arrangements – primarily with the UK coal industry, but also to nuclear fuel.) Up to privatisation UK consumers could be assured of supply as the electricity generators were legally obliged to ensure supply. After privatisation this stricture was removed so freeing the various generators each to produce as much, or as little, electricity as they wished depending on their individual economic circumstances and perceived route to profitability. Instead UK consumers were to be assured of their supply by the 12 Area Boards (and the National Grid): it was they who were to be responsible for supply, though how they would be able to meet demand if the major generators failed to generate adequate amounts was not spelled out. Prior to privatisation Marshall and his senior CEGB management colleagues warned Government of their concerns. At the time Walter Marshall said:

> 'It is well known that my Board of Directors and I did not like the [privatisation] proposals and we said so in a published statement. Nevertheless, I was asked by Mr Cecil Parkinson [the then Secretary of State for energy] and the Prime Minister to implement them and to do my best to promote nuclear power, which was [originally] to be privatised within this framework. I agreed to do that because I do not consider it right to use my position as Chairman of the CEGB to oppose the wishes of a democratically elected Government'[298].

Over the year prior to privatisation the management of the CEGB sought an answer to the 'security of supply versus competition in a free market' dilemma.

However, for reasons such as those given above, they finally came to the conclusion that 'no workable compromise [was] possible'[298].

The 1989 privatisation of England and Wales' ESI demonstrated in real cash terms how isolated a developed nation's energy industry can become from market economic forces; that would make or break most other businesses. That the Government had to remove the CEGB's nuclear component from the privatisation was symptomatic of this isolation. On the face of it – i.e. from the speeches made at the time – the reason was the the true cost of nuclear power was revealed for the first time, but there were other factors operating.

The CEGB chairman's pre-resignation speech to the British Nuclear Energy Society was made in an attempt to set the record straight. He pointed out that the CEGB had accounted for its nuclear stations' decommissioning costs but that being state-owned it was obliged to hand the money over to the Treasury 'on the theory that they know what to do with the money better than us [the CEGB], and that they will let [the CEGB] have it back when [the CEGB] need[ed] it some time in the future'. Prior to privatisation, this money should have been returned.

Whereas the CEGB had an on-going business relationship with the decommissioning contractors (British Nuclear Fuels, BNF, who reprocess nuclear material and 'prepare' nuclear waste prior to long-term storage or disposal a newly privatised ESI would have to forge its own relationships. Given this, the cost of work that the CEGB had set in motion to be carried out at some future, post-privatisation date had to be financially secured prior to privatisation so as not to burden the new power generating companies. This meant, for instance, that BNF had to present its future CEGB work as a fixed price quote in 1988, so ensuring that whatever the price increases might be between then and when the work was carried out (decades) in the future, were allowed for. This pushed up the decommissioning costs, and in turn made nuclear privatisation less attractive. So state-owned Nuclear Electric was formed, leaving only the non-nuclear components of the CEGB to be floated on the stock-market in 1990.

In 1992 Nuclear Electric announced (in its annual report) increased productivity, but many of its critics said that this success was down to the fossil fuel levy – a short-term (to end in 1998) quasi-carbon tax designed to encourage non-fossil generation, and which in 1992 was at about 11%. However, the £1.26 billion (1992) Nuclear Electric subsidy – the 'nuclear premium' – from the fossil fuel levy was no different qualitatively (nor greatly proportionally) from the less visible subsidies the non-nuclear generators and fuel suppliers receive: the only difference was that it was completely transparent. Not only was the Fossil Fuel levy a visible fuel cross-subsidy, but it had to cover a mysterious loss that had taken place during the CEGB privitisation. John Collier pointed to this loss in his Chairman's report at Nuclear Electric's 1992 AGM said that: 'Nuclear Electric had inherited enormous liabilities to pay for reprocessing, and waste management and decommissioning costs of [Nuclear Electric's] existing power stations, particularly the Magnox power stations. Although these were fully provided for in the accounts of the CEGB at the time, when Nuclear Electric was set up it was not provided by Government with sufficient funds to pay off these liabilities'[299] totalling around £8 billion.

The UK electricity privatisation experience, due to the problems in privatising arising from state controls, might at first seem to vindicate the free market theory that only market forces can efficiently manage resource use, but this would be a superficial conclusion. It is certainly true that the 'disappearance' of the CEGB's nuclear decommissioning funds does not inspire faith in central planners (in this case the UK Government of the time): though this is a problem of the central planners and not in central planning itself. Equally it demonstrates that there are limits to the considerations which the market will bear, and that this, and the Gulf War incident, also indicate that the energy supply industries can cope with changing circumstances and great market fluctuation without disruption to supply. Consequently it can be argued that the energy supply industry could absorb greenhouse externalities without undue difficulty.

The problem with the free market with regard to climate change issues is that those operating within it tend to focus their attention on the market as it is today, and its prospects for the future as perceived in terms of today's criteria. To the environmental scientist a resource is a resource, first and foremost, and not a financial variable. The burning of a million tonnes of oil will release 10 000 teracalories of energy, or the electrical equivalent (efficiencies of conversion notwithstanding) of 12 billion kilowatt hours. These are measurable, unchangeable facts: one million tonnes of oil is the same today as it will be tomorrow, or even yesterday. The carbon dioxide released from burning it will have virtually the same effect on the climate this year, as it did last, or for that matter next year. For the market such exactness is rare. From the market's perspective businessmen are more concerned with what the energy released from burning the oil will do economically. Will it be transformed into more wealth, in energising a factory, or will it warm a hospital? Both have markedly different economic implications, but are of little immediate greenhouse concern. That business does find speculation difficult was amply demonstrated by the UK ESI privatisation.

Prior to privatising the CEGB, the UK Government sought advice from the City of London's economic and business experts. The advice sought was of central importance to the privatisation's success and the Government was anxious to receive the best consultations available, and so spent some £50 million towards this end. Yet when, on 11th December 1990, the 12 regional electricity boards were floated on the stock market, the shares were snapped up with such unforeseen enthusiasm that at the end of the first day of trading investors maintaining their holdings saw them increase by just over half as much again. The Government's opposition claimed in Parliament that this underpricing had lost the British tax-payer £3 billion or more[300]. Of course greenhouse forecasts, as we discussed earlier, are also uncertain. Yet natural systems and the laws of physics are such that (incorporating errors of margin) the IPCC is able in its reports' to forecast a far greater time period (decades rather than months) than that required to implement most political energy policies. To date the IPCC has stated that global warming has indeed taken place this century. In just four months since National Power's stock market flotation, despite no major change in its generating plant or material assets and despite a 169% increase in pre-tax profits for the year of £479 million, half a million shareholders (about a third of the total) had sold their shares. This selling of shares was not due to disenchantment with National Power; rather it was to make a profit from the increase in share value. It was for financial reasons of the market, and nothing whatsoever to do with the long-term management of the nation's energy resources and their utilisation, let alone resulting impacts such as those on climate.

The effect of the Gulf War on the oil markets, and the market response to the privatisation of the UK ESI, demonstrate the limitations and uncertainties those in the market have to contend with as they manage energy resources. Energy examples of such market limitations, are not just restricted to nuclear power and electricity privatisation in the UK, or the World's oil markets and the Gulf War, they can also be found in areas of energy resource utilisation usually considered to be environmentally sound. As was mentioned earlier when discussing renewable energy options, some renewable resources are not harnessed in a renewable way. For instance, this has happened when the geothermal energy beneath San Francisco's Mayacamas Mountains began to be exploited by an unrestricted free market. The pool of geothermal energy that was being tapped by an increasing number of individual generators at a greater rate (70 MW extra per year throughout the 1970s) than was being replenished from the Earth's interior, resulted in each generator operating at dwindling levels: it is now envisaged that by the late 1990s if things continue as they are, that power output from the field may slip to half the 1987 level[193].

All of the above might seem far removed from the question of the free market and having to pay for greenhouse issues, but it does demonstrate that real world attempts to place energy supply within the private sector are not easy: they have necessitated multi-billion dollar financial controls and some inventive accounting. Free markets by themselves have little concern for far future, inter-generational issues, including climate change, nor can the best market analysts be relied upon by governments to address even short-term matters. Equally, both the UK privatisation and Gulf War have demonstrated a degree of robustness in the energy industries' ability to maintain consumer supplies, which bodes well for instigating greenhouse controls. Furthermore, as discussed earlier, the free market management of energy resources does not necessarily enable the individual consumer to exert traditional control of price through competition, by fuel-switching; though this is something one would clearly expect in a truly free market. In fact that energy suppliers worldwide do not operate in a completely free market is as true in the USA as it is in the UK today. USA electricity suppliers have also had hidden, and not-so hidden, subsidies; for example, its civil nuclear power subsidised through sharing resources with a state paid military nuclear programme. In 1987 a study by the USA-based World Resources Institute of over 30 countries found that almost all intervened to influence energy prices through tariffs, taxes, subsidies and price controls.

Such market controls are not internationally even. While in 1990 OECD countries oil's pre-tax price lay between $270–350 per tonne (1992 prices), with Greece paying a low $251 per tonne and Finland a high $398, the individual, nationally imposed, taxes vary considerably. At the top end of the range is Italy with its tax of $492 per tonne ($66 a barrel). At the low end we find the USA with $62 per tonne ($8 per barrel).

In short, the ability of an economy to supply and use energy rationally, let alone with a high regard to environmental externalities, in a completely free market is a myth. The centrality of supply, and the timescales on which energy decisions have to be made, preclude energy consumption from ever being completely in the hands of consumers within the market. Equally, while central planning clearly has its faults, Western energy policy has increasingly had to take into account both market and central planning considerations. As such, perhaps the most we can hope for is a green-bounded free market, within a mixed (market and centrally planned) economy; it is in this economic context that greenhouse issues are perhaps best addressed.

THE GREEN-BOUNDED FREE MARKET

Running the free market in a way that has regard for the environment does not necessarily mean that competition is in any way prevented, rather that all those competing are subject to recognised and pre-determined rules. Broadly the regimen is simple, and enshrines a single principle, namely that with human activity dependent on environments and ecosystems there is the need to pay for the environmental impacts of the activity (i.e. the internalising of environmental externalities), and this is essential if the said activity is to be sustained indefinitely. This has most commonly manifested itself in the so-called 'polluter pays' principle.

Ensuring that environmental costs are paid requires a proper understanding of all the processes taking place within the market so that the taxes, financial incentives, limits, targets, tariffs, or whatever actually do the job that they were intended for. Of course such understanding is an ideal, and the best that any economic policy can hope for is that it provides more benefit than cost.

Take natural gas. Since the 1960s, as a primary fuel worldwide it has become of increasing importance. Furthermore, it is a comparatively clean fuel, one that requires little processing, and, importantly from a greenhouse perspective, one that generates less carbon dioxide per unit of energy released than other fossil fuels. Yet worldwide, in common with most other products consumed, the actual quantity a consumer purchases affects the price: the more a single consumer buys, the cheaper the marginal or unit cost becomes. In the traditional view of business this is a perfectly acceptable way of behaving, but from an environmental resource perspective there are still environmental costs, including greenhouse impacts, to be paid for even if the product is being offered at a cheaper rate. Finally, as with all forms of fossil fuel, there is the question of longevity of supply. (In 1994 the World natural gas production to reserve ratio was around 65 years *at the then rate of consumption*[135]). Indeed, while gas' long-term and strategic medium-term supply problems are of concern to many involved with energy policy, they are not to the free market. As a consequence some bizarre situations have arisen where the market has actually stimulated waste. There is the well known example of built-in obsolescence, a not uncommon practice in the 1950s and 1960s, but there are more recent examples within the energy supply industries. To begin with the market and production economics of fossil fuel supply have been such that the natural gas in many oil fields has simply been flared off. Indeed in Earth orbit the brightest lights come from gas flares from oil fields. Fortunately this practice, though continuing, is on the decline as fossil fuel companies seek to recover gas wherever practicable. Profligate gas wastage also takes place at the consumer end of the economic chain. One notable instance was the way gas was marketed in the UK throughout the 1980s. Then, in order to stimulate demand in the time honoured way, as mentioned above, large gas consumers were given marked discounts: consumers who consumed over a certain amount (250 000 therms, or 630 tonnes of oil equivalent, a year) virtually negotiated their own price. This led to some customers deliberately wasting gas in order to save money by getting into this advantageous customer bracket.

Yet in one sense these are unfair examples in that companies like British Gas could easily afford to cease practices leading to waste. On 31st March 1991 British Gas made £3.07 billion pre-tax profits, which represented an increase of 14.2% on the previous year. Furthermore, British Gas (as discussed in the previous section) like many of the specialised large energy supply companies, have more than a little monopolistic element: so industrial and domestic consumers, having committed themselves with an investment in hardware to burn gas, require a significant incentive to make an additional investment in equipment to switch to an alternative, be it electricity, coal or whatever – British Gas can do a great deal to call its own tune. Yet again, if it were not for this monopolistic element, then British Gas' solo effort to reduce the degree of discount-jumping for large consumers might founder. If a readily accessible competitor had a cheaper, but more wasteful, alternative then the consumer would be tempted to take it. In free market economic nations, outside of public services and utilities, monopolies are comparatively rare within the market place, which almost by definition thrives on competition. Here, for such economies to be what is (euphemistically know as truly 'green', environment externalities should be fully accounted and charged for appropriately to individual participants throughout the market, so producing a 'level playing field' for business to fairly compete.

Adopting a 'level playing field' by itself will not straight away make the market responsive to either internalising environmental externalities or maximising efficiency of consumption (minimising the environmental cost per unit of useful work/utility derived from energy consumption). Even if all the environmental costs associated with consumption could be accurately quantified, the benefit is not always realised. The need for the consumer to respond is not always immediately recognised within the market. Assistance may be required. Energy conservation is a case in point in many developed nations – including both the USA[206] and the UK, where governmental policy and infrastructure has focused on supply rather than demand. Here attention has concentrated on the producer, not the consumer.

In the UK in 1991, the Consumers' Association (a voluntary organisation representing consumer interests) published an analysis of three types of domestic dwelling (an apartment, a semi-detached and a detached house) that would benefit from including some basic energy conservation measures. Assuming that the dwelling has central heating, then most basic energy conservation measures pay for themselves within a few years. Depending on the dwelling and type of central heating: ensuring that you have 150 mm of loft insulation will pay for itself within one to two years since a quarter of all heat is lost through the roof; draught-proofing of doors and windows pays for itself within a year; and cavity wall insulation pays for itself between 8 and 16 years. Indeed the only energy saving measure deemed by the study not to be cost effective, was to have double glazing professionally installed: the pay back period is of the order of a century or more – though DIY secondary glazing (fitting a second pane of glass inside the original window) has payback periods of a quarter of this[301].

Yet consumers have been slow to realise energy conservation financial benefits to themselves. Awareness campaigns could help, as would adding global warming costs to energy prices (which would increase financial savings and shorten payback times). The benefits of energy conservation were discussed earlier, but it is worth noting that almost a third of the UK's emissions of carbon dioxide is due to energy consumed within domestic buildings (roughly some 175 million tonnes a year).

Left to themselves, energy producers competing with each other would find it difficult to fairly charge consumers for environmental impacts arising from consuming their products. External intervention is required (which effectively means intervention by government) to ensure that the 'level playing field' is maintained and such intervention affects price. Producers in competition producing the same product would face identical charges, leaving competition unaffected. This green-bounded market approach is a middle road between a purely unrestricted free market and centralised planning. It takes the best elements of the free market system (allowing freedom of competition) and the best from centralised decision making, to ensure that goals are set and met – that environmental, including greenhouse costs are paid for, and that the efficient use of energy is encouraged[302].

Such ideas are neither new, nor radical, especially with regard to the thorny question of the finite nature of fossil fuels and security of supply. For instance, a 1982 report to the UK Secretary of State for Energy concluded that 'the strongest and most necessary part' of any energy conservation programme is 'energy pricing on the basis of long run replacement costs'. And that 'strong clear pricing signals' are required, together with the confidence that such a 'pricing will be maintained' and based on an authoritative view of the prospects of their trends[303]. An earlier Departmental report, in 1979, indicated that savings then amounted to 6% of demand, or 12.5 mtoe a year, rising to 23.5 mtoe or more could be made. Again the top recommendations for Governmental action were: 1) energy pricing for conservation, and 2) the provision of information and motivation[203]. In 1991 the UK Government's Energy Secretary said that 'one of the reasons why the full potential [for energy conservation] is not taken up is that ... the market in energy efficiency does not always work effectively'[304].

Ideally, so as to minimise market disruption, one would not aim to increase the efficiency of resource use soley by pricing. It is possible through reinvesting part of the extra pricing charged (not solely to meet present environmental impact costs but) to avert future impacts as a spin-off to encouraging improved efficiency, and on these grounds sell the energy policy to consumers.

Such integrated environmental policies are rare, but even if only some of the above elements come together in the market place, marked changes in consumer behaviour usually soon follow. In the UK motorists had long accepted the idea that lead in petrol (used as an anti-knock agent) was environmentally undesirable, yet in 1988 only about 20% of retail outlets sold unleaded fuel. The Government introduction of price differential increased demand and encouraged more retail outlets to stock unleaded so that three years later around half of UK petrol sales were unleaded[305]. The UK was not alone within the European Union to see its petrol consumers switch from leaded to unleaded. Between 1988 and 1991, across the 12 member states, unleaded share of the market increased from 13% to 38%[305].

The above use of a combination of pricing and 'environmental awareness' does, at the very least, demonstrate that consumers will switch 'product' brands. At first this might seem a far cry from encouraging an improvement of energy efficiency until, that is, we take the view that consumers in fact do not want energy per se but the services (heating, light, transport, etc.) that energy consumption brings (cf. Chapter 8), and that they will switch brands, and even products, to achieve this goal.

With regard to buying improved energy efficiency, the consumer should *ideally* be choosing between continually paying for a less efficient use of fuel, against an investment in future for more efficient usage. Here, the argument for taxation is that the aforesaid ideal does not always take place since the consumer is unaware of the choice on offer. Taxation helps focus the consumer on the choice by exaggerating price differentials. These taxes have come to be called 'green' taxes.

Sweden has had 'green' taxation for a number of years. However, for many years before such taxes had been proposed there were those who felt that no one should be allowed to 'buy' the right to pollute. More recently it has become apparent that polluters should not be allowed to pollute 'free of charge', i.e. without paying for the environmental impact caused by their pollution. In 1973, Sweden implemented its first 'partly-green tax' of 3–4 cents per non-returnable drinks can. Each year it raises some US$13 million, but it is only 'partly green' in the sense that money raised does not go towards improving recycling technology, or lessening the environmental impacts associated with drinks' packaging, it becomes part of the general central governmental revenue. Even so the small taxation incentive can only have contributed to the situation in 1991 where some 80% of Swedish drinks cans were recycled, compared to 20% of glass drinks containers. Other 'deeper-green taxes' have since been introduced. In 1984, Sweden's 'green' tax on artificial fertilizers and pesticides raised some US$27 million, but this revenue is specifically earmarked for environmental forestry and agricultural work, agricultural advice, and acidification counter measures. Now Sweden has turned its attention to the impacts of fossil fuel consumption. Emissions of both sulphur dioxide and carbon dioxide from installations (power stations and incinerators) have been taxed in Sweden since 1991 at rates of SKR15 000 per tonne of SO_2, and SKR250 per tonne of CO_2. Both these rates were calculated on the basis of the sulphur and carbon content of the fuels burned, but not necessarily by an exact costing of the environmental damage that burning these fuels incurs. Sweden's CO_2 emissions total an estimated 60 million tonnes a year. With the introduction of the carbon tax on oil, coal, natural gas, kerosene and petrol, it is estimated that emissions will fall by 5–10 million tonnes (8–16%), or around 3.5% of its primary energy consumption – indeed, after only one year Sweden's consumption had decreased by over 9%. Furthermore, the US$25 million raised annually by the tax is used for acidification countermeasures, and energy and environmental protection technology[306, 307].

So what would happen if this saving could be achieved throughout the developed world? Sweden's 3.5% saving might not seem great, until one realises that its primary fuel consumption mix is exceptional – Sweden relies as much on hydroelectric power as on both oil and coal combined[1]. The rest of the World is not so blessed with extensive HEP, and relies proportionally more on fossil fuels, which means that across the OECD, the savings from a Swedish green energy tax would be expected to curb total energy consumption by possibly around 8%. It therefore seems that such 'green' taxes have their part to play in helping curb greenhouse gas emissions and so slow climatic change, yet at the same time help eke out and reduce dependency on finite fossil fuels.

MYTHS AND THE GREEN-BOUNDED MARKET

Some members of the business community are alarmed at the mention of green taxes and levys. It is not that they are in principle against the internalising of unpaid environmental externalities (unpaid environmental costs), but that such costs should be paid for in a way that does not disrupt competition within the market place. Equally, others sympathetic to environmental issues view all economic activity as having a detrimental effect on the environment. Both these perspectives can lead to misrepresentation of the goals environmental economists set themselves when trying to identify suitable parameters to the green-bounded market. As noted earlier not all economic activity and its associated pollution levels have a net detrimental effect on the environment. Where economic activity is such that pollution levels are within the absorptive capacity of environmental sinks, then the benefits of the economic activity easily outweigh the environmental cost. For instance, without the management of grouse moors, such ecosystems would largely disappear. Yet such management is associated with heather burning, shooting, the release of lead shot, etc., but have a minimum negative effect on the overall ecosystem being managed.

Another common misconception is that economic growth is ruinous to the environment. This is simply not true! What *is* ruinous is the relentless growth of consumption of non-renewable resources, and the non-sustainable, over consumption of renewable resources. Chopping down a forest without planting new trees as you go will inevitably mean that one day the last tree in the forest will be felled. On the other hand it is possible to have economic growth without increase in the rate of non-renewable resource consumption. It *can* take place without an increase in consumption, and indeed material recycling is set to become a major area of economic growth for OECD developed nations in the early 21st century.

Another common market myth is that wealth is the prime cause of the World's increasing rate of natural resource consumption. This is not strictly true as it is people who consume resources, and not wealth. More people inevitably will consume more resources, so it is population growth that is inextricably linked to the growth in resource consumption. Here one of the fundamental causes of rapid population growth is the insecurity of poor Third World families whose calculations of their future all too often require them to suppose that half their offspring will be dead by puberty, and that their surviving children can not only help them earn a living but will provide security in their old age. Filial regard for the older generation is surely the most primitive life insurance in the World. The position today is that the most polluted cities in the World are in those countries with an average income of about $1200 per head per year. Countries poorer than that produce less pollution (as opposed to environmental impact) as their level of economic activity is so low, and richer countries can afford to run pollution clean-up technology[308]. Furthermore, the rich developed countries are stabilising their populations, and have broken through the so-called barrier of demographic transition, and any real-growth they achieve is experienced on a per capita basis, not purely a national one. If trends continue then by the year 2025 those in the less-developed nations will make up 84% of the World's population, estimated by then to have reached around 8.5 billion, whereas in 1950 when the World's population was only 2.5 billion the less-developed nations contributed only 68%.

The way forward, it would seem, would be to increase the per capita wealth of those in the less-developed nations in such a way that their rate of population expansion slows, combined with an increased efficiency of resource consumption: the latter instigated by the efficient technology-holding developed nations. In terms of greenhouse issues, one of the central questions is how to curb the rate of increase in energy consumption through increasing the efficiency of energy use and generation? As we have seen, when discussing energy efficiency, the energy forecasts and calculations of energy efficiency potentials vary depending on the technical and political biases of those performing the calculations. Conservatively, assuming that North America does everything it can with existing technology, and modifies its mode of consumption more towards that of Japan and Western Europe, then it could conceivably halve its energy consumption. This would reduce the worldwide drain on oil, gas and coal by at least 10%. Optimistically, this is only the beginning since new higher efficiency technology is continually being developed. Using such current state-of-the-art and near-term future technology it would not be difficult to foresee a two to four fold improvement in efficiency within a couple of decades. Some basic calculations suggest that improved efficiency could sustain economic growth through to the early 21st century without increasing energy consumption above early 1990s levels[309]. Even if half the savings of those implied by such a calculation were pragmatically achievable, then this strategy would still be worth pursuing. Yet achieving such efficiencies by establishing a green-bounded market is not straightforward; if it were it would have already been done.

Pitfalls

Economic myths arise easily through the difficulties in translating theory into practice. Past attempts to pay for previously unpaid environmental costs (internalising externalities) have been fraught with both misunderstanding and difficulty in practical realisation; frequently with one feeding off the other. The present attempts in accounting for greenhouse costs (or indeed future scarcity of finite fossil resources) are no exception. When the European Union made its preliminary carbon tax proposals prior to UNCED in Rio (and the signing of the Framework Convention on Climate Change) the House of Lords in the UK sounded an alarm. Its European Committee rejected the EU proposals for a carbon energy tax, preferring instead a direct (centrally planned and non-market) drive to improve energy efficiencies. At the time this came as quite a blow to the EU which saw the proposed tax as being the central element in meeting the twin objectives (for all the reasons discussed in previous sections) of limiting carbon dioxide emissions and improving energy efficiencies. Now, nobody likes to be constrained, and, like those on the House of Lords European Committee, players within the energy supply industries are no exception. Instead of viewing environmental externality accounting systems, such as carbon taxation, as a way of defining the parameters within which the industries are free to operate, they view them with far greater suspicion.

Following the EU's carbon proposals, made in the spring of 1992, six European electricity producers, including Britain's PowerGen and Scottish Power, put forward alternative proposals based on a voluntary code to stabilise carbon dioxide emissions at 1990 levels by the year 2000. The companies in question generate about a quarter of the European Community's electricity, and about 70% of this production is based on coal and lignite. Their offer was to meet the EU carbon emission targets, but only if the EU dropped its proposed carbon tax.

In the UK, PowerGen's principal electricity generating competitor, National Power was quick to respond to the no-carbon tax offer, announcing that it would be able to meet the EU targets with or without the carbon tax. National Power accused PowerGen of putting forward proposals as environmental initiatives, that were commercially desirable in themselves. PowerGen was planning four CCGTs (Combined-Cycle Gas Turbines) to come on line before the year 2000, and carbon dioxide emissions would fall by about 11 million tonnes a year – approximately 5% of the UK industry's annual emissions which would be a major contribution towards meeting the UK goal. With its high hydrogen to carbon ratio, gas produces the least CO_2 of all fuels. Also CCGTs have a high efficiency of around 50%, and this can be increased to over 75% if they are allowed to function as combined heat and power units. However, such was gas' commercial attraction to the various privatised generators that the capacity surplus in the electricity system could reach 57% above peak demand by the late

1990s[310]. If all the 1992 stations proposed had been built. This would have ultimately led to socio-economic disruption.

In terms of disrupting existing socio-economic patterns, at best greenhouse policies will result in improved efficiency in energy production and consumption, so that the continuous growth in energy demand is offset or stemmed, and existing patterns of energy production maintained. The worst that could happen to existing energy-producing utilities is that the fuel mix changes from those that generate considerable greenhouse gas emissions (such as coal and oil) to those that generate less (such as gas), or none (the renewables and nuclear power). Users too will initially have to pay more, investing in energy efficiency so consuming less, or continue existing consumption patterns, settling for paying for energy at a higher rate due to some greenhouse externality accounting mechanism, such as a carbon tax. Clearly, effective greenhouse policies will disrupt the existing pattern of energy resource use; after all it is the existing, high greenhouse gas emitting, pattern for which change is sought.

Another pitfall, in introducing a greenhouse policy that accounts for environmental externalities, stems from the uncertainties of the costs associated with global warming, and the purpose of the policy. Clearly if the costs of global warming are large, then the sums to be internalised into the greenhouse gas generating economic activities will also be large: if global warming costs are small, then so will be the amount to be internalised. Unfortunately, here we do not have an accurate figure. Supposing that the mechanism for internalising the costs is a carbon tax, we are then left with the question as to how much to charge? We also have to address the question if the aim of the exercise is not necessarily to pay for externalities, but to stem the increase in greenhouse emissions.

The 1992 EU proposals for a greenhouse carbon tax were for $1 a barrel of oil in 1993 rising to $10 by the year 2000, but is this enough, or too much? Should the tax be imposed at the primary production level (when the oil comes out of the ground), or at the end-use level when the electricity is metered in the home, or petrol put into the tanks of cars? The implications of the EU 1992 $10 carbon tax proposal were clear. Petrol prices would have been increased by about 3.5 p (UK) a litre; electricity prices by about 1 p per kWh, and gas prices by about 0.3 p a kWh. For UK end-users, this represents an increase of 7.6% on garage bought petrol, a 13% increase on domestic electricity, and nearly 20% on domestic gas consumption.

Two researchers in the USA from the Environmental Protection Agency, addressed just this problem of how much and where to tax. They found that a 125% tax imposed at end use is likely to achieve stabilisation of USA carbon dioxide emissions, whereas a 200% tax would be required at the primary production level[311]. (End use taxes proportionally need not be so high as at primary production level where the unrefined and/or undelivered product costs less.) Such levels (100–200%) of taxation may seem high, and indeed they are, but they are not unknown. USA citizens largely enjoy tax free automobile gas (petrol), whereas Western Europeans are already used to paying around 100% tax on their petrol. Such views can only be broadly considered, such is the inexact nature of economic science. Even so, researchers in Europe too had their own thoughts on carbon taxes.

In December 1992 Cambridge Econometrics, a private company split off from Cambridge Universities' Applied Economics Department, announced its analysis as to the effects of the EU carbon tax proposals. They assumed that the tax would be introduced in 1993 at $3 a barrel, and rise $1 a year to $10 a barrel in the year 2000. Their analysis concluded that such a tax would only be likely to reduce Europe's carbon emissions by about 3% to a level close to that found in 1990. However if the tax were to be implemented beyond Europe, throughout the developed (OECD) world, then emissions would be reduced by 4%. In short, they concluded that the EU carbon tax proposals were not strong enough to completely stabilise emissions, that such taxes should be as international as possible, and that the revenue from such taxes should be spent in a non-inflationary way: for example, they should not be used to reduce income tax[312].

Finally, even analyses from bodies with oil connections (who presumably wish to see the oil industry flourish, and so for carbon taxes to be low) agree that the 1992 EU carbon tax proposals did not go far enough. The Centre for Global Energy Studies, established in 1990 by Sheikh Yamani, the former Saudi oil minister, reported in 1992 that most European countries require oil increases of $30 per barrel (not $10) to stabilise CO_2 emissions[313].

Such understandings of the way proposed carbon taxes interact with the economy are in a rudimentary stage and many questions need to be answered. Following the UNCED conference in Rio, the UK government approached industry for answers to a number of key questions (see Table 11.1[314]). However, for carbon taxes to succeed it is important that competing economies account for greenhouse impacts in a similar way (or to ensure that the combination of state and carbon taxes are at least similar, notwithstanding an allowance for national growth[313]). If they fail to do this then one economy will have a short-term competitive edge over the other: the nation not choosing to implement the tax, will be able to produce goods more cheaply. Keeping this macro-economic international dimension in mind, the signing by World leaders of the Framework Convention on Climate Change at the UN Conference on Environment and Development in Rio was a key step forward, though at the time USA reticence towards carbon tax accounting of greenhouse externalities was a major stumbling block.

These pitfalls of ensuring that the environmental impacts of greenhouse costs are paid for, ensuring that charges do not disadvantage one country compared to another, deciding in what way and how much to charge etc., ultimately incur a cost on the average citizen. Here it is important that the paying of greenhouse costs are in line with the reduction of greenhouse emissions and importantly

Table 11.1 List of main carbon tax issues the UK government identified as requiring further analysis[314]

Economic consequences

- The macro-economic impact of the tax
- The impact of the tax on specific industries and other sectors
- The impact of the tax on each industry's competitiveness, in particular with (a) OECD countries and (b) non-OECD countries
- The impact of the tax on companies in particular markets where there is keen competition
- The impact of the proposed industry exemptions
- The effect on industry of similar exemptions in other countries
- Issues arising from the application of the destination principle, including treatment of energy intensive exports (export products which require a great deal of energy in their manufacture) and refinery emissions
- Response of non-EU industrialised countries to imposition of (carbon) EU tax

Administration of tax

- The administrative costs for industry of compliance with the tax
- Clarification of the issues arising from the definition of the tax – the chargeable event, tax point (i.e. where between primary production and end-user to charge the tax)

Timing of implementation of tax

- Implications of the (then) proposed early date (1st January 1993) for implementation and its phased introduction to 2000
- The implication of conditionality: that the tax will only be introduced if other OECD countries adopt similar measures

Base and effectiveness of tax

- The impact of tax on UK (greenhouse) emissions
- The impact of proposed exemptions on UK emissions
- The detailed specification of the fuels liable for tax and the effects for industry on choice of fuels

Fiscal issues

- The yield of the tax, taking into account exemptions and changes in rates
- An analysis of the effect of tax on investment incentives

Basic principles and rationale of tax

- The rate of the tax
- The carbon/energy split
- The rationale and proposed method for taxing electricity

Legal/procedural issues

- The issues arising from implementation of the tax – timing, scope for periodic reviews of tax rate, factors to be taken into account in change of tax rate, etc.

Social issues

- The distributional impact on households (the domestic sector) done in a morally acceptable way by society. (The last concern listed in Table 11.1 – the way the costs are distributed among households.) While this is more a political decision than one for environmental scientists, there are human biology dimensions that must be recognised. For instance there is the link between energy consumption to keep warm, and healthy.

It is well established that in temperate countries such as in northwestern Europe and the north of North America, deaths and hospital admissions for coronary heart disease and stroke are higher in winter than in summer. In some severe winters mortality has been as much as 70% higher than summers, with the size of this winter excess being related to the difference in environmental temperature. Indeed seasonal fluctuations in cardiovascular events are greater in Britain than in some other countries with wider temperature variations and colder winters which suggests that the excess winter deaths may be preventable. In the UK excess winter mortality is greatest in the poorer socio-economic groups who have the worst domestic heating, hence consume less fuel[315].

When in 1994 the UK government introduced an 8% Value Added Tax (VAT) ostensibly as a 'key element of [its] programme' for climate change[280], the potential effect on the health of UK citizens was outlined in an editorial article in the *British Medical Journal*[316]. Of course the UK's tax on fuel was not a true carbon tax as its revenue did not go towards improving the efficiency of energy use, or ameliorating energy resource-use environmental impacts, but towards reducing the Government's own budget deficit. Consequently this tax is a good example of what can go wrong if (so-called) greenhouse taxes are improperly applied.

Those affected most by the tax included the poor and the elderly. At the time almost three quarters of UK pensioners paid no income tax, but nearly all had consumed energy that attracted VAT. Furthermore, although low income households spent less on fuel than those with high incomes (spend being related to size of dwelling, hence wealth), the proportion of household income the UK poor spend on heating is higher: 13% in the lowest fifth of income as opposed to 4% in the highest fifth. An even finer comparison of the top and bottom tenth household incomes reveals burden of VAT fuel cost are proportionately seven times greater for the poorer families[316]. This unequal situation is compounded in that poorer families tend to use more expensive fuels as they cannot afford to install gas central heating. The situation is worse for pensioners who spend a greater proportion of their time at home. While British shops can be closed if they do not have an internal temperature of 16°C, some 37% of households of the elderly have winter temperatures below this level.

The consequence of the UK's introduction of VAT on fuel would have been to increase the seasonally related illnesses described above. However, as with a number of public health issues, the greatest problem is not with a small number of people on which effects will be large, but a large number of people on whom effects will be small. The health conclusion[316] is that VAT on fuel per se is 'unfair' and 'should be opposed on principle'. Furthermore, 'no mystery concerns [fuel poverty's] cause or solution. Improved insulation creates warm, dry homes that cost less to heat thereby increasing the occupant's disposable income'. From the greenhouse viewpoint, allocating (part of) the revenue from carbon taxes to providing insulation for disadvantaged households, and protecting the elderly as well as those in long-term rented dwellings (as landlords have no motivation to invest in insulation), means that ultimately greenhouse gas emissions will be lower.

While pitfalls such as those cited above do exist for those wishing to implement a proper greenhouse policy, it is possible to have such a policy. The danger, though, is that by instigating policies in the name of greenhouse (and other environmental) concerns for ulterior reasons, voters will become disenchanted with greenhouse and environmental policies when they are properly formulated. All of which demonstrates that the way environmental and greenhouse issues are perceived is of great importance.

ECONOMY, CLIMATE CHANGE AND SUSTAINABILITY

One key theme of this text so far has been the need to pay for the environmental costs associated with climate change, and that those human activities that contribute to such change should be where remuneration is found (i.e. the need to pay for energy consumption's greenhouse externalities). Yet, while we can identify ways of accounting for such costs, the economies of developed nations have (one might say) functioned perfectly well since the industrial revolution: which incidently is when the atmospheric build up of anthropogenic greenhouse gases started. So why then, the need to modify the present economic system? In other words, as laudable as fully accounting for greenhouse costs might be, the present system is at least proven so to all intents and purposes there is no pragmatic need for change.

This counter argument to internalising greenhouse externalities was rebuffed by the then British Prime Minister Margaret Thatcher in October 1988. She said: 'No generation has a freehold on the Earth. All we have is a life tenancy – with a full repairing lease'. The implications of that distinctive turn of phrase 'full repairing lease' are profound. To date no generation has paid for the environmental costs resulting from its economic activities. These were passed on to subsequent generations. Today we inherit from our parents a world not only less well endowed in tropical forests and species, but with less potential agricultural top soil, less fossil fuel, and fewer other resources upon which future economic activity depends.

It may seem that the problem is one of reconciling large short-term gains (at the expense of paying for externalities), as opposed to making smaller gains per unit time, but sustainable over a far longer period and providing an overall greater return. In fact this is somewhat of a simplification. The only people to make large gains through not paying externalities, are the immediate consumers of whatever the resource involved in the economic activity. Society as a whole does not gain, it instantly starts paying whether or not it realises it. When a tonne of a finite resource is consumed, then straight away society is deprived of options for its use in the future, and straight away the pollution released counts as a cost. The real choice is not just between large gains now, and smaller but more repeatable gains into the future: the real choice is quite different – it is about who benefits, the immediate consumer or society.

Consumers, of course, make up society, but the consumers of a specific resource consumed by those involved in a specific economic activity are a subset of society – they are not synonymous with it. Equally, those consuming more of the resource are not synonymous with those who consume less: compare the average Western European with the average Ethiopian, or the average North American with the average Indian. The choice is really between whether those involved in economic activity should benefit at the expense of the longer-term benefit to society (which includes benefit, albeit smaller but more sustainable, to those conducting the economic activity). Very often the potential quick gains that an entrepreneur could expect to make are dwarfed by those that society could realise straight away if externalities were paid.

The above might seem counter-intuitive. Supposing a widget producer either started to pay a tax to meet the cost of externalities (or alternatively paid in some way other than directly in cash, say by building a pollution control plant) then you would expect the cost of the product to increase slightly, and the level of demand to decrease, with the producers profits being squeezed. You might think that the net result in terms of economic activity would be for a decrease. This is not so. The reaping of an environmental tax could (if spent properly) provide jobs *elsewhere* – for example in ameliorating the environmental impacts of widget production. Or alternatively, the building and running of, a pollution control plant as part of the widget production process also provides jobs and is just a real an economic activity as widget production itself.

In terms of the greenhouse issue, the above choice was behind the thinking of the USA Committee Science Engineering and Public Policy report on greenhouse warming[232]. It concluded that the USA could cut emissions of greenhouse gases by up to 40% for little or no cost if 'the proper policies were implemented'. It specifically called for regulations and price incentives (including taxes) to encourage energy conservation and develop new energy technologies. The adoption of such a policy will obviously have an effect on a nation's economy. With higher fuel prices goods will become more expensive and there would be a negative effect on manufacturing industry – though if fuel prices worldwide increased then at least one country's manufacturing base would not lose out to others. On the other hand, less energy would be consumed which in turn would mean that less fossil fuel needed to be imported: a plus point in terms of balance of payments. Industry produces more from less, and if the resources it uses are finite, then it is able to increase its viability lifetime. Then again, there would be jobs created (helping to offset those lost from manufacturing itself) in monitoring environmental impacts, helping reduce them, in developing and applying new energy technologies, etc. This then is the theory, which sounds good but what of reality?

While strict greenhouses measure have yet to be developed and observed worldwide, many countries have adopted environmental legislation covering general pollution concerns. Furthermore, industry has seen the writing on the wall, and in most developed countries is beginning to recognise that it has to clean up its act voluntarily or have legislation imposed; meanwhile environmental legislation has been increasing. Throughout the 1980s industry, along with society, became increasingly aware of environmental concerns and correspondingly spent more to reduce its environmental impacts. By 1992 the OECD estimated[317] that the annual global market for equipment and services in the environmental field was worth some $200 billion, and that this was expected to increase to $300 billion by 2000. Here money is actually spent on dealing with environmental costs, and not some other purpose. In comparison to this 1992 World environmental spend of $200 billion, the aerospace products sector had an annual market value of $180 billion, and the chemicals sector $500 billion. Not all of this growth in environmental business was through industry voluntarily spending to ensure a cleaner activity; legislation has provided a major impetus. About 25% of the environmental market in the USA is in the area of waste management, whereas in Europe it is only 15%. The USA also produces relatively more contaminated land clean-up equipment which reflects the USA 'Superfund' legislation (which provides a pool of finance to reclaim contaminated land). In Japan the big spend is on air pollution which probably reflects 1970s Japanese legislation for acid gas abatement and the introduction of catalytic converters for cars – converters which someone had to build, install, and monitor the air where traditionally there have been high levels of pollution.

Paying for environmental protection is not just good for our quality of life or enhancing our prospects for a sustainable future but it is, in the here and now, big business in its own right. As a business, it can operate alongside other enterprises in the green-bounded free market. There, business can buy and sell goods and services at whatever price they want provided no externalities remain – free markets being one of the most efficient price setting mechanisms. In this way environmental resource preservation can be optimised so that resource consumption stands the best chance of sustaining economic activity into the future. However for this to happen we have to change our view of the way we consume the World's resources. We have to modify our perceptions.

CHAPTER 12

PERCEPTIONS AND RESPONSES

'To err is human; to forgive, divine.'
Alexander Pope, *An Essay on Criticism*

PERCEIVING IS BELIEVING

To the scientist, isotopic hydrogen and oxygen ratio changes in ice cores, dendrochronology, pollens in peat, geology, and other biotic and abiotic evidence all provide pieces in the picture of natural climatic change across the ages which is accompanied by fluctuations in the Earth's atmospheric greenhouse gases. Direct contemporary measurements, backed up with recent ice core analysis, reveal that our atmosphere is once again experiencing an increase in these gases, particularly in carbon dioxide which we are releasing into the atmosphere. Finally, direct meteorological measurements reveal (even allowing for growing urban island heat effects) that the Earth is warming. Even without a decades-old Worldwide meteorological monitoring network we can see for ourselves the glaciers retreating all over the globe, and can measure the sea level rise.

The international scientific community has seen this evidence and has stated with certainty that there is 'an additional [human-induced] warming of the Earth's surface[20].'

An informal report from the 1992 Rio Earth (UNCED) Summit highlighted differing perceptions on greenhouse effects. The USA emits eight times as much carbon dioxide as India, despite having less than a third of the population, but when recalculated as emissions in tonnes of carbon dioxide per $1000 of GNP, India's clean image is swept away and the formerly heavily polluting USA appears less selfish. This of course, comes as no surprise, considering that the USA has only one third of India's population, but a GNP roughly 19 times larger[318].

Science may be an extremely useful tool but the scientists are human: they can, and do, make mistakes. Two British scientists, Farmer and Gardiner, using an old fashioned, inexpensive instrument discovered the Antarctic stratospheric ozone hole. Lovelock was disturbed to learn that so confident were those who programmed one particular satellite that they knew all that mattered about the stratosphere, that they programmed the instruments aboard the satellite to reject data that was substantially different from model predictions. The ozone hole was seen by instruments, but the instruments were not to be believed[319].

Lovelock considered that, until the 1980s the list of global dangers in order of public priority were: all things nuclear; ozone depletion; and chemical pollution. The UK Prime Minster's 1988 Royal Society speech stressing the danger from human-generated atmospheric 'pollutants' such as methane and carbon dioxide moved these compounds toward the top of the priority list of concern. Lovelock has his own priority list, at the head of which he places the (greenhouse-related) issue of the clearance of tropical rainforests.

For a human to be greatly affected by an environmental issue, the object of concern (or a resulting instance of it) must be perceived both as great in magnitude and (related to this) close at hand. One well documented instance of perception affecting a population was that of the 1979 Three Mile Island PWR nuclear reactor accident. Nearly eight years after the accident, and a year and a half after electricity generation at the site was restarted, 21% of residents examined by psychologists reported depression, a quarter claimed raised anxiety, 36% scored well above average for hostility, and all say that they are not convinced of the safety of the reactor[320]. Parallels can be drawn with the anthropogenic greenhouse issue.

One example manifested itself for the Swiss village of Saas Balen in the mid-1990s. Wilfried Haeberli, Director of the UN's Glacier Monitoring Service, linked the threat of the village being swamped by mud and water to climate change predictions. The 423 inhabitants live below the 3 kilometre Gruben glacier which has been melting for a century, and recently has been losing 60–70 cm in height annually. The village was flooded in 1968 and 1970 and in 1995 one of the melt water lakes above the village was drained as part of a £5 million engineering project to protect the community. While the older citizens accept that they have to live with danger, the younger inhabitants leave seeing no future in the valley[321]. This clearly illustrates that those affected by marked expressions of warming – for instance populations on low-lying coastal land – will be most concerned and have the way they live their lives affected. Furthermore, perceptual dimensions to human life and belief that superficially appear to have little immediate bearing on the greenhouse issue may in actuality affect the human response of many.

In January 1990 the USA (under President George Bush) and the coalition forces went to war in the Middle East to oust the invading Iraqui army from Kuwait. The motivation for this costly international force was not just a desire to uphold international law but to protect the oil reserves for the developed West, Kuwait has over 9% of the World's proven reserves[134]. Yet President Reagan (Bush's predecessor) heavily cut the US Department of Energy's research and development in energy conservation and renewable energy – the very technology that could lessen their dependency on imported oil. The Reagan administration in particular has been criticized[322] for arguing that government intervention during the 1970s oil crisis caused market distortion and disequilibrium. The

1981 National Energy Policy Plan articulated this: 'Free markets will not work perfectly during severe disruption... [but] they will work smoothly, with greater certainty, and ultimately more fairly than complex systems of price allocation controls managed by Government'. This contrasted sharply with the Department of Energy earlier view expressed in 1979:

> 'The World oil problem will be resolved, either uncontrollably, through severe economic shocks to national economies, or controllably, through concerted early action to ease the transition that must be made to more abundant sources... faced with growing dependence on imported oil and what it is doing to our economy, this country does not have the luxury of waiting for leisurely market processes to work'.

In short the Reagan and Bush administrations perceived it necessary to promote a *purely* market economy and a market led energy policy despite there being other (at least) as informed views at the time. This is just once example demonstrating that beliefs and perceptions can help shape human activities.

The problem with looking at human perceptions and their effect on environmental issues, is that the perceptual processes are multi-dimensional, and frequently vary considerably from group to group, if not, with complex energy–climate issues, from individual to individual. Indeed, a human appreciation of environmental perception is itself dogged with perceptual bias. Yet just because a factor affecting the way we relate to a problem is very complex is no reason to ignore it. Furthermore, the perceived usefulness of solutions to environmental problems itself relates to the solution's acceptability and presentation. This will be considered in the following sections when a number of ways in which the antropgenic greenhouse issue has been and is, commonly perceived will be examined.

HUMAN PERCEPTION AND RESPONSE AND THE GLOBAL COMMONS

Human beings, by their very nature, are limited: as such human perception is imperfect, and the human perception of complex phenomena, such as global warming, frequently fails to result in an optimum response (or sometimes even a predictable one). Human psychology gets in the way so that perception fails even a basic understanding of many phenomena. Indeed, human perception and psychology on occasion even conspire so that falsehood is believed with conviction. Consider the difficulty even of trying to get across to a school class (who's very *raison d'être* is to learn) such a multi-faceted issue as global warming, compared to the limited success attained in communicating simple scientific laws, e.g. the conservation of energy. One recent study highlighted just how limited our ability was to do this[323]. Involving 1130 pupils, aged 11–16 from 15

UK schools, the study came up with what at first might seem some counter-intuitive results. The researchers asked whether the law of conservation of energy pertained to one of the following:

(i) a parliamentary law,
(ii) countryside conservation,
(iii) energy saving,
(iv) that energy cannot be created nor destroyed but can change form,
(v) a law that said that energy could only be used when really necessary.

Having chosen an option, the pupils were then asked whether they were:

(a) sure they were right,
(b) only thought that they were right (i.e. not certain),
(c) did not really know,
(d) thought that they were wrong,
(e) were (subsequently) sure they were wrong despite choosing the option.

As one might suppose, the older the pupil the more likely they were to give the correct answer. The percentages of those who chose *correctly* and who said they were sure that they were right, rose from 8% to 33% between the ages 11/12 to 15/16 years. For this *correct* group, the proportions giving all other categories of answer (options 'b' to 'e') decreased with age, including, strangely, those who just thought they were right without being certain. However the surprise comes with the age of those who chose an *incorrect* definition for the law of the conservation of energy.

Intuitively, one might think that older children would acquire greater powers of discernment so that, in giving *incorrect* answers, they would be less certain that they were right. Further, thinking that one was right (as opposed to being sure) – when really wrong – would increase with age group: in other words the recognition that one might be wrong when one really was wrong would increase with age. It turns out that this is not so! In fact, counter-intuitively, in all categories of answer (including the correct one) the numbers of those 'thinking they were right' decreased with age, whereas 'being sure' in all but one class (option 'ii') increased with age. It seems that while the older pupils were more likely to choose the correct answer, and with greater certainty; if they did choose incorrectly they were more certain about that too. The pupils, whether right or wrong, became more entrenched with age.

Of course this is only one study and, while the number of pupils surveyed was large, the pupils had all been educated within the UK secondary school system, and one could argue that pupils who had been educated differently would respond differently. Yet this does demonstrate that human nature, albeit when moulded by a developed nation's educational system, can bring forth bigoted beliefs. Not surprising, a number of public understanding of science surveys carried out in both France and the UK reveal that about 30% of the population thinks that the Sun revolves around the Earth[324]!

Against such a backdrop one can begin to see that those concerned with getting greenhouse issues (and other subjects relating to the use of the global commons) across to the public face a formidable challenge. If bigotry is not enough, those trying to enhance human perception of greenhouse issues face the problem that tackling global commons issues may result in the compromising of individual benefits for the good of many. Indeed sometimes individuals will cling to personal benefits at the expense of the greater public good. This situation is one that some psychologists refer to as a 'social trap'[325]. Garrett Hardin provided an example of a social trap[21] in 1968 when he likened the tragedy of the commons to the over-grazing of common land when it is not in the individual's interest to withdraw their animals from the over-grazed common even though the community as a whole would benefit (see Chapter 2, p. 24). The trap is that each individual animal owner continues to do something for their own advantage that is collectively damaging to the group.

Very often social traps are self-reinforcing, and include a positive feedback loop. For instance inflation is a self-reinforcing social trap, with every increase in inflation producing new demands for higher wages which in turn drives inflation ever upwards. Another example, which is related to greenhouse policies, is that of a deteriorating rail service. Rail, and indeed other public transport systems, are more energy efficient than individual use of cars. Yet as public transport systems erode more people switch to private transport. The process continues to the point where most people are daily stuck in traffic jams on the way to work and would prefer to go by public transport, but the public transport systems have decayed to the point that they do not provide a realistic alternative. Furthermore, the system has inertia built into it. With investment made in a car, there is less of an individual's money left to spend on public transport.

Illustrating the conflict between individual versus collective benefits, social traps have been caricatured as a game by Merrill Flood and Melvin Dresher in *The Prisoner's Dilemma*. In this game, devised in 1950, two 'prisoners' have been caught by the police and are suspected of a crime. They are held incommunicado and individually questioned. The justice system has the following pattern of rewards and penalties. If they both talk or defect on each other they are given the standard sentence; if the prisoners 'cooperate' with each other, so that neither talks, they get off lightly; but if one talks and the other does not, the first gets a reward while the second receives a severe sentence. It can be readily seen that the pay-off matrix is such that individual rationality is at odds with collective rationality. The same is true of many greenhouse issues as they relate to individuals. Why should an individual forgo a fast, energy inefficient, fossil-fuel powered, car, or even a car at all, for the good of the community, when those who do not forgo a fast car (or car at all) have greater personal freedom and quality of life?

The analogy with social traps in global commons exploitation, be it ocean, tropical forests or air, is that consumers (the World's citizens) continually spend to a point beyond resource sustainability so that the resource is eroded. We are using fossil fuels up, and are having to spend more in extracting some of them (oil from Alaska and the North Sea for example) as well as beginning to see the Earth's climate change. However, a most striking example is that of exploitation of the world's fisheries. Today the World's fishing fleets are spending more per tonne of catch than they were ten years ago; indeed the total marine catch is now even beginning to decline! Another striking social trap analogy is with the super-power nuclear arms race, a race in which no-one wins and which until recently the participants continued to up the stakes.

A simpler game to play is the sell-a-five-pound-note game, or the sell-a-£5-note game as described by American psychologist John Platt[325]. The rules are that a £5 note is auctioned but that the money from the two highest bids goes to the highest bidder, that bids should increase by multiples of at least 5p, and that the total bid should not exceed £50. The bidding may start with a 5 pence bid, followed by a 10p bid. Should the bidding stop then the winner gets the £5 note making a £4.90 profit. The loser loses 5p, and the seller loses his £5 less the 15p he receives from the bidders. Of course the bidding does not stop there as the loser raises the stakes to 15p and so the bidding continues. The game's first key point comes when the bidders respectively bid £2.50 and £2.55. From here on in the person selling the £5 note will make a profit. The next interesting point comes when one of the bids reaches £5. At this point, should bidding cease the winner wins the £5 but buys it for the same amount so making no net gain, whereas the loser loses their last bid £4.95. However more often than not the loser will bid £5.05p to win, where upon his competitor will increase their stake to £5.10p so that both are bidding more than £5. If the game were to stop here, then the winner would get the £5, but spend £5.10 in the process so in fact lose 10p. It should become apparent why there is the £50 total bid rule, because each competitor wants to minimise their net losses by getting back the £5 note being sold.

The analogy with social traps in global common exploitation, be it ocean, tropical forests or air, is that consumers (the World's citizens) continually spend through utilities, industry whatever, to a point beyound resource sustainability so that the resource is eroded. With regard to climate change we are at the point where bidders have just exceeded the £2.50 mark so that the total bid is greater than £50. We are using fossil fuels up, and are both currently having to spend more in extracting some of them (oil from Alaska and the North Sea for example) as well as beginning to see the Earth's climate change with the increasing costs that this will inevitably entail. However a most striking example is that of exploitation of the World's fisheries. Today the World's fishing fleets are spending more per tonne of catch to catch than they were from the oceans ten years ago; indeed the total marine catch (the £5 note if you will) is now even beginning to decline! Another striking social trap analogy is with the superpower nuclear arms race, a race in which no-one wins and which until recently the participants continued to up the stakes (if you like to the £50 bid total).

The burning question is, of course, are there ways of escaping these social traps? Fortunately history reveals indications that there are. In spite of all our serious problems and traps today, the majority of the World's poor are more or less fed, the number of children educated increases, and garbage is disposed of. We have learned to convert long-range social goods into daily wages for farmers, teachers, and garbage collectors. The human race has managed tremendous feats through channelling resources through a complex web within the population. Tasks such as placing a man on the Moon, or a satellite into orbit to monitor resources and environment, including the weather. And individually, you (assuming you are from a First World country) can buy coffee in your local supermarket without having *directly* motivated the person who cultivated the original coffee beans.

Equally the tragedy of the commons is essentially a problem of the allocation of scarce resources. Humans have managed resources in a variety of ways including by force, by tradition, by inheritance, or by election. So marketing boards are established. Hunting seasons are legally bound and harvesting quotas determined. It is because these have been so successful that we are almost lulled into a sense of security. For instance, it is now possible to make the best agricultural land a few fold more productive than in the last century. Hardin's tragically over-grazed common land was only, almost by definition, badly managed, and his commoners therefore were fairly stupid. The problem is not one of thoughtless competition but one of establishing and maintaining the discipline of getting out of the trap. This can be done in a number of ways[325].

- *Change timescales* to convert long-range sequences into more immediate ones and bring the consequences of our actions home to us sooner. Warnings on cigarette packets attempt to do this as public health education and, in greenhouse terms, what is needed is public environmental education.
- *Dissuade* If pointing out the benefits of the carrot do not work (i.e. stop smoking and gain health), then use a stick. Taxation is one and a carbon tax would be a greenhouse dissuader.
- *Move the goal posts* People continually move goal posts. At one time smoking was thought of as being trendy: today both in the USA and in many parts of Western Europe smoking is increasingly viewed as anti-social. In a greenhouse aware culture, travelling by energy efficient

public transport wherever possible might be considered as virtuous, and the ownership of high-performance, low-energy efficiency sports cars as a lowering of status. Statutory regulations as to the thermal insulation of new buildings might be strengthened.

- *Providing rewards for trap-escaping behaviour* In other words, providing carrots for 'greenhouse friendly' activities. Energy efficient goods might attract less sales tax than low energy efficient goods.
- *Finding external help to get out of the social trap* For the individual this might be by accepting society's, or the government's help. For government's this would be by arriving at international agreements and for greenhouse issues would mean arriving at evermore.

Given that social traps exist, and that erosion of the global commons exist and even that there may be ways out, it still might help to consider why such traps are allowed to come into being. Identifying the reason why humans behave the way they do, especially when such behaviour is against the common good, may help us fine tune the ways out of such traps. Of course it does not help when frequently traps involving the global commons are nested: nested much in the same way that increased unemployment leads to increased crime, which itself makes an economy function less efficiently. Similarly, late twentieth century developed nations while being the worst offenders (on a per capita basis) in terms of anthropogenic climate change forcing, they do have the technology to develop energy efficient fixes, they also have greater population stability. Finding some common *raison d'être* for such behaviour might itself point to solutions.

Unlike the harder sciences, such as biology (and more so physics), psychologists have yet to establish exactly what makes people act in a specific way in response to certain conditions. They have though made some broad inroads into the human mind. Psychologists have, for instance looked at why some humans are more able to delay initial, low-gain gratifying behaviour in favour of long-term, high-gain gratification, while others are not. The greenhouse analogy here would be to invest first in energy efficiency, and then in supply so as to minimise energy wastage. An alternative analogy (in a world of ideal greenhouse policies) would be the choice of making do with less personally convenient, but energy efficient and cheaper, public transport to commute to work, and so make a financial saving to enjoy at the weekend.

Psychologists have looked at this ability to delay gratification and, for school children at least, have even managed to correlate psychological profiles of those who are best at delaying gratification for greater long-term rewards, and of those who cannot[326]; such studies are thought to reveal relationships in core psychological dimensions and so are relevant to adult behaviour. Those who are unable to delay gratification tend to be rebellious, unpredictable, self-indulgent, and even sometimes hostile. Those capable of the greatest delay for the greatest long-term reward are described as being responsible, productive, ethically consistent, interested in intellectual matters, and overly controlled. Differences in delay behaviour are found among both sexes, and in both sexes psychologists have repeatedly found that it correlates positively with IQ. But, while this indicates that those most likely to accept greenhouse, delay gratification, policies are most likely to be more intelligent, productive etc., it does not tell us *why* these types of people have this ability, or indeed why others of different character do not. Equally, while one could say that some criminals have been disadvantaged in society in one way or another, this does not let us know why some people turn to crime and others do not (not all socially disadvantaged people turn to crime).

It is not difficult to view those who impair the sustainability of the global commons as indulging in criminal behaviour. Many psychologists have examined the psychology of crime and have asked how such behaviour arises. Theories have been put forward involving: individual morality, social setting, economic circumstances, the prospects of avoiding punishment, etc. That psychologists have not come up with a clear-cut definitive answer to crime as opposed to any other behaviour, is obvious in that not only does crime still abound, but that arguments as to its prevention and punishment still prevail. Perhaps the closest we have come to such is the explanation presented by two criminal psychologists, Richard Hernstein and James Wilson, in the mid-1980s[327]. Their theory (not being the clear-cut answer we all seek) rests on a fairly broad assumption, that people, when faced with a choice, choose the preferred course of action. In fact this assumption is so broad that it is almost tautological. However, when they use the word 'choose' they do not mean that people consciously deliberate about what to do. What they are saying is that behaviour is determined by its perceived consequences. At any given moment, a person can choose between committing a crime and not committing it. The consequences of committing a crime consists of rewards (possession, or receipt of all, or part, of the value of what is stolen) and punishments. The consequences of not committing crime also entails gains and losses. The larger the ratio of net rewards of crime to the net rewards of not committing crime, the greater the tendency to commit crime. Clearly then, with this view and most of the early theories expounded by early psychologists, criminologists, sociologists etc., not to mention many still publicly discussed today, that relate crime to social background, economic factors, or other causes, there is in common the feature that they contribute to the reward–punishment (or gain–loss) equation.

There is, of course, more to criminal theory than this. Human behaviour is shaped by both primary reinforcers (for instance, hunger and sex), and secondary reinforcers which are learned. This is further complicated in that the line dividing innate reinforcers from learned is blurred (for instance is true altruism possible, and if the need to satiate hunger is innate, then why is the World's cuisine so diverse and some dishes extremely unappetising to members of other cultures?). Delay and uncertainty also add their own dimensions to behaviour: ones that are most relevant to greenhouse issues. Millions of cigarette smokers ignore the (possibly) fatal consequences of their actions because they are distant and uncertain. If smoking one cigarette caused certain death tomorrow, it would be fair to anticipate a sharp contraction of the tobacco industry. Similarly, if future climatic effects of current and future greenhouse gas emissions could be experienced now, and the pain of the transition from finite liquid fossil fuel dependence felt promptly, then one could easily suppose that energy efficiency and the development of greenhouse friendly and renewable energy resources would become a matter of urgency.

This matter of choice and resource use as a reward, also manifests itself in medicine. Just as we are using fossil fuels to the point where it is to the detriment of the environment, so patients who are given medicines to administer to themselves can easily overdose unless precautions are taken[328]. This happens when patients in pain require continual low doses of pain killer: allowing the patient to control the flow of drugs in a drip negates the necessity for a nurse to be continually present. The danger, of course, is that patients can give themselves more pain killer than is good for them. Doctors overcome this by ensuring that there are limit mechanisms in place. The analogy is that we need something similar to regulate fossil fuel consumption.

We can escape environmental social traps by adopting rational management, dissuading people from damaging the commons, encouraging people to preserve the same, and coming together hierarchically (from individual to local council, to government, to international, to worldwide frameworks within which good management of the commons can occur). We can use human psychology to further such resource management goals by making people more aware of the issues, by enabling them to recognise long-term benefits and making these relevant to individuals. This means investing (i.e. providing and recognising the opportunity) for those indulging in such behaviour, investing in generating awareness, and in greenhouse taxes.

In terms of psychology and human behaviour, the prognosis is not clear. The elevation of environmental issues on the political agenda, the increased membership of environmental groups (in the UK a 100% increase from 1.5 million in 1960 to 3 million in 1980), and internationalisation of environmental and energy-related policies[329] is surely to be welcomed.

The Prisoner's Dilemma is a good game to research in that it easily lends itself to computerization. Martin Nowak and Bob May arranged players on a two-dimensional computer grid. They programmed the players so that they either continually cooperated or defected. Furthermore, after each round with each player's neighbours, the score was totalled and the player either stayed where it was or moved into a neighbour's space who had achieved a lower score. What took place was that mathematically chaotic, and 'beautiful' patterns emerged[330]. Given the correct circumstances a dynamic stability occurred, which gives rise to the question as to whether this applies to real world analogies?

In 1993 Martin Nowak and Karl Sigmund designed a computer programme incorporating memories of recent moves. It quickly became apparent that a - tit-for-tat strategy tended to achieve high scores. However, players programmed for tit-for-tat (TFT) behaviour did stumble with the introduction of unconditional cooperators who were taken advantage of by habitual defectors. In real life, populations dominated by central planning are softened and subsequently exploited by entrepreneurs (both legal and illegal). TFT also can result in continual back-biting after an initial mistake. However, the best strategy appears to be one played by a Pavlov player (the animal behaviourist who made dogs drool at the sound of a bell). They only cooperate if in the previous round both they and their neighbour make the same move; i.e. if they both cooperated before and won or they both lost together and so learn from the error. Such lessons would seem to indicate that it is indeed best for nations of the World to move together, and to promptly punish (economically) those who do not cooperate in global policies. Mid-1990s events in the former Yugoslavia, China and Nigeria for instance, demonstrate what happens when aggressors (defectors) are allowed to get away with defection without prompt sanctions.

Throughout this text there are numerous examples of how we can lessen our effect on climate. Nonetheless, it is perhaps worthwhile citing one example here that does recognise psychological and perceptual barriers to adopting a 'greenhouse friendly' stance. The USA (currently the country whose citizens contribute most to anthropogenic climatic forcing) is perhaps the best place from which to take an illustrative example. Since the late 1970s, house-builders and homeowners in many parts of the USA have been able to have their dwelling appraised in terms of energy efficiency through schemes known as Home Energy Rating Systems (HERS). Where successful, HERS have been paid for by the authorities, or, frequently, the energy utilities, and not the homeowner, so that in terms of the immediate cost–benefit, gain–loss psychological equation, the balance is in favour of the home-owner accepting a HERS audit. In some areas hierarchical policy control have made HERS mandatory for new houses. So successful have HERS been at bringing the energy efficiency message across that it has been estimated that where they have worked some 40% of the new construction market has been penetrated, and 20% of existing buildings, with the realisation of energy savings of between 10–50%[332].

INDUSTRIAL AND COMMERCIAL PERCEPTIONS

That the human psyche can sometimes present perceptual barriers to energy issues is nowhere more evident than in industry. In the USA between 1978 and 1982, a Georgia state-wide survey of over 150 energy extensive industries revealed that senior managements were reluctant to instigate expensive energy saving measures even though they were demonstrated to be fully cost effective. Low cost measures were readily adopted where top management involvement was not required[333]. A decade later perceptual barriers are still the principal stumbling block for instigating energy efficient measures.

In 1989, in West Germany another survey was conducted of over 500 companies in eight areas of activity: bakeries, butchers, dairying, food production, the wood industry, brickworks, concrete manufacture, and textile finishing[334]. For each of these activities, a list of technical and organizational energy saving measures was drawn up. The researchers found that of these measures between 40% and 60% were being used; indeed, some low energy-using companies (especially those small firms with less than 50 employees) used only between 30% to 40% of energy-saving options. When asked why more energy efficient measures are not instigated, over half the companies claimed that their new production plants were already energy efficient, or that money was needed (or perceived as being needed) for more important investments. Less than half the companies said that the managers could not ascertain [perceive] future energy costs hence cost effectiveness, or that they wanted to wait for new technical solutions. Nearly a quarter of the firms claimed that they lacked the right personnel to implement the necessary measures. Importantly, less than half of the firms systematically calculated the profitability of energy saving measures. The question that follows is what is it that prevents firms from saving energy? In the USA the Office of Technology Assessment concluded that 25% of the energy consumed in federal buildings could be saved with no sacrifice to comfort or productivity[335]. So what prevents organizations from implementing these savings?

Economist Stephan DeCanio has presented an answer based largely on perceptions and psychology[336]. First he says that we have to recognise that firms do not behave like individuals. The firm makes choices and decisions that are generated through its rules of procedure rather than a single individual's action (especially at middle management level). Second, management and shareholder have interests that do not necessarily coincide. Often managers will be deterred from making decisions that may not result in instant payback and so are risk adverse, whereas shareholders are more risk-neutral. Here DeCanio points to managements' desire for large returns over short time scales, whereas energy saving measures frequently have a low rate of return over longer time scales. The reason for this is that management compensation (salary rises and promotions) are often tied to recent performance. Furthermore, many firms rotate their managers every two or three years in order to preserve motivation, prevent ossification, and to ensure that they have a broad perspective on the company's activities. A manager who expects to be in a post for two or three years has no motivation to promote a project with a more distant payoff. Thirdly, top management gives low priority to relatively small cost-cutting projects.

To overcome such problems the researchers have put forward a number of solutions, in addition to making the less visible costs more visible (or the internalization of the environmental and social externalities associated with energy production/consumption). These include: information, motivation, training, and consultation measures. One comparatively new measure is that of Energy Performance Contracting where a third party – be it a consultant, manufacturer of production plant, or an energy utility – takes on the financing and the risk of an energy saving investment. In return the third party receives a fixed proportion of the saved energy costs. Sixty-one percent of the 500 firms in the West German study above[334] thought this to be a good suggestion (10% gave no reply), but only 29% were planning investments for which contracting (in their opinion) could be the right solution.

At the end of the day no one can force firms to adopt such non-legislative measures to improve energy efficiency. Measures must, from the business perspective, clearly help the business turn a profit *within* a timescale that continues to attract investors (shareholders), and which maintains competitiveness. One way parallels the need for human economic activity to recognise and pay for the hidden costs (our environmental externalities) by adopting some 'environmentally friendly' measures, not solely to pay for some of the externalities but also for promotional reasons. Customers may favour the environmentally friendly firm over its competitors. This factor should not be taken lightly. Firms wish to be seen to be responsible, and this can have other, non-customer, spin-offs. What is more, the advantages of being seen to be environmentally friendly are becoming increasingly important in many firms' public relations policies. For instance, in the year of the first IPPC report (1990) ICI made considerable play of its responsibilities to the environment and even produced a 42 page brochure[337] extolling its environmental virtues. ICI was (before splitting in 1993) a UK chemicals multinational, and a major player within the UK economy with a 1989 turnover of nearly £13 billion. In addition to the special brochure, ICI's 1990 Annual Report led with an article on its activities and the environment, which announced a four-policy strategy, 'Objective 3' of which was to establish a rigorous energy conserving programme. If major companies, such as ICI, need to show that its key documents are environmentally aware, then there is a real chance that the coming decade will see a significant change of attitude and behaviour throughout industry and commerce.

Having a clean image helps to sell products, but it is not the sole motivator for industry to recognise hidden environmental costs. Such externalities ultimately bite back, for in essence if industry is harnessing resources from the environment, and if such processes 'damage' the environment, then at the end of the day they are harming the source of their wealth. In the case of global warming, if the warming is rapid, or if it affects climate in a way not suited to existing industrial and commercial patterns, then industry will have a cost it must pay: it will no longer be an externality.

From a scientific perspective, the risk of climatic costs being imposed on industry is still unclear. Will, for instance, the Broecker salt conveyer stall shortening hot Western European summers and deepening its winters in the space of one or two years? No one knows. However there are a few recognised possibilities. There is a theoretical link between a warmer Earth having more energy available for atmospheric water latent heat transfers that are central to storm and storm track formation[338]. But this is theory and has yet to be scientifically validated to the level of practical utility.

Yet scientists are not the only people to concern themselves with the risks associated with climate change – insurers are too. The insurance business makes its money from accurate risk assessment: it must charge premiums to cover claims, but not such that a competitive insurer can steal its business. So the way the insurance business looks at 'natural disasters' (both climate/weather related such as hurricanes and floods, or earthquakes) can be quite revealing. Over the three decades prior to 1990 the World decadal average real cost to insurers of damage from natural disasters has risen more than three-fold to US$3 billion, whereas the overall cost of damage has risen from US$3.7 billion in the 1960s to US$11.4 billion in the 1980s[339]. Part of this trend is due to the real growth in World population and economy, but natural hazards themselves have not yet shown any significant increase in number. Though the insurance profession recognises that the evidence for climate change is circumstantial, they take the issue seriously. This can only serve to encourage others in the commercial and industrial sectors to do the same.

We have looked at perception within industry at the middle management level but the view from the top may be quite different. In October 1995 the Chairman of Shell was invited to share his insights into the future of energy use and its sustainability with delegates attending the World Energy Council Congress[340]. He pointed out that the energy surplus of the late-twentieth century would probably quickly disappear early in the next century and that meeting this growth in demand would require a massive investment – possibly of the order of US$200 billion a year (a little over 1% of the World's GDP.) He said, 'over--zealous taxation in the name of environmental conservation could jeopardise energy supply security'. It is important to note that those in business are also trying to modify other people's perceptions. So we must question carefully such warnings. Perhaps coherent international policy is required. Whatever, such expert views certainly should not be dismissed out of hand.

ECONOMIC PERCEPTIONS

The *Oxford English Dictionary* defines economics as: 'a) the science of the production and distribution of wealth, b) the application of this to a particular subject'. It also provides an alternative definition – one that is perhaps more broadly recognised: 'The condition of a country etc. as regards material prosperity'. What economics is not, is the simple way money changes hands as measured by 'profit', 'losses', 'surplus', 'turnover' etc. These are accounting terms and accounting in its various forms is a tool that economists use to identify the ways money flows.

From an environmental perspective, the measure of the way 'wealth' is currently 'produced' and 'distributed' in the free world, is not entirely environmentally sustainable. For instance, the fundamental national measure of a nation's wealth-producing capability used worldwide is the Gross Domestic Product (GDP) – the sum of all the earnings of each and every citizen of that nation. Yet just as the earnings of all those involved in, say an oil spill accident, are included in the GDP, so paradoxically are the earnings of those involved with the clean up. The way around this, so as to reflect the economy more as a measure of the environmental sustainability of that economy, is to subtract from the GDP the costs, be they those of environmental damage or clean up, that impair or are diverted from the creation of future wealth. Whereas those diversion of future wealth would be, the 'opportunity costs' associated with the clean up. David Pearce has produced a modified Gross National Product (GNP) – 'green GNP', which includes overseas investment income[341].

Modified GNP = Conventional GNP + Value of environmental service
– Value of environmental damage.

Here, while the value of environmental damage is more readily ascertained, the value of environmental service is difficult to determine, it being the net gain in natural capital stock. A good greenhouse example would be that of deforestation. While the salaries of those causing the deforestation contributes to a nation's GNP (and the costs associated with, say, the extra siltation of rivers due to increased soil erosion from deforested areas contributes to the value of environmental damage), the value of the environmental service would be negative, it being the cost of not being able to replant the forest (through soil loss), loss of biodiversity, etc. On the other hand where forests are replanted (even if they may not have the same value as before) the environmental service would be positive since this activity would add to the country's future ability to produce wealth*.

* GNP may be modified with environmental considerations, Joel Cohen suggests that the term 'carrying capacity' can similarly be modified when applied to humans[342]. An individual human's action (life) can increase or decrease the human carrying capacity of the Earth. It follows that our behavioural response towards global warming (our policies) will affect the Earth's carrying capacity for humans.

While such modified GNPs, in theory, provide a more accurate reflection of a country's economic performance in terms of how it manages its natural resources in a sustainable way, problems arise when trying to put numbers to the 'value of environmental service' and 'value of environmental damage'. This problem is the same as those encountered in cost–benefit analyses; indeed the 'value of environmental damage' would be a grouping of costs in a global warming cost–benefit analysis, and the 'value of environmental service' a grouping of (environmental) benefits. It is comparatively easy to quantify some components of the 'value of environmental damage', but some are virtually unquantifiable. For instance, a good greenhouse example would be that of deforestation. Loss of top soil from a deforested tropical forest, costs of fishery loss from siltation, the loss of future income from the forest and other such costs can be determined, but the loss of a unique plant species is unquantifiable. That species *might* have contributed a new medicine, or a new insight into plant ecology with possible benefits to agriculture. Equally, in cash terms, what is the intrinsic value of a species?

There is also a second problem in bringing the above elements together in determining either the 'value of environmental service' or the 'value of environmental costs' (or the costs and benefits of global warming) in addition to quantifying components such as the intrinsic value of a species. This arises from the uncertainty of the degree of global warming, and the effects of adding an additional tonne of greenhouse gas to the atmosphere. If both effects are large, then the associated costs will be high, but if the effects are small then so will the costs (and benefits) of control. Climatologists and environmental scientists are trying to ascertain the likelihood and degree of climate change and its environmental impact, while economists are trying to work out ways to translate these into hard cash terms. As yet, there are no answers, all economists have to offer for the moment are a range of opinions.

Economists' views vary, from the belief that the net costs associated with global warming are fairly low (Nordhaus' view), hence so are the benefits of curbing such costs, efforts should therefore be focused on reducing CFCs[343], to proposals that we must keep the warming below 2.5°C above pre-industrial levels[283] and introduce controls to keep the rate of warming to within 0.1°C per decade, a rate that would allow many natural systems to adapt and species to naturally migrate. The economists in both the absolute-limit-to-warming and the limiting-the-rate-of-warming camps suggest that the costs of exceeding these limits would outweigh the benefits, so that nations' GNPs have to be 'modified' by valuing 'environmental services' and 'environmental damage' in such a way that it is not worth exceeding these limits.

Whatever limits are adopted as goals for greenhouse policy, economists have to find a way of encouraging nations to keep to them. One could (in theory) arbitrarily restrict fossil fuel production, and hence CO_2 emissions. But the option increasingly favoured by economists and governments is to modify the market with a carbon tax. This too has its own problems, which are discussed by David Pearce in *Blueprint 2*[344] and include the need to ensure that carbon taxes are applied to both developed and less-developed nations in an equitable way that reflect their respective abilities to pay, and that no nation shirks its obligations to charge the tax. However a carbon tax has advantages in that it is a way of internalising externalities which can generate revenue to use to improve the energy use efficiency and diversification of energy generation into non-fossil, greenhouse-neutral, modes of supply.

This last point is important, for a carbon tax can be either a way of internalising costs (all or in part) or just another tax, from which the revenue goes towards non-carbon related economic activity. The establishment of an energy tax does not necessarily mean that positive measures are taken to minimise 'greenhouse' impact, e.g. by subsidising installation of energy efficient devices or non-greenhouse energy-generating technology. Governments could use revenue from a carbon tax to meet other (political) goals, such as lowering income tax, or even paradoxically encourage fossil fuel consumption to bolster an ailing industrial sector (e.g. by subsidising construction of coal or oil-fired electricity generating plant). Whatever a carbon tax is, it is not just a mechanism for increasing the price of fossil fuel so as to allow market forces to push energy efficiency to the fore; true, this is one effect of a carbon tax but the other is to provide revenue. *Blueprint 2*[344] cites some 11 so-called carbon tax studies published in the late 1980s or 1990s that assumes that the revenue raised from the tax is lost from the economy! However, it is now virtually universally recognised by energy economists that the revenue from a carbon tax should be used to address greenhouse energy issues, since it is coupled to the greenhouse issue.

Carbon taxes also offer an additional benefit: they are a way of making intergenerational transfers (i.e. provide funds for a future that is climatically changed and short of fossil fuel). In 1929 the economist Pigou noted that people frequently irrationally distribute their resources between the near-future and far-future[345]. When there is a choice between two satisfactions the larger of the two is not always chosen, in that frequently an immediate small gratification is preferred to a large one some years into the future. This is because the human perception of time is at fault: we see future events, needs and wants on a diminishing scale. By choosing short-term options at the expense of greater long-term benefit we not only diminish our own welfare, but also those of future generations (in which we have a genetic and mimetic (cultural) interest).

This faulty perception does work on the level of self interest. The harm to future generations from realising short-term benefits will be much greater than to those living in the present. Yet as the human life-span is finite, the benefits accruing from long-term investment may well not be realised by those making the investment. Therefore it is not surprising that effort directed towards distant future goals is frequently much less than the efforts invested in the short term.

The consequence of such behaviour is one of reduced saving, hence loss of the ability to create large amounts of new capital so that existing resources are consumed in a way that sacrifices future advantages. Pigou illustrated this through finite resource depletion whereby such resources, abundant in the present, are squandered on comparatively trivial purposes. To avoid such mismanagement Pigou recommended policies such as:

- legislative resource protecting measures to prevent frivolous squandering,
- introducing incentives for long-term investments,
- and the abolition of taxation that discourages saving.

This last is of particular relevance in the current carbon tax debate. If one were to abolish taxation on long-term saving then the lost taxation revenue could be made up from part of the revenue of a carbon tax (and other taxes on extraction of finite resources). In other words, taxes could be used to dissuade short-term squandering, and provide tax relief to encourage long-term and even medium-term saving. Large projects of national and greenhouse import could be encouraged in this way; such as investment in a new energy-efficient infrastructure.

Taxes are currently the preferred method of the domestic regulation of fuel prices, and so it should not be surprising that they are one of the most talked about mechanisms for recouping the costs to pay for environmental externalities. Yet there are other mechanisms, and these have already been employed to regulate the market. Fuel stocks can be purchased and bunkered, and indeed such 'price fixing' could be used to offset environmental externalities, despite the one argument frequently used against carbon taxes is that they constrain the free market. The counter to this is that when applied correctly (and internationally) they should enable the market to reflect the 'true cost' of fossil fuel (i.e. to internalise some of the externalities).

Finally, whatever the economists' perception of the economics of climate change, it rests with the politicians to sell the policies arising out of such perceptions to the public, and to determine the level of tax. Here, US economist Galbraith has pointed to one major stumbling block. If the majority are happy to ignore externalities (i.e. prefer cheaper fuel for short-term gain) then they will vote for politicians backing such policies. Galbraith calls this characteristic of our society 'the culture of contentment'[346]. It is exactly for this reason that perceptual barriers must be surmounted so that the electorate is aware of the significance and import of the policy packages. The alternative is for politicians to positively lead, something notoriously difficult within a democracy and which can run the danger of undermining democracy itself. The classic example here being Yeltsin in the 1990s who led Russia along the painful path towards a free market economy with all its long-term benefits, but who nonetheless had to struggle against the old guard communists who represented price stability for the poor and elderly.

This idea of politicians determining taxation level, and not economists, is an important one. Economists provide mechanisms by which goods, costs and benefits are accounted (or not), both within and between societies. Economists do *not* have absolute control, or in all cases the necessary day-to-day facts required in setting national budgets – that is done by politicians. Grubb (Royal Institute of International Affairs) criticizes the economist Nordhaus on just this point. Nordhaus says that 'his hunch is that the overall impact [of anthropogenic global warming] is unlikely to be larger than 2% of the total [World economic] output'. But Grubb asks why should 'an economists hunch about climate change command more respect than the warnings of hundreds of scientists about the risks associated with human interference with the atmospheric heat balance, and of social scientists about the possible human impacts[347]'.

To sum up, economists have introduced the concept of unpaid costs – externalities – associated with the selling of goods. Anthropogenic climate change is one such externality, an environmental externality. While some economists believe that externalities arise due to free market failure, it is widely recognised that whatever their cause, they exist and so need to be accounted for. To date, taxation has been the obvious choice to balance the books, but the debate continues as to the level (accounting) and method of implementation (economic policy) of such taxation. Debate also continues with regard to the fate of the derived tax revenue. Currently such monetary issues are not solely the concern of economists: politicians, who are ultimately responsible for the implementation of policy, are increasingly recognising the need to address the way environmental externalities are met.

POLITICIANS' PERCEPTIONS

If anyone had thought that anthropogenic climate change was not relevant to politics, or politicians, then the 1972 UN Conference on the Human Environment in Stockholm firmly demonstrated the contrary. Not only was that conference attended by the representatives of 113 governments, but political rows dominated its proceedings for the first few days. But importantly the Stockholm conference and its precursor (Founex Report) focused on whether environmental concerns would hamper development? Twenty years on and the 1992 United Nations Conference on Environment and Development (UNCED) in Rio attracted over 150 representatives and heads of state.

Anthropogenic climate change is, in terms of international politics, a particularly thorny problem impinging on many other international issues, not withstanding World development. Not only do the effects (hence costs and benefits) of anthropogenic climate change vary from one part of the World to another, but so do greenhouse emissions. As we have already noted, the low population growth, high GNP per capita countries release the most carbon dioxide per citizen, and in the process power much of their economic activity. On the other hand the high population growth, low GNP per capita countries produce the least CO_2 per citizen. Politicians from these less-developed countries naturally want improved standards of living for their people, and this will inevitably include higher energy consumption, but it is difficult to see how this can be done without a massive increase in fossil fuel consumption. The only obvious solution would be for these countries to develop energy efficient, and non-greenhouse energy technologies. However, these are expensive and any extra expense will hinder their economic growth. Politicians from less-developed nations are therefore calling on their developed nation counterparts for the extra funding required, but the developed nations are reluctant to pay. This poses such an intractable dilemma that one political response that scientists have faced, is a denial as to the magnitude of the problem. One memorable instance took place at the 1990 World Climate Conference. While Margaret Thatcher was warning of the need to combat climate change, US President Bush refused to accept the seriousness of the problem as presented by the IPCC Report and said that it was 'not what my scientists are telling me[348]'. The USA later signed the 1992 UNCED Framework Convention on Climate Change and by 1997 US President Clinton accepted that climate change was an issue that needed addressing. And so the international debate continues...

Individually, many politicians worldwide have found that expressing a view on anthropogenic climate change can be politically risky. Those who derive their livelihood from the fossil fuel industry are often (though not always) worried when politicians, in airing greenhouse concerns, seem to be threatening the way they make their living. And of course where one politician may say one thing (and make political capital), another may wish to make political capital out of this. Margaret Thatcher herself experienced such an attack by her UK political opponents. In the summer of 1990 one of the leaders of her opposing (Labour) party said, projections had 'been deliberately inflated to make her position look radical... no one else believes this far-fetched forecast[349]. To put this into the context of the time, the forecasts made by UK Department of Energy experts were on the high side.

While politicians perceive issues such as global warming both as real concerns to tackle as statesmen as well as ammunition for political in-fighting, they also suffer the same perceptual problem as the rest of us in that we do not have all the facts, let alone the ability to process them. Indeed, given that politicians have to consider so many wide ranging issues, it has been argued that not only are they unable to process data optimally, but that their daily information load is so great that some of it is passed over. Faced with such an overload, the temptation is to ignore material that conflicts with their beliefs[350]. One example illustrates the point, in May 1992 Britain's Environment Under Secretary, during oral questions in the House of Lords, was asked to elaborate on what measures the Government had introduced to curb carbon dioxide emissions. He replied, 'We have encouraged the use of unleaded petrol; we are introducing catalytic converters on new cars from next year, and we have also introduced the use of sulphur scrubbers at power stations'. Unfortunately for the Minister, the switch to the production of unleaded petrol (while reducing lead pollution) resulted in a small (about 2%) increase in energy consumption at oil refineries, while catalytic converters (while lowering NOx emissions) result in a car increasing carbon dioxide emissions by 9–23%. Finally, power station desulphurisation equipment reduces the efficiency of power station electricity generation by about 3% – hence more carbon dioxide is emitted for the same electrical output of a non-desulphurised station (though desulphurisation reduces sulphur gas emissions and hence acid rain).

The following year the UK Environment Secretary provided another example of erroneous greenhouse perceptions. In a speech outlining the government's strategy for sustainable development he said: 'In recent years we have come a long way towards making our energy policy more sustainable. In particular, we have made great strides in decoupling economic growth from growth in energy consumption'. He was correct only in a long-term context – the energy needed to produce a real-term cost unit of UK GDP has fallen by about 40% since 1950. However between 1989 and 1992 it rose by 3.7%. At no other point since 1950 was there such a sustained three-year increase of this kind[351].

Politicians not only 'inadvertently' affect their perception of issues by passing over information purely as a response to information overload, but because they may be tempted to accept material that confirms their own ideas of policy, and bias can be introduced. One 'greenhouse' related area where this abounds is that of energy forecasting. Energy forecasts, by their very nature, are both inexact and complex. Consequently, failing to understand their limitations, or even failing to forecast, has a detrimental impact on the formulation of an 'optimal' energy policy. Forecasts are created using a set of assumptions, e.g. a national energy consumption forecast might assume real-term increases or decreases in the price of various fuels, a certain rate of population growth, population mobility, etc. Being assumptions they are not facts, so that failing to understand assumptions is also a failure to understand the forecast. It is also possible to take this further in that forecasts can be created with assumptions favourable to a preferred policy, e.g. the instigation of a particular fuel programme. Indeed the trend was that, through to the early 1980s, the forecasts steadily lowered their estimates. Such a trend was observed in many countries, including the USA, UK, and Western Germany[352].

It is easy to blame politicians for their failings, but it has to be remembered that in the majority of countries (and certainly throughout the OECD developed world) politicians are subject to democratic constraints. Not only do politicians have to deal with the wide range of issues of this complex late-twentieth century world, but they have to cater to the needs and wishes of their electorate. Just as we (the electorate) are governed by politicians, so their fate is sealed by our vote and somehow society has to chart a way between these two poles. One navigational aid can be found in the guise of the media. Politicians note what the press has to say as they are, in many democracies, *an* indicator of public feeling. Equally newspapers note what politicians say and do. The media's perception of issues is a significant factor in shaping both political and public opinion.

MEDIA PERCEPTIONS

Virtually all of Western society receives news through the media, be it newspapers, radio, or TV. Even those who feature in the news only make a part of it, and so even they rely on the media for the bulk of their own perception of the news. So powerful is the media in influencing events and society that virtually all major political parties, firms, and organisations have press liaison offices. In addition to national newspapers and TV channels, there are their regional counterparts. Finally there are the trade newspapers, magazines and journals for virtually every aspect of business.

It goes without saying that these last specialist publications are the ones that can most fully cover specialist topics. Global warming and energy policy issues are most completely reported in scientific journals, and less so in other vehicles. However some issues, including global warming, are so complex that so specialist analyses are required that even respected journals devoted to one aspect of science (and/or its application) can inadvertently misrepresent issues just outside of their usual remit. One such instance (and there are many) can be found in one of the World's leading medical journals. In this case they were publishing a series of articles on 'Health and the Environment', yet its first page misreported figures by a factor of ten, misquoted reports (by one word) so as to change their meaning, and introduced uncertainty where certainty had been declared. Fortunately, as often is the case with respected scientific publications, this leading journal was anxious to maintain its high standards, so that when the paper[353] was subsequently reprinted in a collection, as part of a book[354] these errors were corrected.

Clearly if scientific publications covering science outside their area of editorial expertise can make slips, then it is not surprising that general newspapers and magazines can completely misrepresent an issue. This is not to say that such misrepresentation is deliberate: it is frequently a consequence of other pressures, such as the need to get stories out quickly and in an understandable form for the general public (even if over-simplification leads to error). It must also be remembered that while science journals exist to educate, newspapers and magazines exist to inform entertainingly, and so to sell. Therefore if one science team reports a maverick result (which the rest of the scientific community recognises and will probably be conventially explained on further investigation) newspapers will have no qualm in reporting this (even though there may be a body of evidence pointing to a more mundane explanation waiting in the wings). Again, as scientists are human, one can invariably find a challenger to the majority view. While sometimes such challengers do have insight, or extra evidence, into some scientific concern, frequently they do not – their individualistic stance may be borne of a pet theory, or a pedantic interpretation of some technical point, or even for purposes of devil's advocate. Newspapers are not concerned with this, they want exciting copy that sells their publication. Out of this misreporting perceptual barriers are created.

There are numerous illustrations of this. A review of the US and German media's reporting of science[355] has shown that coverage of nuclear power stories is driven by the size of public protest demonstrations not the science, and that coverage of AIDS in the USA is not influenced by the number of deaths, but 'events' such as the death of public personalities.

With its *raison d'être* of informing entertainingly (but not to educate) scientists' expectations for the general media to raise the level of the public's understanding of science should be set against the existing level of such understanding from which, at best, one might expect a marginal improvement. Compare for yourself the complexities of as complex an issue as anthropogenic climate change with the scientific understanding extant in most developed nations. For instance, a recent French magazine's survey of its readers revealed that 25% did not know the temperature of boiling water (at S.T.P.) and that 30% thought that the Sun went round the Earth[324]. According to those concerned with the public understanding of science, there is a sizeable adult minority (typically well over 5%) of those in most developed nations with such misconceptions! Though it must also be remembered that, while such a low level of the public's understanding of science seems abhorrent to many scientists, scientists and non-scientists alike share a common *mis*understanding of science.

Be it 'information' or 'education', the one thing the media is good at is providing a channel of communication, hence its potential for altering perceptions. Indeed, in this frantic fact-feeding, dream-selling, modern World of ours, there is so much material to hand that the media not only entertainingly informs the public, but scientists themselves (as a subset of the former). This happens to such a degree that the media directly affects what is discussed by the scientific community!

In 1991 a study showed that *New England Journal of Medicine* articles publicised by the *New York Times* received 73% more academic citations in each of ten calendar years after their publication than articles from the same section of the journal that did not benefit from such coverage. To negate the obvious criticism that the *New York Times* only covered those papers that would have received heavy academic citation due to their inherent merit, the researchers had an experimental control. In 1978 there was a three month strike during which time the *New York Times* was produced but not distributed: during this period scientific articles were still selected for coverage but, without the paper's dissemination, they received no extra citation in scientific papers[356].

Given that the media provides some feedback into the academic community, the question arises as to whether this is a beneficial service? One could argue that no publicity is bad publicity. However, this perhaps begs a more fundamental question as to whether scientists can rely on the media to obtain, and disseminate, accurate information. Though a newspaper's primary motive is to provide information in a way that sells, and not to deliberately misinform, the likelihood of misinformation is significant. This is quite understandable if, as discussed above, specialist journals operating outside their area of expertise have problems in relaying information accurately.

However it is not all bad news. While a survey of scientists at Germany's Mainz University revealed that 48% of those whose work had been popularised noticed that journalistic accounts were at least partly inaccurate, and a similar survey of scientists at the Research Centre at Julich revealed even more perceived errors, many of these concerned technical detail not broad meaning. It appears that the scientists' perceptions of the media's perception showed that the scientists were being fairly critical, with many of the errors being errors of omission and most were coded in the surveys as 'not serious'. It would seem that scientists may perceive coverage of their own speciality to be more flawed than that of other fields simply because they are more competent to detect errors, or that scientists may scrutinise coverage of their own fields more closely since they are more involved[355]. Similarly a study of New Zealand's news coverage of climate change showed that 80% of stories were no worse than slightly inaccurate[357].

Yet while many of newspaper's and the media's science reporting of science subjects is *broadly* fair, the media does tend to both 'filter' news in its own way (which may not be optimal for society) and fail to check whether its copy is accurate[355]. The media sets its own agenda in that it uses non-academic criteria (such as what competing newspapers are publishing, or whether there are catchy quotes or personalities associated with one story over another). At the same time it fails to check accuracy, and why should it for it is merely reporting what 'it' sees, albeit through its own perceptual bias. Furthermore, while in an ideal world journalists should report the 'truth' there is rarely time for a reporter to become sufficiently expert in a technical issue in order to assess conflicting claims. Even an undergraduate scientist reading this text will have his or her work cut out validating much of what is said. Furthermore, the academic references have their own references, so that the student has to undertake more work to home in on any part of the global warming debate. Nonetheless, such a critical analysis is relatively straightforward, even if a chore. Notwithstanding this the student has academic tutors to turn to for help. On the other hand reporters do not have such experts on tap, if they absolutely need to they have to spend time looking them up, and even then they, as non-specialist scientists, themselves do not have the background to properly comprehend a quick and easy answer. For the media, with time being money, balance is a surrogate for validity checks. If a journalist is confronted with a group of scientists who say that we are probably currently living in the greatest mass extinction of species in the history of the Earth, against another that says we are probably not, then who does the journalist believe? And so we have the real life position in which, for instance, the media reporting of an early 1980s US court case challenging the State law that creationist principles be taught in school, transforms competing counter-arguments of differing strengths into equivalent ones[355]. One survey covered 6 months of news coverage provided by three American television networks. It concluded that 49% of science stories broadcast were 'completely accurate', 34% generally accurate and 14% somewhat inaccurate. In short 83% of the coverage had no or few flaws[357]. Even so, much damage can be done when an issue is misrepresented as it serves to confirm misconceptions.

With regard to global warming, in 1990 there was one TV programme[358] broadcast in the UK that epitomised the above, in that it took as its 'angle' the counter position (i.e. that anthropogenic global warming does not exist). Furthermore, it sensationalised this view by inferring that there was some 'conspiracy' among scientists. The programme presented scientists discussing 'catastrophic' warming without strictly defining catastrophic. It reported (from a news clip) that in 1988 most scientists were convinced 'that average [World] temperatures will rise by 2–5°C over the next one hundred years'. But it immediately went on to say: 'But is there any real evidence of a forthcoming disaster', leaving the viewer to make up his or her mind as to whether 2–5°C over a century is 'catastrophic' or not, or whether climatologists were predicting a worse fate! The programme then quoted a respected climatologist as saying that he did 'not think there is any evidence for a catastrophic change in our climate at the present time'. Again, given the meaning of catastrophic this is a perfectly rational statement, but equally it does not deny the majority science view that global warming, as discussed in this text, may very well be taking place. The programme continued in such a vein examining four areas of climatological evidence revealing the weaknesses in using them (by themselves) to 'predict' (rather than forecast) the future. The programme did not say that environmental scientists and climatologists were quite aware of these weaknesses (let alone that no-one can *see* into the future) which is why so much evidence was being gathered, and examined together, as possible. The programme ended with the allusion that scientists were in a coalition of interests, promoting greenhouse theory to obtain research funds!

Stephen Schneider, a leading climatologist at the National Centre for Climate Research, USA, has had numerous encounters with the media and personally experienced its frequent misrepresentation of his presentations of academics' then current understanding of climate change. So often has this happened that in his book *Global Warming*[360] (published just prior to the first IPCC report[20]) Schneider devotes an entire chapter to the way the US media reported global warming – particularly coverage in 1988; a year which saw a very warm North American summer. Against a national backdrop of water shortages, withered plants, reduced river flow etc., the media flocked to the Conference on Climate Change in Toronto in such numbers that extra press rooms had to be provided. They had been primed a few days earlier by a quote from atmospheric physicist Jim Hansen who said that he was '99%' sure that the warming of the 1980s - evident in his calculations was not a chance event. Immediately, Schneider says, that '99%' was everywhere. Journalists loved it. Hansen appeared on a dozen or more national television programmes, and was quoted in a front page story in the *New York Times*. But Hansen had inadvertently shot himself in the foot in that one person's colloquial opinion that there is a 99% likelihood of something happening does not mean that there is a 'statistical confidence' of 99%, in the mathematical sense of the term. Consequently Hansen received flak from other scientists, which the press, of course, equally loved. However the public, reading all of this, were as likely than not to perceive undisciplined academic wrangling, than any sense of what the true debate was about.

Science rests on the trinity of hypothesis, experimentation, and conclusion (from which may spring validation). Take away any one of these and you destroy the science. It is this incomplete coverage of science that results in non-science, if you will, being reported which happens all too often and of which Schneider fell foul.

Is there anything about the way reporters work that encourages errors? A survey of news reports on climate change, carried out in New Zealand, suggests that there may well be[357]. Reports covered by the survey that were journalist initiated, as opposed to source, or press officer initiated, contained an average of three errors compared to one. Reporters using a press release similarly made fewer errors. However, the greatest number of errors (nearly 7 per story within the survey) came from stories based on a public speech, as opposed to an interview or document. Surprisingly, stories based on an interview (lasting more than 20 minutes) contained more errors than those based on shorter interviews or no interviews at all. At first this may seem counter intuitive.

TECHNOLOGICAL IMPERATIVE AND PERCEPTION

So successful has science been over the past few centuries that it has enabled a technology to be spawned which has transformed the human condition planet-wide. Today, a quarter of the planet can speak to a quarter of planet within minutes. Today there are more human beings alive than there has ever been before. In the past few centuries the vast majority of the World population has seen their life expectancy dramatically increase: indeed, between the latter half of the 1960s and the latter half of the 1980s the UN estimate for the World death rate decreased by 25% to 9.9 deaths per thousand, with nearly all of this improvement occurring within the less-developed nations[362].

Associated with this growth in science, technology, and population has been an associated growth (but on a far greater scale) in information required to service human activity. Not surprising then that politicians find it increasingly difficult to assimilate the necessary information required for policy-making (see section on political perception). This information myopia can be exploited by the less scrupulous. Such exploitation is something John Tyme in his book on motorways versus democracy[362] called the 'technological imperative'. The technological imperative arises out of four key components converging:

(*1*) the technology itself.
(*2*) an industrial/financial complex associated with the said technology,
(*3*) a 'lobby', or pressure organization for the industrial/financial complex,
(*4*) an 'interest section' within government or the state apparatus which will cooperate with the lobby.

John Tyme examined the way the technological imperative works towards the ends of the UK automobile and road lobby[362]. Aside from road building, it is also found in other areas of today's technological society including energy policy – hence its relevance to greenhouse issues. In terms of public and political perception of these issues, perception can easily become biased in the face of a technological imperative doggedly lobbying one side of the case (sometimes to the point of being economical with the truth).

Of course it has to be recognised that not everyone in any industry contributes to the technological imperative. Furthermore, industry not only has the right, but should be expected, to blow its own trumpet, promote itself and market its goods. However, this does not mean that a radical part of industry cannot go to extremes to secure its position and future prospects. With regards to the anthropogenic greenhouse effect and the technological imperative such extremes are found primarily in the energy supply industries.

The nuclear industry, for instance, properly promotes itself by stating facts such as 1 tonne of uranium in a fast reactor is worth more than one million tonnes of coal. This does not mean to say that nuclear fission alone is the answer to our energy problems. In 1982 the Director of non-nuclear research at Harwell addressed the UK Energy Industries Club on long-term energy options. While he cited nuclear power, he also referred to the potential of the extensive use of renewables, and the possibilities for 'very large improvements in energy efficiency[363].' There was nothing that new in what he was saying, he was painting a useful picture, the point is that these non-nuclear options were clearly recognised by UK energy industries well over a decade ago. Yet while the industries recognised these options, quantifying them (for purposes of energy policy formulation) has not proved easy. One major (1990) study into possible US energy consumption in the year 2000 (assuming 100% market penetration of the most efficient end-use technologies) calculated savings anywhere between 24% and 44%. Another calculation for Europe forecast potential savings of between 15% and 30%[364]. While the less-developed nations have even further to catch up, expressed as energy efficiencies of their economies; on average they consume 40% more energy to create a dollar of wealth than developed nation's[365]. It is not untypical for conservation-savings calculations to be couched in an error safety bracket of 100% or more above the lower estimate. Such a margin of error, while prudent for energy policy formulators and commentators, does leave open a flank for some sharp comment within the energy debate. Some, associated with the nuclear industry have, for example, said that introducing savings into the future energy supply-demand debate presents 'conservation as equivalent to a new supply or fosters the illusion that energy use can be continued unchecked by making existing supplies go further', and that this is 'the comforting illusion of energy conservation[366].' While such an argument is technically true, such is the historic and current magnitude of waste in the energy supply/end use system, it is largely irrelevant.

In the 1970s and 1980s much of the energy debate centred around the pro–anti nuclear fission issues. It could be seen that when public opinion was hostile to any of the well organised energy players, rather predictable arguments were heard. For instance at the 1989 World Energy Congress in Montreal, it was concluded that:

> 'The worldwide problem of adverse public opinion on nuclear power needs to be addressed by all. The lack of understanding... as well as feeling there is no reason to take any additional risks... remains an obstruction to its orderly development'.

These risks were, and are, involuntary ones for which many see alternatives[367]. Nonetheless, under attack from such 'adverse public opinion' the nuclear industry understandably sees the need to defend itself. Where the technological imperative comes in is where this is done so enthusiastically that matters become distorted, and the discussion of new issues at the top of the energy debate agenda (such as global warming from 1989) become confused.

Examples of some nuclear-power advocates distortions were remarkable, indeed audacious. One lavishly illustrated booklet, *Understanding Electricity – the need for nuclear energy*, was produced by the (UK) Electricity Council in 1979 specifically for 'school and college use'. Not content with stating that 'nuclear energy... is cheap... Electricity produced in nuclear power stations costs less than electricity from power stations burning coal or oil', (the antithesis to the 1990s view) the same page had a diagram of a 10p piece with three segments labelled 'oil 1.31', 'coal 1.29', and 'nuclear 1.02' yet the segment for coal was over *twice* the size for that of nuclear[368]!

Promotion emerges out of the technological imperative as a consequence of wanting to further the use of the technology being promoted in society. Its function is to persuade of a need which may (or not) be there, but which is nonetheless overrated. There is therefore the danger that policy makers and their electorate not only may have the technology's long-term credibility placed in doubt, but the public may have their judgement side tracked. As the various energy technologies are inexorably linked to greenhouse issues, the perception of these may also become distorted.

'If the Earth is getting warmer, why is Minneapolis getting colder?' was one message presented to the public by the inappropriately named Information Council for the Environment, which is funded by a US consortium of electricity utilities[370]. Minneapolis *may* over recent decades had been getting cooler, as may some other regions of the Earth, but this does not contradict the IPCC's announcement of an overall warming Earth. Another, in 1991 (after the IPCC's confirmation of anthropogenic climate change), asks: 'How much are you willing to pay to solve a problem that may not exist?'

Erroneous information campaigns equally have been mounted for quite laudable purposes, including 'greenhouse friendly', energy efficient technologies. Again in 1991, but in the UK, the Department of Energy mounted an energy efficiency campaign under the banner, 'Global warming. We have been warned', accompanied by six photographs of the great storm of October 1987. In March 1992 the Department of Energy was reprimanded by the Advertising Standards Authority who ruled that the advertisements gave the impression that there was a causal connection between the 1987 storm and global warming, and that in doing so it failed to convey adequately that no scientific consensus existed on the matter. This type of misinformation represents another facet of the technological imperative – albeit a different technology. While the above nuclear examples presented that technology in a *favourable* light over its competitors, the DoE greenhouse poster takes the opposite approach of showing the fossil fuel energy generating technology in a *worse* light compared to the energy efficient end-use (consuming) technology. As for the success of such campaigns, it is difficult to come to a definitive conclusion: perception being such a multi-dimensional phenomenon. Even so, whether or not such campaigns do impact significantly on perception, what is certain is that the public in most developed Western nations feels less hostile to nuclear power than in the 1970s and early 1980s. In the USA (the World's greatest consumer of fossil fuels), between 1983/84 42%–46% opposed nuclear power with 48%–52% being in favour, by 1993/94 only 35% opposed with 54%–57% being in favour. Indeed the only time between the mid-1980s and mid-1990s when this trend in the US public's increased favour in nuclear power was a temporary reversal in 1986 and 1987 presumably as a result of the USSR Chernobyl incident[371]. In the UK part of the nuclear lobby has continued its campaign, and at times aggressively. Of its targets, the case for wave power (a potential clean competitor) in particular was severely undermined.

A consultative paper commissioned by the UK Energy Technology Support Unit in the 1980s reportedly calculated the cost of energy from a wave energy device, known as a Salter's Nodding Duck, at nearly twice what it should have been. Salter's Duck was one of eight short-listed out of some 300 ideas to generate electricity from waves. Each Duck would be about 38 m long, and some 400 (very roughly four miles long) strung together would generate around 1000 MW. However the report was still being referred to by policy-makers in 1990 when the European Commission dropped a £9.5 million research programme into wave power. Stephen Salter, the Duck's inventor, said 'I really can't help feeling that there was a conspiracy among a small number of people so that the facts and figures were distorted?'

All this is not to say that energy interests, be they nuclear, energy efficiency or whatever, are completely unethical: it only takes an over enthusiastic PR department, key managers, or 'innocently' misinformed personnel for the presentation of an unsubstantiated case (or subversion of counter cases) on behalf of an energy lobby. Most of those involved in the energy industries (and at the other end of the spectrum, the lay environmental movement) are sincere well-meaning individuals. However this does not stop misrepresentation from taking place.

At, arguably, one of its grossest levels, it is not that difficult to hire out the scientist with the views of your choice, such is the spectrum of opinion. The problem with unresolved scientific questions (such as the degree to which climate change is taking place) is that there will be those who believe it worse, or less serious, than others. Uncertainty provides a lucrative opportunity for a few outspoken proponents or sceptics who are willing to go beyond the usual caveats and cautions of accepted science practice, and into outright advocacy. Reportedly in the early 1990s the potential profits of public scepticism over the anthropogenic climate change was put at $10 000 a month for US researchers with contacts in the fossil fuel industry. While this figure was itself no doubt an exaggeration, scientists on either sides of the debate can collect reasonable fees (over $10 000 a year) for offering their versions of the truth – be it pro-fossil fuel or the so-called environmental (green) lobby. US climatologists have given a mixed response to such selling of services. Stephen Schneider (of the National Center for Atmospheric Research) says that both industry and green groups are simply ensuring that their side will be heard, and that 'money [does not make] these guys say what they do; they were saying the same thing before the money came along'. He is of the opinion that researchers sometimes have to become advocates, and if it means making some money on the side then so be it[372].

Not only do some US scientists (and indeed scientists from other nations) receive financial rewards from lobbyists, but they also are involved in lobbying in their own right, and spend some considerable sums in the process. In 1994 a US congressional committee was told that US universities spend some $60 million a year to Washington lobbyists, helping them to bypass the scientific peer review process and attract funding for research projects and facilities[373]. The motivations for such lobbying are obvious, but increasingly environmental issues are being sucked in to lobbying, both incidently and as the direct subject of the lobbyist's attention.

Because of the ready availability of threats arising from the future supply or use of energy – the lights going out *vs* climate change – energy policy is easily related to environmental policy. The political activities of the 'green lobby' rely on environmental threats as much as the energy industry does on threats to living standards (or even counter environmental threats) if this suits their commercial objectives. And so regulation and consumption changes, and switches between fuels, with serious consequences for major investments.

Considering the rise of a politically potent environmental lobby in many developed and less-developed nations is in part an expression of organised public opinion no longer satisfied with traditional policy, and short-term based decision-making systems, it might appear that the time has come when we should develop new systems rather than rely so extensively on existing technocratic evaluation and planning. Wide consultation with information seen to be obtained from independent (non-governmental and non-commercial) sources might help validate, in the public's eyes at least, long-term policy-making in areas such as anthropogenic climate change where an exact scientific perspective is difficult to obtain.

PERCEPTION BY PROXY

Phenomena that are elusive to direct experience on a day-to-day basis have to be learned from others. Anthropogenic climate change with its few degrees average global rise in temperature over a century or two is a case in point. For many, not being able to sharply experience climate change on an immediate and personal basis means that they have to trust their sources of information. Bad news if there are technological imperatives around, or that the issue is fraught with subtleties and complications. It is in just such an absence of direct experiential information that misinformation thrives (albeit from innocent cues, ignorance, or those actively seeking to impose their own views).

As a social sentient species, we humans share our perceptions with each other, and the way we share these – the words we use – tend to reveal something about the way we consider our views. Furthermore, as the advertising business knows full well, the words we use to convey information can themselves influence our own view of things. Various group perceptions have already been explored in previous sections, but all have language in common in the way these perceptions are mooted and shared. The day-to-day use of language can itself (subliminally) affect the issue being discussed. For anglophobes, discussing greenhouse issues two instances readily spring to mind.

The word 'greenhouse' itself has an innocuous connotation to the point of being mundane. A 'greenhouse' is where things grow, surely it cannot threaten you and your way of life. Greenhouses are harmless! Of course the situation is further complicated in that the greenhouse effect is fundamentally a natural phenomenon without which our planet (and other worlds) would have a far cooler surface. One has to identify, and then differentiate between, the natural and anthropogenic greenhouse effect. So the 'greenhouse effect' though one term, colloquially represents two concepts. The result is that the majority of the population, when asked to define the 'greenhouse effect' would restrict it to the anthropogenic warming due to carbon dioxide, and not recognise either the natural role it plays, or for that matter many gases, including water, contribute to the overall effect. It also means that the public are less likely to consider natural climatic changes when discussing what is the anthropogenic greenhouse.

Our regular use of the English language interestingly betrays the way we perceive our environment. Typical of this is the use (particularly in the UK) of the word 'earth' and the 'Earth' as a proper noun. Similarly the world 'world' means 'planet' or sphere of human interest, such as 'the world of stamp collecting. But 'World' – this is synonymous with 'Earth'. Both World and Earth are proper nouns denoting the special status we should afford them.

Leaving aside astronomers, most of whom seem to recognise a planet when they write it, look at common usage. The 1991 London launch of the IUCN, WWF, & UNEP World Conservation Strategy II report, *Caring for the Earth* presented press pack sheets which kept switching between 'Earth' and 'earth'. Language is complicated, dynamic and evolutionary. A living language should be expected to change, but when that change gets in the way of communication then surely the change should be challenged.

As we have seen, fundamental to World development and anthropogenic climatic change is human fertility. The larger the World population the more carbon dioxide is emitted. As with the confusion of 'Earth' and 'earth', the term 'fertility' is all too frequently used incorrectly. Many books on the current population problem on both sides of the Atlantic refer to how the developed nations are successfully lowering their fertility. This is news to me, and worrying. Am I really less able to have children than my father? Topical concerns such as falling sperm counts spring to mind and it really is all very worrying. In fact what is really happening is that the word 'fertile', which really relates to the '*potential* to reproduce', is used to relate to the 'reproduction *rate*'. This is quite nonsensical because it is surely obvious that someone who is fertile is capable of having offspring (whether or not this capability is realised). In fact there is a word that relates directly to the number of offspring an individual (or couple) give rise to, and that is 'fecundity'.

To sum up, the way we use words can (unintentionally) affect the perception of others. We need to take more care lest complex issues (such as global warming) get misrepresented.

MANIFESTATIONS OF PERCEPTIONS

The previous pages have summarised the principal factors involved in shaping the way issues of the day, with specific reference to threats to the global - commons, are perceived. How, then, are these perceptions manifested? What do individuals think?

Environmental perception, as with many of the greenhouse connected issues covered in this text, can be and is covered as the principal subject in its own right of other texts. Evidence has been presented that suggests that environmental concerns have only recently rivalled top issues in the public's mind in developed nations. However, I propose to disabuse the myth that somehow there was a huge wave of public concern over the environment in the late 1980s and early 1990s, so illustrating the triviality of media coverage and aspects of formulating the political agenda of the day. Turning to the USA, whose citizens generate the most carbon dioxide per capita in the World, we will look at the acceptability of a number of solutions to greenhouse issues by both informed and uninformed members of the public.

First, in the UK the government began a series of general UK environment reports. In 1992 *The UK Environment* included a substantial section on public attitudes[374]. It noted that in 1989 about 25% of those in England and Wales said that 'environment/pollution' was one of the most important problems the government should be dealing with. As such it was of similar concern to health and social services, but more than concern relating to pensions/social security, crime/law and order, inflation, housing/mortgage rates, and education, all of which rated less than 20%

Ministry-funded surveys, covering England, Wales and Scotland in 1989 and 1990, reveal that of the top environmental issues of the day, concern was greatest for pollution in the rivers and seas, sewage on the beaches and nuclear waste, with over 50% of those surveyed indicating that 'they were very worried about the problem'. The depletion of the stratospheric ozone layer and nuclear waste were very worrying problems to 42% to 55% of the population. Global warming at 41%–45% led the next cadre of issues along with concerns over the quality of drinking water, with acid rain and factory fumes in the 38% to 45% bracket. Wildlife protection and road traffic came next, with access to parks and derelict land only worrying less than 20% of the population. In short, at 41%–45%, global warming was just over half way up the top league of environmental issues that the UK public found very worrying.

The report took pains to point out that the public's citing 'pollution/environment' issues generally as one of the most important, or the most important concerns facing Britain, varied considerably with time fluctuating between 5 and 35% of the British public. Between December 1988 and the summer of 1992 there were two peaks of public interest in the summer of 1989 (over 35%) and 1990 (over 30%), with lesser peaks (at about 20%) in the summers of 1991 and the end of the survey run in 1992.

But do these polls paint an accurate picture? Not entirely. It is important to treat opinion polls with a certain amount of latitude. To begin with most national polls are accurate to about plus or minus 3%. Second, it all depends on the exact nature of the question asked.

In 1983, as part of the UK World Conservation Strategy, the national Conservation and Development Programme was launched. The World Wildlife Fund (now Worldwide Fund for Nature) commissioned a national poll. This revealed that 21% of those aged 15 or over would seriously consider switching their vote from a major party to another they would not normally support if it included in its manifesto a commitment to reduce wastage of natural resources. Eleven percent would seriously consider switching to another party should the commitment be one of further protecting wildlife and the environment. Armed with such hindsight, the six-years subsequent, 14.9% Green Europarliamentary vote in 1989 seems hardly surprising – certainly not news, but it made the headlines then. More recently, between 1990 and 1992 a series of UK national polls revealed that concern for the environment – as represented by that 'the government should give a higher priority to environmental policy even if this means higher prices for some goods' – was fairly constant (see Table 12.1). This despite a deepening economic recession over the same period[375].

Table 12.1 Responses to the proposition that 'the government should give a higher priority to environmental policy even if this means higher prices for some goods'[375]

	1990	*1991*	*1992*
Agree strongly	33	31	31
Agree slightly	36	38	34
Neither agree nor disagree	12	9	10
Disagree slightly	8	10	9
Disagree strongly	5	5	6
Don't know	6	6	9

So what is going on? It is difficult to say, but I suspect that what we are seeing is a broadly consistent level of concern with time. However depending on how the various poll questions are actually phrased, what other concerns (that do vary greatly) currently compete with the poll's subject for the environment, the media agenda, and indeed all the perceptual factors, social traps etc, that operate at the time, we end up with an illusion of widely varying public concern for the environment with time. This is not to say that concern for the environment is static, just that it does not vary as greatly as is made out by media fashions. Given this, it would seem logical to build on this core of public backing for the environment and to try to stimulate those less concerned about environmental issues. But if we did this would we have success? In America some interesting social research indicates that we might.

The USA is just the place to address this success in perceptual change question. Solve the problem of getting greenhouse issues understood here and there is a good chance of beginning to tackle seriously the problem of greenhouse gas emissions. In 1995 Doble published the results of a study[376] about the way public opinion related to issues characterised by technical difficulty and scientific uncertainty. He chose two issues: the disposal of solid waste and global warming. Four hundred people were chosen to be representative of the US population but were in actual fact from four urban areas: Chicago, Hartford, Los Angeles and Nashville. Each filled in a pre-test questionnaire that asked them to rate the urgency of the two environmental issues among a list of national problems and then to evaluate a variety of possible actions to deal with each issue. They then watched an informative 15 minute video on one issue, broke for discussion before watching a second video on the other issue and have a second discussion. Finally they filled out a post-test questionnaire. Doble then mailed the questionnaire to 418 of the country's leading scientists (not necessarily experts in the two environmental subjects) and found that the responses from the subject experts and the other scientists were nearly identical. The importance of Doble's research is that it addressed the critical question of whether the public could be meaningfully included in science issues of complexity and uncertainty, or whether such issues should be removed from the public agenda and left for elites to address? What Doble found was that the public panel, whose pre-test questionnaires were markedly different from the scientists' perceptions, did become very much closer. Doble summed up the effect of showing the informative video and holding the discussion saying that: 'After the educational intervention and a chance to deliberate, there was a shift in people's thinking about both [global warming and solid waste] issues, especially global warming'. Furthermore, 'in the post-test, the panel laid out clear guidelines about what was, and was not acceptable in terms of policy'. In the post-test the greenhouse proposals that proved acceptable to over 65% (and up to 75% the greatest degree of support that any proposal received) were:

- Tax breaks for home-owners who improve home insulation *even if* that means a tax rise of $50 per taxpayer per year. A move in support from the pre-test proportion of the public panel of +4% to 65% at post-test.
- Requiring business to improve fuel efficiency *even if* that means higher prices. A move of +11% to 70% at post-test.
- Increasing mpg standard to 40 by the year 2000 *even if* that would increase car prices and decrease performance. An increase of +20% to 75% at post-test.
- The USA should lead international efforts to address global warming *even if* that means they must sacrifice more than others. A move of +15% to 70% at post-test.
- Requiring communities to plant new trees *even if* that means a tax-rise of $50 per year per tax payer. A move to 68% post-test, an increase of 13% from pre-test.
- Reducing the number of trees that are cut *even if* that means higher paper prices and some logging companies go out of business. An increase of +21% to 69% at post-test.
- Increasing funding for solar energy research *even if* that means a tax rise of $50 per tax payer per year. An increase of +24% to 67% at post-test.

On the other hand, proposals that the panel found unacceptable, as represented by those proposals that only had the support of 30% or less of the panel at post-test, included:

- Rationing gasoline (alternate day buying) *even if* this would cause hardship, especially for those who need their car for work. Which had only the support of 7% of the panel at post-test, up +1% from 6% at pre-test. This was the least liked of proposals.
- Discouraging two car families by taxing a second car *even if* this would be a hardship for many people. This had the support of just 11% at post-test but still an increase of +3% over pre-test proportion.
- Raising the gasoline tax by 25 cents a gallon *even if* this would burden truckers and others who need their cars for work. This also had support from 11% of the panel at post-test, but up only +2% over the pre-test proportion.
- Banning cars that have low gas (petrol) mileage *even if* that means most sports cars and station wagons. This received just a quarter of the panel's support, 25% both at pre and post-test.
- While making exceptions for the sick and elderly, requiring people to keep their thermostats no higher than 65°F *even if* this would be hard to support. A measure which, though only approved of by a minority, received a substantial 30% of the panel members support – up +7%.
- Raising gasoline tax by 10 cents a gallon for each of the next ten years (to $1 in total) *even if* this would seriously burden truckers and others

> who need their cars for work. This received just 11% of the panel's support but interestingly this was 1% *less* than the level of support at pre-test.

Interestingly, other than the 25 cent extra gasoline tax all the above proposals, to which the public panel did not give substantial support, also failed to secure much support from the scientists questioned. Where the scientists and the panel parted company was in the two proposals acceptable only to the panel: increasing mpg to 50 by the year 2000, and impose a strictly enforced 55 mph speed limit. Conversely. acceptable to the scientists, but not the panel, were: imposing a tax on 'gas guzzlers'; building more nuclear plants; increase aid to Brazil in return for forest protection; and raising the gasoline tax by 25 cents.

With regard to measures welcomed by scientists, and supported by the panel, all those listed above supported by over 65% of the panel were also supported by between 57% and 79% (i.e. over half for each measure) of the scientists questioned.

Though Doble's work does indicate that the public can grasp complicated issues that involve uncertainty, it also reveals something about the overall human (or North American) reaction to global warming. Both the scientists and the lay panel were more reluctant to pay for direct taxes, such as: imposition of a 10 cents a gallon tax on petrol (gas) rising to $1 after ten years; taxing home owners' heating oil; taxing two car families. However, both scientists and panel accepted tax breaks for home insulation, tax breaks for industry improving energy efficiency, government funding for solar energy, and tax breaks to builders using passive solar power. After a bit of thought you will realise that this is rather strange in that giving tax breaks, or increasing government spending does mean that extra tax revenue has to be raised! Could it be that the scientists (who were not necessarily experts in energy, climate, or policy) were behaving similarly to the public in wanting carrots (rewards) and not sticks (penalties)? Equally, what this means is that the climatic environmental externalities of - consuming fossil fuel would not necessarily be paid for by those doing the actual consuming. Whether or not this is right, proper and fair is down to society, though it is worth noting that it does depart somewhat from the 'polluter must pay' principle that lies at the heart of much developed nation late-twentieth century environmental policy. Equally, turning this all on its head, it does show that people generally do not like 'sticks', preferring 'carrots', and that as a policy tool the motivation is there for sticks to work. A logical conclusion would be to have a policy in which the sticks paid for the carrots, though there are many other options.

Given that it is possible for the lay public (in at least one developed nation) to appreciate global warming issues, all the perceptual barriers, social traps and the like, described in this section still need to be overcome. In addition to which policy makers need to pull together all the strands connected with climate change issues. Palaeoclimatological, ecological, geophysical, developmental, economic and social perspectives therefore need to be put into focus; a difficult but not an impossible task.

CHAPTER 13

FUTURE CLIMATIC AND HUMAN CHANGE

The relationship between human civilisations and climate is fundamental, so fundamental that it can make or break a culture, just as climatic change can alter entire landscapes. Modern Man evolved out of East Africa between two million and five million years ago. Some 12 000 years ago, as the World left the last glacial and entered the current (Holocene) interglacial, East Africa saw climatic changes as significant to life there as anywhere else. During the previous glacial–interglacial cycles the giant buffalo, *Pelorovis antiquus*, and a number of arid-adapted species, particularly an impala-sized antelope, were common[377]. Given that there were over a dozen previous glacial–interglacial cycles it seems unlikely that the termination of the last glacial in itself would be as damaging as it was. More likely something else was happening, either the current glacial–interglacial cycle is deeper than previous ones, or the additive effect of a newly evolved intelligent hunter, *Homo sapiens* (as opposed to the other species of early Man), resulted in extra pressure on the region's ecology, or both. Whatever, the significance of climate, and the human-climate relationship remains.

PAST HUMAN AND CLIMATIC CHANGE

Even within the current (comparatively stable) Holocene interglacial, climatic changes have been so great as to intensely stress human culture. These have either adapted or perished. Evidence of one of the earliest known instances of human response to climate change can be found in Texas, USA where wells have been found dug by the Altithermal people to combat a drought that began 7500 years ago and lasted (according to analysis of a core from the North American Quelccaya glacier) some 2500 years. The drought was of such severity that it is thought that the landscape turned to dust and access to groundwater must have provided a very real lifeline.

Around 2200 BC (4200 years ago) climate change was probably the key factor that resulted in the fall and disappearance of the Akkadian Empire which centred on an 800 mile stretch of the Euphrates River. This happened around the same time as the Bronze Age ended and the World became much drier, however, this last might be coincidence. What is known is that after about 2225 BC the pharaohs of the Old Kingdom lost their grip on Egypt and both palaeobiological and cultural evidence suggests that this might well be linked to climate change; though whether this link is participatory or causal remains debated[378].

Another dry period is known to have affected the Yuccatan Peninsula in Meso-America, especially between 150 BC and AD 250[378]. However, sediment cores from nearby Lake Chichancanab, Mexico reveal that AD 800–1000 was the driest period of the middle to late Holocene. This period coincided with the collapse of the Mayan civilisation[379] sometime between AD 750 and 900, although it was severely climatically stressed beforehand. Many Mayan cities declined or were abandoned in the AD 150–250 period. Some, such as Coba, Yaxuna, and Oxkintok persisted despite being located in the most arid parts of the Yuccatan. That these sites continued to thrive while others failed is quite telling about the relationship between human civilisations and climate change. Faced with the onset of climatic change, effort can either be put into adaptation, such as the construction of irrigation systems, or there can be a continuation of business as usual[378].

Of course while such incidents are illustrative of the way human societies and climate can interact and change, much of our historical knowledge of climate change lacks the detail of how this change affected human beings. For instance, in the first century AD Flavius Josephus, a Jewish historian, noted that when King Herod built the mountain fortress at Masada by the Dead Sea he reserved the top of the hill for cultivation. Today the area is quite arid, so clearly the climate has markedly changed between now and then, though the *exact* human consequences as far as Masada goes are not known. Even so, there are a number of examples, like the Mayan collapse, that do illustrate the importance of climate to human societies and we ignore this relationship at our own peril. Another telling example is that of the Norse farmers of Greenland. In early medieval times their farms were fairly extensive and their living good enough for their settlements to sport solidly built churches. But by AD 1500 the Norse farmers were gone and Greenland was left to the Inuit seal hunters. The message here is again one of adaptation and failure to adapt. The Norse farmers could have taken up fishing and seal hunting (i.e. they could have adapted), but they did not. On the other hand the Inuits survived.

If all this seems far removed from Western culture, it is important to remember that Western Europe and North America have both had their share of climatic changes influencing history and culture. The Medieval Warm period, with temperatures half a degree or more warmer than today's, saw vineyards in northern Britain and the Vikings colonize Greenland. This warm period helped spread vermin species that led to one of mankind's most horrific periods in history: the wave of Bubonic Plague, or Black Death. This plague spread throughout the known world lowering the global population by an estimated third. However, there were other climate-related factors, indeed other factors, the Medieval Warm Period in the temperate zone increased agriculture, and throughout this period Euro–Asian trade, conquest and colonization increased in intensity. It was therefore only a matter of time before such activities, enabled organisms to be transported between biogeographic zones, and bubonic plague (or something like it) was almost an inevitable result. The organism concerned was *Pasteurella pestis*, a bacterium transmitted by fleas on black rats, *Rattus rattus*, that were introduced into Western Europe through trade caravans returning from China.

Yet just as climate change may have assisted the spread of the Black Death, so it helped eliminate it. From 1500 the cooler climate that had wiped out the Greenland Norse farmers provided harsher winters that lowered the population of vermin and so helped bring the Black Death to an end after London's Great Plague in 1665. Indeed, this cool period troughed (1°C to 2°C cooler than the peak of the Medieval Warm Period) around the sixteenth century, a time known as 'The Little Ice Age' (though a more accurate a title would be 'The Little Glacial'). It was a time that saw huge cuts in the human population, without a corresponding decline in infrastructure (houses, roads, ports and such) and allowed some countries to re-organise their economy favourably. Britain was one which, aided by the land closure legislation, saw a growth in its urban population at the expense of its rural peoples. A large work force was therefore on-hand for when the industrial revolution took off. This tale is grossly simplified, but shows that climate change may have been a causal factor in bringing about the wave of plague, a key factor without which history might have been very different.

The climate warmed in the eighteenth century, although half a degree cooler than today, until the mid-nineteenth century when warming resumed. It was from this time that significant quantities of carbon dioxide began to be anthropogenically released and began to affect the climate. It also was the beginning of the end of the traditional Western European white Christmas; itself a theme arising out of the earlier Little Ice Age when each year northern European rivers froze, even those benefiting from the warm North Atlantic drift, such as the Thames.

Today, as discussed earlier, we are undergoing another period of climate change. If the IPCC 1990[2] Business-as-Usual scenario is right then, before the middle of the next century, the Earth's climate will have warmed by about a degree Celsius, returning to that of the Medieval Warm Period. Shortly after the middle of the next century the Earth's climate will be as warm if not warmer than the current interglacial (Holocene) maximum that began some 8000 years ago and ended 4000 years ago. As the Earth further warms, it will be experiencing climates not seen since the height of the last interglacial some 150 000 years ago. Right now, therefore, we should not be surprised to be experiencing occasional extreme weather conditions; extreme, that is, in terms of modern human experience. Examples of extremes range from the UK hurricane of 1987, the droughts in California, UK and Australia, to the 1994 record-breaking warmest November for the UK since records began in 1659, and a succession of warmest years on record for global average temperatures from the mid-1980s onward. Some of these extremes are dramatic, others less so. The coolest May since records began was experienced in the UK in 1996, and the warmest August in 1997, but these cannot be *proven* to be related to climatic change. Furthermore, cool examples are mild and not unexpected in a temperate zone given with overall warmer World with higher evaporation (and hence prospect for cloud and regional cooling). However when weather extremes occur with sufficient and discernibly increasing regularity they can begin to be linked with climate change. One such weather-related phenomenon is the flooding of the Mississippi tributaries in North-central America in 1993, causing many millions of dollars worth of damage and disruption to the regional economy. The area itself is geographically noted for two things, first the large catchment, and secondly that it lies at the border of two climatic zones. Ironically, also in 1993, James Knox presented a sediment analysis that covered some 7000 years and showed that some 3300 years ago floods were quite common. Still larger floods occurred between about AD 1250 and AD 1450 as the climate slipped from the height of the Medieval Warm Period into the Little Ice Age. He found that throughout this 7000 year geological record large Mississippi floods were associated with changes in climate of only about 1–2°C and changes in mean precipitation of 10–20%[380]. Similarly between AD 99 and 900 in northern China, storms and large floods were rare and seemed to be negatively correlated with temperature.

As shown, there are ample examples of indicators of current climate change, and impacts of climate change on human activity. But are changes in human activity the only effects climate change has on our species? The health implications of climate change were discussed in Chapter 11, but there are also other factors of greater significance. It has long been known that environmental change (not just climate change) is important for speciation (creation of new species) such as a mountain range rising or continental land bridge flooding, so dividing populations. Evolutionary pressures (perhaps spurred by genetic drift) then act on the two populations creating new species, with the original species being a common ancestor. Another instance is the use of antibiotics and pesticides (chemical change to the environment) leading to the evolution of drug and pesticide resistant strains. Climate change too can lead to speciation: the creation, or flooding, of land bridges due to water being locked up (or released from) ice caps may force some members of a species to adapt to the new conditions, while others migrate or die out; either way, separate genetic pools within the population are created and diverge, finally to such an extent that the sub-populations become species[381]. Is this what happened with the evolution of *Homo sapiens* in Africa?

Peter deMenocal (Earth Observatory USA) examined palaeo-records from marine sediment cores from seven sites off the coast of East and West Africa.

He also looked at a range of climatic indicators so as to build up a detailed picture of how Africa's climate north of the equator changed with time. He then compared this record with that for hominid evolution and found that there was, not just one but, a series of significant developments in human (and other vertebrate) evolution that coincided with changes in climate. With regard to African climate, he noticed that over the past 4 million years there were roughly three step-like increases in African aridity and in the variability of the African climate at about 2.8, 1.7 and 1.0 million years ago. With regard to climatic variability, before 2.8 million years ago the African climate (not *necessarily* the global climate) varied over a cycle of 19–23 ky (ky = 1000 years). After 2.8 million years ago a 41 ky cycle dominated, while after 1.0 million years ago the 100 ky cycle dominated (that broadly corresponds to the current glacial–interglacial cycle). These three kiloyear cycles may be linked to the various periods of the three Milankovich orbital parameters. It is also known that shortly after these three periods African ecology changed markedly. With regard to human (or to be exact 'hominid') evolution, between 3.0 and 2.5 million years ago (i.e. about 2.8 million years ago) two distinct lineages emerged from *Australopithecus*; one became the *Paranthropus* genera, and the other the *Homo* lineage. Between 1.8 and 1.6 million years (i.e. about 1.7 million years ago) ago, the climatic records indicate increased regional aridity that coincided with the extinction of *Homo habilis* and the rise of *Homo erectus* (thought to be our direct ancestor). It also saw the final extinction (by 1.4 million years ago) of all species of the *Paranthropus* lineage. By 1.0 million years ago *Homo erectus* had expanded its range beyond Africa into Europe and Asia. In short, the picture that seems to be emerging is that climate change has played a key role in the evolution of mankind[381]. It certainly seems strange, given that evolution has repeatedly given rise to key developments (such as limbs or eyes on four or five independent occasions) that intelligence capable of generating technology (which has obvious evolutionary advantages) has taken so long to manifest itself (within 0.1% of the time life has been on the planet). Could it be that while climate change has taken place in the past, that this Quaternary period of time (the past two to three million years) is unique in that it has been a sustained period of climate variability?

The fat hairy people of Pompeii may well provide an insight as to how a - sustained period of cyclical climatic change provides the necessary evolutionary hoops leading to the rise of intelligence. According to archaeologist and anthropologist, Estelle Lazer, a substantial but significant minority of those who died when Vesuvius erupted in AD 79 were on the hairy side, suffered headaches and had a form of diabetes[382]. Possibly about 10% of the city's women suffered from these symptoms because they had the hormonal disorder, hyperostosis frontalis interna (HFI). Today HFI occurs mostly in post-menopausal women which indicates that the people who died at Pompeii lived well into their 50s and 60s. Other features of the 300 skeletons she examined indicate that the people who perished might be closely related, for instance, their tooth roots were unusual. This finding challenges the common view that Pompeii was a cosmopolitan city and is supported by that of the vulcanologist Haraldur Sigurdsson who hypothesised that the area around Vesuvius had been subject to the occasional earthquake for a couple of decades before the cataclysmic eruption. Even when the fatal eruption took place, the volcano spewed ash and pumice for 18 h beforehand, and this obviously did not scare those who remained in Pompeii. Lazer is of the opinion that the wealthier residents probably left the city years earlier leaving behind the poor and those with nowhere to go. Further, from an analysis of 30 houses there is clear evidence of squatters, and that when the final eruption arrived Pompeii's population may have been quite small[382]. Could this be analogous to how intelligence arose? Could the brightest of the tribes of the early species of man have adapted to the ever-changing (in evolutionary timescales) climate, whereas those not so endowed were at a comparative disadvantage every time the climate altered sufficiently to warrant a behavioural change (be it migration, developing new hunting skills in a new ecology, or whatever)?

Clearly climate change is of intrinsic importance to our species both on a mundane day-to-day basis, at the level of entire settlements and societies, and, even possibly of evolutionary consequence. Equally clearly climate change today is of at least as great an importance to our twentieth century global culture as at any time in the past. This last is a result of demographics if nothing else. Today there are about 6 billion people on the planet and nearly all the large areas of potentially fertile agricultural land are already under the plough, levels of marine fish catches worldwide are stagnant, if not declining, so indicating over-fishing and the number of megacities is increasing. So, faced with climate change, what will we do if our agricultural systems falter while we adapt? Where will the people displaced, for example, by desert expansion or sea level rise, go? A third of Bangladesh is below the 5 m contour, one sixth of the land area is less than a metre above sea level and has a population of a million people. The latter is likely to be flooded by the rising seas over the next century[20] while the former will not escape unaffected; water tables may become saline. And not withstanding all of this, the world of the twenty-first century will have more mouths to feed even without the added burden of climate change. Some of the currently wealthier countries may, with their cars and washing machines, also face unpleasant surprises. The IPCC, in 1990, stated that it was 'certain' that 'emissions resulting from human activities are substantially increasing the atmospheric concentration of greenhouse gases'[20]. Further, it calculated with 'confidence' that the 'relative effectiveness of greenhouse gases can be estimated'. What it did *not* say in 1990 was that it was 'certain' that greenhouse gases from human activities were causing the planet to warm. So what have we learned in the seven years since 1990? Further, if there are still uncertainties, do we have enough information to prescribe with surety a course of action?

DETERMINING SHORT- TO MEDIUM-TERM CLIMATIC CHANGE

Of all the uncertainties of climate change, arguably those that need the most urgent clarification relate to the short to medium-term future of a few years to decades. If investment is to be made now to meet the challenges of climate change then those making the investment are going to want to know what they will be getting for their money. Furthermore, there is little point addressing longer-term concerns if those in the more immediate future still have considerable uncertainty.

Looking at the IPCC 1990 and later projections, you might be forgiven for thinking that the Earth would slowly but steadily get warmer in a linear way. The IPCC projections are, after all, nearly straight line graphs (see Figure 5.5). Yet the climate does not behave like this, it is more chaotic with numerous attractors, the greenhouse gases being only part of the story. Of course the complexity of climate (change), even over the short term can be easily seen by looking at a graph of global surface temperatures over the past few decades (see Figures 3.7 and 3.8). In addition to the anthropogenic and natural greenhouse gases affecting climate these are: albedo (ground, sea, cloud surface-reflective) effects, solar cycles with variations in the Sun's output, ocean current flips such as the El Niño (and possible changes in the Broecker salt conveyor), and even cataclysmic events. Major volcanic eruptions have been implicated in some of the greatest instances of extinctions in our planet's history, including that of the dinosaurs. This may come as a surprise as the asteroid impact theory put forward by the Alveraz brothers has attracted considerable credibility, but there is some fairly substantial evidence to suggest that prolonged volcanic activity 'softened up' the dinosaurs; indeed the K/T (dinosaur) extinction did not just happen overnight[383]. If this is the case, then for the volcanic activity to result in global extinctions its effects could not be put down to the physicality of the immediate eruptions themselves, rather some other planet-wide factor: the answer could well be global climate change. It is known that large volcanic eruptions can inject sulphate into the stratosphere and that this reflects the Sun's heat back out into space, rather than warming the lower atmosphere, and so cools the climate. This cooling effect has been known since the time of Benjamin Franklin when the eruption at Tambora in 1815 was followed by the 'year without summer'. More recently, evidence has come to light that during the last glacial period 73 500 years ago the World was further cooled by a massive volcanic eruption at Toba, Sumatra, in the Indian Ocean[384]. The eruption was so large that it is thought to have injected about a billion tonnes of ash and sulphur gases, 27–37 km into the atmosphere. This matter would have shielded the lower atmosphere and ground from the Sun resulting in a climatic cooling of about 3–5°C. It has also been suggested that the sea level changes, or the change in weight of sea water on magma chambers (which decreases during glacials when water is locked up in ice caps) might trigger eruptions. Super eruptions are fairly rare, but even smaller more common eruptions, such as that of Agung (1963), El Chichón (1982) and Mount Pinatubo (1991), can cause cooling. These three each resulted in temporary cooling of about 0.4°C (when the effects of other factors such as *El Niño* current swings are accounted for).

This cooling effect is not new, and in the 1970s climatologists were aware of the potential that sulphates and aerosols have for cooling, but thought these outweighed greenhouse considerations. They should not be blamed for getting the balance wrong, anymore than climatologists at the end of the twentieth century should be blamed for not having as perfect an understanding of the climate as their successors in the twenty-first century may have. Nor, just because aerosols and sulphates cool the climate, should we (on an increasingly greenhouse planet) be reassured. After a sulphate-rich, stratospheric-piercing, volcanic eruption, the Sun's energy is prevented from reaching the surface by sulphates so the energy content of the upper atmosphere increases at the expense of the lower. On the other hand currently, due to human emissions, greenhouse gases prevent energy leaving the lower atmosphere. So, overall, the energy within the atmosphere is increasing, even if the temperature at the surface hardly changes.

Clearly this situation is far from satisfactory, but it is only now that the magnitude of greenhouse warming and aerosol and sulphate cooling are being estimated with any useful accuracy. Sulphate aerosols can act as nuclei on which cloud droplets condense and, for a fixed availability of water, the average cloud droplet is smaller and smaller droplets are more reflective. The net consequence is that, with sulphate aerosols, clouds become more reflective. This is known as the *indirect* aerosol effect and it is hard to quantify.

The bottom line is that, though molecule-for-molecule the warming effects of the principal greenhouse gases are greater than the cooling effects of sulphate, their effects in space and time are different. Greenhouse gases have an atmospheric lifetime of many years, whereas the lifetime of tropospheric (low atmosphere) aerosols is only of a few weeks. (Unlike the stratospheric (high atmosphere) aerosols which require a volcanic explosion to carry them into the jet stream above the lower weather clouds that would wash them out.) Therefore aerosols generated by human activity have their greatest effect near that activity and the result is that the anthropogenic aerosols cool mainly in the northern hemisphere, over Europe and Western Russia as well as the Eastern part of the USA and parts of Asia. The scale of this effect is broadly to be able to *directly* cool (reduce the solar energy input to the lower atmosphere (troposphere) by about -0.3 to -0.9 Wm^{-2} and *indirectly* by as much as 1.3 Wm^{-2} [385, 386]. These figures do not mean much by themselves but when one considers that the (long wavelength forcing) anthropogenic greenhouse gas effect which is estimated as being between 2.0 and 2.5 Wm^{-2}. Taking these aerosol (and factors such as

Pacific *El Niño* current) effects into account have improved modelling studies and were used by the IPCC in 1995 to slightly modify its 1992 forecasts. One way to see how good a model is, is to run it backwards in time, as well as forwards into the future and climate modellers at the Hadley Centre in the UK have done just this, with quite impressive results. By 1995, their models were agreeing with what actually happened historically to within a quarter of a degree Celsius. Indeed, in sophisticated models from 1995, the range of variability of the actual year-on-year weather records and that of the models overlapped[387]. These models are still improving through either the inclusion of additional factors affecting climate and a better understanding of how known factors impact on climate. For instance in 1996 it was discovered that mineral dust from North Africa has a significant climatic impact and that about half the dust from that part of the continent was due to soils being disturbed for agriculture in arid and sub-arid regions[388].

Given this gradual but steady improvement in understanding of how the global climatic system works, it would appear to be only a matter of time before the caution of the IPCC in 1990 was replaced with greater certainty. In fact the landmark statement came at the end of 1995 and the US journal *Science* Richard Kerr superbly caught the mood on September 22nd with the headline, 'Scientists see greenhouse, semi-officially'. It went on to say, 'The greenhouse warming is now official – at least that was the unofficial word last week[389]. A draft IPCC report had been circulated that suggested that it was unlikely that the global warming of the twentieth century was entirely due to natural causes. Two months later, the IPCC concluded in an *approved* report that 'the balance of evidence suggests that there is a discernible human influence on global climate'[390]. From here on in the priorities and goals in determining the nature of short- to medium-term climatic change shifted, from whether or not human action was affecting the climate, to exactly how these effects would manifest themselves.

DO WE NEED TO PAY FOR CLIMATE CHANGE?

Given that there is so much evidence that human activity *is* affecting climate change the key question now facing policy makers whether this human-induced climatic effect is large enough to worry about? In other words are we going to have to spend money on climate change policies? Of course the ways of answering the question 'do we need to pay for climate change?' depends on what is meant by 'we,' 'need' and 'pay for'.

From the politicians' view the way we address (or fail to address) environmental issues significantly depends on the electorate, which in turn is affected by the relative urgency of competing issues. The average man in the street, be he from the affluent parts of New York or poorer parts of Rio, is less likely to be concerned about remote issues than immediate matters. So the news that analyses of 50-year meteorological records had revealed warming on the Antarctic Peninsula, or that the five outermost Antarctic Peninsula ice shelves have retreated dramatically over the same time period (which appears to be highly indicative of climate change[391]), is going to be of less concern than something more immediate, e.g. a relative or friend dying as a result of a record-breaking heatwave. This is a serious point, as, in mid-July 1995, over 400 people died in Chicago as a result of a severe heatwave[392]. Such local *versus* global perceptual differences have been quantified. A Gallup *Health of the Planet* survey in 1992 showed that between 11% and 29% more people in developing countries, as compared to industrial nations, viewed the local state of poor air quality and inadequate sanitation as 'very serious'. Conversely global problems, such as global warming, were viewed with more seriousness in developed nations[393]. Similarly, just as there is an economic difference *between* rich and poor nations, there are also differences in affluence *within* them. If political leaders truly care about issues that, though major, will not have an impact until many years from now, or in places far away, then they will take steps to ensure that their electorate is sufficiently aware to make an informed decision. Politicians rarely truly encourage an informed electorate, and it can lead them and their nation into a social trap. Remember Galbraith's 'culture of contentment'[346] discussed earlier where provided a politician keeps 51% of the population happy then it does not matter about the remaining 49% was of great relevance in the early 1990s in US politics when politicians had to grapple with reducing a massive federal budget deficit. The poorest 40% of the USA population account for 20% of the vote and the richest 40% account for 56%, so it was difficult to get the rich to vote for more taxes – the sufficient numbers of 'altruistic' voters simply did not exist. So the deficit remained and further pressure was put on the poor as social spending was cut. Given this, what chance then for seemingly remote and long-term global warming issues?

The answer to the question, 'do *we* need to pay for climate change?' must include the point that the '*we*' that is detrimentally affected by climate change would like their climate costs to be paid for, but there is nothing to absolutely compel fossil fuel consumers to meet climate costs; they do not 'need' to make climate cost payments per se. Of course most climate change costs in the short term are negligible for the majority of the World's population. Only through informing those that have to fund climatic measures of the scientific evidence of climatic impact manifestations can politicians hope to win the electorate's confidence for greenhouse policies and the payment of climate costs: hence the need to break through the perceptual barriers.

The answer to the question 'do we need to pay for climate change costs?' might also focus on the 'pay for' dimension. One can pay for climate change in three ways. Strictly speaking, externalities (unpaid pollution costs) are paid, but not by the polluter in business-as-usual. For instance large drought, such as that which affected Texas and its neighbouring states in 1996, did have costs in terms of loss of crops and related economic activity. The costs were of the order of $1 billion so that, if the drought was a manifestation of climate change, then this loss could be considered a climate change payment, but not one paid by fossil fuel users but by farmers and their economic dependants (and/or insurers). The other ways of paying for climate change relate to payment for the implementation of a climate change policy: either by paying for adaptation to a new climate (for example in costs for the introduction and marketing of drought-resistant crops), or by paying to prevent climate change (for example, reducing greenhouse emissions). There is also the option of a mix of paying for some adaptation to climate and some for the amelioration of climate change.

Finally, given that the question 'do we need to pay for climatic change?' depends on the meaning of 'we', 'need' and 'pay for', one could further distil the question to its bare essentials. At this fundamental level the answer depends on whether there are net benefits to funding a greenhouse policy or to simply allow market forces to operate (assuming perfect customer information).

COSTING CLIMATE COSTS

We have already looked at the various types of cost likely to be incurred as climate changes under the IPCC 1990 Business-as-usual scenario in Chapter 6 and we will continue to use this scenario (B-a-U) for reference as the subsequent IPCC 1992 refinement falls within the IPCC 1990 high and low B-a-U estimates. Furthermore, discussions within the IPCC, prior to its 1995, assessment reflected a wide range of warming estimates and only a slight cooling modification due to regional short-term sulphur air pollution. With regard to the big picture, it does not matter whether warming will be 2.4°C (IPCC, 1992) or 3.3°C (IPCC, 1990) over the twenty-first century. Either way the question of how we cost climate costs remains.

In terms of a proactive (as opposed to a passive) response we can either spend resources to adapt economies to a new climatic regimen or spend to reduce greenhouse emissions by developing renewable resources, improve energy efficiencies, etc. The decision depends on what costs the least and on what is practically possible. If we are to rely totally on amelioration then the level of human-generated greenhouse emissions needs to be reduced by some 60% on their 1990 levels, if continued growth in the atmosphere's concentration of greenhouse gases is to be stopped. Given that World population is continuing to grow, and given that there are developing nation pressures to increase per capita energy consumption, it is unlikely that a reduction in emissions of 60% is achievable. At the very least, some of any expenditure resulting from implementing a greenhouse policy will need to be spent on adaptation and not just amelioration.

Estimates have been made, both as to the cost of individual, as well as the total global, climatic impacts (see Chapters 4, 6 and 11). Some individual costs can be fairly well determined, others, especially the more global ones cannot. Take a maize field. It produces x tonnes of corn per year with an economic value of \$$y$. With climate change and its associated drought, sea flooding, whatever, the value of the crop is wiped out. So how much money should the polluter pay (assuming internalised costs) either to compensate the farmer or to persuade him or her undertake a less climate-sensitive business, relocate, or whatever? Unfortunately x and y are not fixed, they are variables that are dependent on a host of factors. While it is possible to say that the value of a specific crop in a specific place at a specific time in the past was a certain amount, it is not possible to be so exact without all that information, or to predict the future. If nothing else, the problem is that goods and services are valued very differently in different parts of the World. Witness the variations in a price of a loaf of bread in northeast rural India compared to rural Eastern Europe, compared to suburbia in southeast England or Washington, DC.

It was this problem that the IPCC came up against in 1995, what is the value of a human life? They needed to know because disrupted, or likely disruption of livelihood through climate change, or death by heat wave, drowning, starvation, whatever, would require compensation or cost to prevent this loss of life/livelihood. The answer was, 'it depends'. Their economists concluded that a human being was worth between \$100 000 and \$1 500 000, depending on where they lived. Those in the wealthiest countries were valued at about \$1.5 million, while those from poor nations were thought to be worth as little as one fifteenth that sum. Clearly if the value of human life was reduced in the Third World (which is relatively climate-sensitive) compared to the industrial nations then the policy odds are stacked in the industrial world's favour.

Their economists also noted that the IPCC scenario projections assumed that the less-developed nations would develop in time, so contributing more greenhouse gases. If this was so then the value of life in the less-developed countries would rise. Such gyrations in formulating internationally agreed scientific greenhouse perspectives were common in the summer of 1995, notwithstanding that many developing-nation representatives at the time viewed this economic analysis as 'discriminatory'.

If it is so difficult to ascertain greenhouse costs, what of the costs associated with business-as-usual? It might be that by turning the greenhouse climate change question on its head and looking at the costs of climate policies in the absence of climate change, we can obtain a better insight into the scope and nature of the problem. The question is, what would the costs be of imposing greenhouse policies in a climatically stable world?

GREENHOUSE POLICY COSTS IN A GREENHOUSE STABLE WORLD

Suppose that either the climate is not changing, or that changes are either part of the natural variation in weather or that, if changes are taking place, their cause is due to some agent other than human activity (e.g. solar variability). Now ask the question, what are the costs of imposing greenhouse measures?

Greenhouse measures fall into two energy-related categories and a carbon cycle, afforestation category. The energy-related categories consist of reducing fossil fuel consumption and increasing non-carbon fuel consumption. Related to all three is the increase in human population and one greenhouse measure might be to reduce or lower population growth. Clearly this has a cost, but the Earth cannot sustainably support more than a population which most demographers put at somewhere between 8 billion and 15 billion[342], a level of population that we are likely to reach by the end of the twenty-first Century. We should, therefore, halt, if not reverse, our current trend of population increase. There are signs of this and we are spending some money on it. There is doubt such aid expenditure will increase.

Another fact is that estimates of the number of environmental refugees today is of the order of 10 million, but the estimate of the number of environmental refugees in the globally-warmed year of 2050 might be about 150 million out of a total World population of about 10 billion[394]. This figure is equivalent to 1.5% of the World's population, which compares with the current (early 1990s) estimate of 10 million environmental refugees or about 0.2% of the global population. A third of the 150 million refugees in the year 2050 might result from the movement or collapse, of old and generation of new agricultural zones, while the remainder are largely at risk from sea level rise. The costs are considerable and estimates made in the early 1990s for coastal defence costs resulting from rising seas have been in the range of \$2.5 trillion to \$5.0 trillion (trillion = 10^{12}), with coastal land loss of \$15 trillion[394]. Of course land (in cash terms) can only be lost once, and coastal defences, once built, can last for decades with only moderate expenditure required on maintenance. We might well be looking at a total cost of \$17.5 trillion to \$20 trillion spread over 50 years (i.e. \$350–\$500 billion a year); although this ignores resettling refugee costs and so the estimate might be conservative. These figures compare with a global World product (the total of everyone's earnings per annum in the early 1990s) of about \$23 trillion a year. Consequently, a \$400 billion annual greenhouse cost bill would be roughly equal to a 1.7% tax on income, or alternatively, 60% less than the estimate of that spent on arms in 1986[15], or approximately a 22c an imperial gallon equivalent fossil fuel tax (on gas and coal, as well as oil) or 18c a US gallon equivalent given that the World consumed about 7000 mtoe a year in the mid-1990s.

Another approach is provided by the World Conservation Strategy organizations which costed the major aspects of the world strategy for sustainable living at US$161 billion a year (1991 money). Here the money is envisaged as being spent on switching to renewable resources, reforestation, increasing energy efficiency and stabilising World population, as opposed to the greenhouse impact costs of combating sea level rise, etc. Of course if the human population in the twenty-first century was less than expected in a business-as-usual scenario due to World Conservation Strategy spending, then conceivably the refugee problem would similarly be less as there would be fewer displaced and fewer already occupying fertile lands in competition. Given $161 billion is much less that the previous $400 billion estimate, the fossil fuel tax equivalents would also be proportionally less, so our 18c a US gallon equivalent of fossil fuel tax would become a little over 7c a US gallon equivalent of fossil fuel.

The bottom line of all these estimates is that though the initial sums are huge, when divided by the equally huge consumption of fossil fuel, they become quite manageable and broadly equivalent in magnitude to the $1–$10 per barrel carbon tax being proposed in the early 1990s by the European Community. It therefore does not seem unreasonable to conclude that it would be possible to impose an affordable carbon tax that would go a considerable way towards paying for many of the World's developmental costs, while at the same time internalising many of fossil fuels externalities (such as from acid rain) and, through the marginal raising of fossil fuel prices, provide fiscal deterrence of continually increasing World consumption of this finite resource.

Even if climate change through a human-induced greenhouse effect is not taking place, as a species we need to become less dependent on finite resources and switch to renewable ones. Of course there are those who will rightly question the need to switch to more renewable resources after all, nobody should pay large bills without question, and there is a group that has repeatedly challenged the IPCC and its findings. In 1996 the IPCC was lambasted for its suggestion (in its 1995 review) that 'the balance of evidence suggests a discernible human influence on global climate'[390]. One vanguard of such efforts has been the Global Climate Coalition, a body which the leading interdisciplinary science journal *Nature* has described as 'the voice of fossil fuel producers, primarily in the oil, coal and petroleum industry'[395]. But what of countries, such as the USA, who are heavily dependent on fossil fuels?

From the early through to the mid-1990s oil imports represented about 43% of US consumption, while by 1995 oil imports had steadily increased to amount to over 47% of total US consumption. In terms of overall US primary energy consumption (including non-fossil fuels), oil (both imported and home produced) in 1995 represented some 39% of total energy consumption and 52% of all fossil fuel consumption. In short, the USA is dependent on imported oil to make up about half its oil consumption, while its oil consumption itself accounts for well over a third of all its energy consumption. With US domestic production of oil down some 23% in the decade up to 1995, and oil consumption up by over 11% over the same period, the US is caught between two opposing trends, neither of which by themselves would cheer an economist while together they make for a non-sustainable trend. The US is increasingly dependent on imports to maintain its industry's and citizens' supply of oil. Should anything happen to halt the flow of oil into the USA then a major restructuring of energy supply industry would be required. In common with most Western nations, the majority of passenger-miles travelled in the USA are by car. Without oil the US might increase its non-fossil electricity supply and correspondingly increase its electricity-powered modes of transport: trams, rail, etc. Electricity can be generated from a number of primary fuels and would provide greater energy flexibility, hence national security, through reducing future dependence on overseas oil.

After oil, natural gas is the fuel accounting for the next biggest slice of the US energy budget. Natural gas accounts for about 30% of US fossil fuel consumption and around a quarter of its mid-1990s primary energy consumption. Consumption has also grown since natural gas became a significant energy resource in the 1960s, with overall growth close to 25% for the decade up to 1995. However, as with oil, the USA has had a net dependency on natural gas which goes back to the 1970s. In the mid-1990s demand for natural gas outstripped its domestic supply by over 15%, with the shortfall being made up by imports.

Coal is the only fossil fuel making a significant contribution (about 24% in the mid-1990s) to US primary energy production that is not dependent on imports. Indeed the USA is a net exporter of coal, producing about 10% more than it consumed. Without imports there would certainly be an immediate shortfall which would have major economic repercussions, and most likely entail some lights going out across the nation. To sum up, the USA lacks energy security. Ignoring greenhouse considerations for the moment, the obvious way out would be for the USA to increase its domestic production of fossil fuel. It can certainly do this with coal, the mid-1990s estimate of reserves is equivalent to nearly a quarter of the World's total proven recoverable reserves (under existing economic and operating conditions), or some 258-years worth of reserves at its then current rate of consumption. If it did increase its coal production then energy would be lost in conversion to gas or oil, and so lower the production to reserves ratio.

As indicated above, unless there is a change of policy, the direction the US is currently going is most certainly not sustainable, it cannot continue to consume energy resources in the way it has in post war years. So what can the US do? The obvious answer is to increase its domestic production of non-fossil resources (nuclear, HEP, and the other renewables) and second, to improve energy efficiency. Yet if this answer is so obvious, then why has it not already been done? Here the answer is simple, money! Not the shortage of money, rather money to ensure the market responds in a way that encourages energy efficiency, so the price of fossil fuels will rise. Customers (which effectively means all adult citizens) do not like price rises. Given that all adult citizens equate with a nation's electorate, politicians are loathe to be seen to back increases in energy prices. There is a great irony here: while the average US citizen (if not environmentally aware or unconcerned about their children's future) may not want to pay a bit more for their energy now, in encouraging cheap imports they are in fact paying a long-term higher price.

Consider it this way, either a country saves a tonne of oil through increased energy efficiency, it produces a tonne of oil itself, or it imports a tonne of oil. If the country improves its energy efficiency then jobs are created in designing, manufacturing and distributing energy-efficiency technology. If oil is consumed instead, less jobs are created as oil production supports less than three direct jobs and one indirect job per million US$ (1990 prices) of revenue, - compared to nine direct jobs for the industrial sector and 18 direct for commerce. Importing oil provides even less jobs, but only marginally fewer as much of the workforce is employed transporting oil to refineries, refining and then getting it to the customer. Once the well has been drilled few jobs are required to keep it flowing to the well-head. One estimate[396] for the USA is outlined in Table 13.1 and shows that about twice as many jobs (both direct, in provision of the goods and services, and indirect, support jobs for direct workers) can be expected by improving the efficiency of energy technology, rather than providing more oil.

Table 13.1 Relative labour intensity of supplying or saving oil[396]

	Number of US jobs supported by supplying or saving 1000 barrels a day		
Option	*Direct jobs*	*Indirect jobs*	*Total jobs*
Oil imports	114	42	156
Domestic oil production	126	50	176
Energy efficiency	207	120	327

Assuming that the USA had continued the energy efficiency trends it adopted after the 1973 and 1980 oil crises, it could, have now become more energy-efficient, importing some 2.1 million barrels a day less oil. On this above basis, the total number of jobs would have increased by about 475 000. Even if this estimate is inaccurate, we are probably still talking about an order of hundreds of thousands of jobs. In short, the selfishness of the individual, and for that matter the politician, is actually costing America jobs! On the other hand, assuming a figure of 20% improved efficiency on 1990 figures, then the jobs created in *lieu* of savings for oil alone would amount to well over half a million.

Then again, the USA has other externalities to pay in relying heavily on oil imports (greenhouse gases and climate change impacts are not the only form of externality). In paying for imports the USA is transferring wealth out of its territory to those who (primarily through OPEC) have a certain monopoly on oil, hence limited control over its international price. One estimate has been that the US has transferred to the oil producers some US$1.2 to US$1.9 trillion dollars between 1972 and 1991. Furthermore, because oil prices were high (and the US lacked the energy efficiency technology) the economy suffered and incurred costs of a further US$1.2–2.1 trillion over the same period[397]. To put these figures in perspective, the extra cost to the USA each year is approximately equal to about 5% of Gross Domestic Product. In 1991, The US Office of Technology Assessment (OTA) published a 'moderate' scenario in which the USA could tackle greenhouse problems at no net cost. Under this scenario increases in motor efficiencies are modest, no non-fossil power stations would come on line, and nuclear power stations would increase the proportion of time that they generate power from 50% to 75%. Carbon dioxide generation would increase by about 15% (to 22%) by 2015 as compared to the 50% increase anticipated in the OTA business-as-usual scenario[398]. Another study, In 1997, by the US Department of Energy echoes this conclusion. It looked at 200 technologies to improve energy efficiency in four economic sectors: industry, buildings, transportation, and electric utilities. These could cut the nation's fuel bill by $50–$90 billion a year. This would be enough to offset the cost of developing new greenhouse technologies[399].

The above economic benefits are estimates only. Even if they were accurate only within an order of magnitude they demonstrate the considerable scope for savings in the US. But what of elsewhere? Western Europe consumes roughly half as much energy per capita as the USA but even here there is scope for improvement (see Chapter 8). Eastern Europe consumes much the same as Western Europe but without the economic benefit that Western Europe derives from that same energy expenditure and, now freed of a centrally planned economy, could now see considerable economic growth without any increase in energy consumption; it *could* do this but whether it *will* remains to be seen. Finally, the less-developed nations and China are developing and their economic growth could have a tremendous impact on fossil fuel consumption and CO_2 emissions in the short to medium term. If the developed nations, and especially the USA, are not seen to be curbing their own CO_2 emissions then what motivation will the developing nations have to give energy efficiency priority? Furthermore, if they do not give energy efficiency priority then the World's fossil fuel reserves will dwindle at an even faster rate and that, surely, cannot be good news for any nation in the medium to long term.

In the last century many countries were energy self-sufficient but, as energy demand increased, it was simpler to buy more seemingly plentiful cheap energy: oil has in real terms been cheaper since 1885 up to the early 1990s, with the exception of two decades, the 1890s and 1915–25. We can now see above the non-sustainability of the way we use fossil fuels and we have the technology to obtain our energy from elsewhere and to improve efficiencies of use. Today we could change to a more sustainable economy if we wanted to. What has not changed is that some costs would rise even if unemployment (for instance) fell and that the more sustainable economy requires a longer term view that could be greater than one or two terms of political office so that the political leadership has less stake in success of such policies (see Figure 10.1). We have saturated our global and regional commons and are closer to the edge over which lie all the demographic problems previously described (see Chapters 1 and 6). Of course the human-induced greenhouse effect may not be real even though sustainability concerns are. Even though the IPCC says that the 'balance of evidence' indicates a human impact on the climate, this indication is not absolute proof. We need to decide whether to act.

TO ACT OR NOT TO ACT?

There is a problem in statistics known as type one and type two errors. Most observations, be it of height of adults, size of industrial product, whatever... fall in a range between a maximum and a minimum acceptable, or useful, value. Those values falling outside this range might be discarded. However it is possible to err one's decision as to what is acceptable one way or another. Indeed different people may well err differently. Take, for example, the size or amount of product. It might be that legislation dictates that a certain sized product must be of a certain minimum size (*ex:* a bar of chocolate), or have a certain minimum number of contents (*ex:* a box of matches), however the manufacturer would not want the product to be very much larger than the minimum requirement; after all the manufacturer *does not want* to give away product. On the other hand the consumer *may well want* more chocolate, or more matches in a box. In short, it is possible to deviate from the desired range in two ways. The same is true of deciding whether or not to engage in climate change policies. We know that the weather, which is the local day-to-day expression of climate, varies. June 5th this year will not necessarily have exactly the same weather where you are now as it will June 5th next year, or did have June 5th last year. But there is a range of variability within which we might say indicates climate stability, and outside of which indicates climatic warming or cooling. Now those who say that global warming is taking place could be wrong. It might be that through a series of very unlikely events warming took place in recent decades but that this was within natural variability and that next year's average global temperature will be more akin to that typical earlier this century. Alternatively those that say that global warming is not taking place could be wrong, and that what we are seeing is not part of natural variation but due to a discernible trend.

Statisticians refer to these ways of being wrong as the afore-mentioned type I and type II errors. Another example of type I and type II error is perhaps clearer, that of capital punishment. Should we send to the electric chair ultra-violent criminals? If we decide to then we have to balance our judgement system to err either towards a type I or a type II error. We can either bias the system so that innocent people are let off, but that will invariably mean that some guilty people will get let off too, or we can bias the system towards all those who are guilty getting punished, but that will mean some innocent people will get punished by mistake.

With regard to global warming, should we indulge in global warming measures that have a hundreds of billions of dollars price tag per year, or do we not undertake mitigating measures and spend the money on other things? In fact the global warming problem is even more complicated: in addition to whether or not we can be 100% certain (0% error) that human-induced climatic change is taking place, we equally do not know whether all our counter-measures are for the best, there may be side effects. For instance some of the economic controls suggested in this text might undermine the free market to such a degree that we would be better off without the controls.

With the greenhouse issue there are two sets of questions. First, is the planet in all probability experiencing human-induced climate change? Second, whether or not we should pay for counter-greenhouse measures, as these themselves might have certain unforeseen costs? This dual problem, each with two broad answers, can be addressed using a two-by-two matrix, see Table 13.2.

This reveals the four principal outcomes as to whether human induced climate change is real and whether greenhouse measures should be taken. Clearly the worst outcome is if global warming is real but we have not invested in greenhouse measures: in this option the full costs of climate change are not paid by those causing it. On the other hand, the best outcome is not necessarily the opposite: the best option is not necessarily that global warming is real and greenhouse measures are adopted. Depending on the actual measures and the way greenhouse policies are adopted, it might well be that even if global warming is fictitious adopting greenhouse measures is the best policy, there would be none of the climatic costs but all the possible benefits. The question therefore becomes not whether or not anthropogenic climatic change is taking place but that we should adopt greenhouse policies, regardless of climatic change. The only question that now remains is what greenhouse measures are required and to what extent can they be expected to drive energy efficiency and greater sustainability.

It is interesting to note that *even if* climatic change were not taking place, adopting greenhouse measures does result in non-greenhouse spin-off benefits that can result in a potential net benefit. There are those, such as the US-based Global Climate Coalition, who disputed the IPCC 1995 conclusion[400] and so it is

Table 13.2 Two-by-two matrix for hypothetical global warming and global warming counter- (both adaptive and mitigating) measures.

	Business as Usual	*Undertake greenhouse measures*
Human-induced global warming is real	Cheap energy prices Fossil fuel runs out more quickly Population rises faster (by comparison with greenhouse measures) Lower energy efficiency/higher international energy trade Less use of renewables/higher international energy trade Greater shifts in climatic belts with health and agriculture costs Greater rise in sea level, flooding, salination of groundwater, etc. No economic gains through manufacture and no implementation of energy efficiency and energy sustainability measures	More expensive energy prices Fossil fuels last longer Population rises slower (by comparison) Higher energy efficiency/lower international energy trade More use of renewables/lower international energy trade Slower/lesser shifts in climatic belts/lower impact costs Slower rise in sea level. Protection of low-lying coastland Economic gains through manufacture and implementation of energy efficiency and energy sustainability measures
Global warming is fictitious	Cheap energy price Fossil fuel runs out more quickly Population rises faster (by comparison) Lower energy efficiency/higher international energy trade Less use of renewables/higher international energy trade No shifts in climatic belts, no health and agriculture costs No rise in sea level No sea defence costs No economic gains through manufacture and no implementation of energy efficiency and energy sustainability measures	More expensive energy prices Fossil fuels last longer Population rises slower (by comparison) Higher energy efficiency/lower international energy trade More use of renewables/lower international energy trade No shifts in climatic belts, no health and agriculture costs Agriculture contingency plans costs (but not implementation) No rise in sea level Sea defence contingency and some implementation costs Economic gains through manufacture and implementation of energy efficiency and energy sustainability measures

worth identifying the non-climate related benefits to greenhouse policy and it would be foolishly myopic to ignore this option.

Notwithstanding that there are those who disagree with the IPCC, the evidence that carbon dioxide we are adding to the atmosphere is affecting our World is continuing to mount. In 1996 the results of a fundamental piece of research were published[401] that looked at the way CO_2 was building up in the atmosphere, not just on a year-on-year basis but how it changed within each successive year. Carbon dioxide concentration varies within each year in the

Northern and Southern hemispheres in opposite directions. As each hemisphere moves towards its respective spring, so plants begin to grow, trees put out leaves etc. The plants absorb CO_2 through photosynthesis and emit oxygen, thus decreasing CO_2 concentration. As each hemisphere moves into winter the reverse happens: plant growth declines, vegetative matter decays, and atmospheric CO_2 increases. In the far north, and south, this annual variation in CO_2 is between 15 and 20 ppm but diminishes to about 3 ppm near the equator as the differences between seasons declines. Since the early 1960s the annual amplitude of the cycle (as measured in Hawaii) has increased by 20% (and 40% in the Arctic). This increase has been attributed to the increased photosynthetic assimilation of CO_2, which could be expected with plants growing in an atmosphere richer in CO_2 compared to three decades ago. However, in addition to this increased amplitude, the researchers noticed that the declining part of the annual cycle, which corresponds to Spring growth, has advanced over this time by about 7 days. Though CO_2 concentration is not in itself a measure of climate (in that rainfall, temperature and other climatic elements were not being measured) this directly corresponds to global climate change. The climate warning signal is there!

Despite the evidence continuing to mount, it would be foolish to think that the international scientific consensus on climate change is equivalent to all scientists holding this view. A number of scientists with expertise on climate change hold different views as to the magnitude of anticipated climate change, and a very small minority even disagree that human-induced climate change is taking place. Nonetheless, the IPCC does reflect the broad consensus and its assessments are reference points for both academia and politicians. Furthermore, we have shown that when non-climate-related scientists are appraised of the core facts their appreciation of climate change issues does largely fall in line.

POSSIBLE SCENARIOS FOR 2025

We have covered much ground in this text, and have enough information to begin to form a coherent picture. Using the mid-1990s as our starting point and looking ahead three decades to the year 2025, we can outline each end of a spectrum of possibilities. One is the IPCC 1990 business-as-usual scenario (which is only slightly different from its subsequent IPCC 1992 and 1995 scenarios) and the other might be if *all* the greenhouse policy options discussed in earlier chapters were simultaneously implemented, that is to say that forests were planted rather than felled, renewable energy supplemented fossil fuels, and energy efficiency was enhanced within the limits technically and economically available today. The results of this very rough assessment are outlined in Table 13.3.

Table 13.3 Approximate (~) practically achievable savings on anthropogenic carbon emissions (from energy and land-use) due to implementing greenhouse policies for 2025 using existing technology, compared with Business-as-Usual for 2025 and the 1995 estimated actual. Given that it is unlikely that *all* these savings will be achievable, the actual 2025 carbon emissions are likely to lie somewhere between 7 and 14 Gt C/yr; but the closer to the 7 Gt C/yr figure the more successful we are in implementing greenhouse policies. The estimate below assumes existing technology. New technology (subsequent to mid-1990s) will undoubtedly be available so enhancing energy conservation and efficiencies of use, however savings from the employment of new technologies are likely to be offset by those nations (or those within nations) failing to fully implement greenhouse policies.

Greenhouse qualifier	*Manmade CO_2 emissions GtC/yr*		
	2025 B-a-U with greenhouse measures	*2025 B-a-U without savings*	*1995*
Business as Usual (B-a-U) C emissions	14	14	8
Decline in deforestation and some	save~2 GtC	–	–
forestation	save~20% energy	–	–
Improve energy efficiency	save~2.2 GtC	–	–
Increase use of non-fossil carbon energy			
Human carbon emissions GtC/yr	7 low – 8.5 high	14	8

The picture portrayed in Table 13.3 (based on the figures, and discussion, given in the earlier chapters) is not one comparing the business-as-usual (B-a-U) forecast with that of the *maximum* theoretical savings, rather savings that might practicably be achieved with today's technology. Greater savings would be possible but to achieve these might require an unrealistic re-direction of resources. For instance, it would theoretically be possible to devote cropland (as opposed to woodland or marginal lands) to generate biofuels rather than food. Yet feeding the larger population of the twenty-first century is surely a priority. Table 13.3 also assumes use of existing technology. No doubt new (post-mid-1990s) technology will become available so further enhancing the potential to save energy, however implementing such improvements in efficiency are likely to be (partially) offset by technological use and political inertia.

The B-a-U estimate for the year 2025 is commensurate with the range of high and low estimates used for the 1990 IPCC B-a-U scenario[20], and between the IPCC IS92a and IS92e scenarios used subsequently for 1992[93] and 1995 (published 1996)[402] reports. The IS92a scenario is roughly equivalent to the 1990 B-a-U scenario, while the IS92e scenario assumes a decline in nuclear power and slightly higher economic growth than the IS92a. As the B-a-U scenario in Table 13.3 is broadly in agreement with the IPCC scenarios, so too the greenhouse policy savings identified for 2025 tie in with many of the alternative more energy-efficient IPCC scenarios. However, while the forecasts are for the year 2025, the IPCC scenario forecasts run to the year 2100 and they do significantly diverge from each other after 2025. Nonetheless the year 2025 is a useful year to examine as by then greenhouse policies instigated at the turn of the century will begin to make themselves felt.

Assuming that the emission targets identified in Table 13.3 are achievable, what will they mean in terms of the World's climate? As discussed in the 1990 IPCC report[20] even if all anthropogenic emissions of CO_2 were halted, the atmospheric concentration would decline slowly and would not approach its pre-industrial level for many hundreds of years. Even if, in the late 1990s, we decided not to exceed a CO_2 concentration of 420 ppm, which is 50% above the pre-industrial level, annual anthropogenic emissions would have to be reduced continuously to about 50% of their present value by the year 2050. Alternatively, in order to stabilise concentrations an immediate reduction of 60–80% would be necessary! Clearly, even with the theoretically realizable targets outlined in Table 13.3, we are going to see some increase in global temperature. This means that, of the revenues raised through greenhouse policies to meet current fossil fuel (and land use) externalities, spending on adaptation to a warmer climate *as well as* some spending on abatement will be required.

With regard to the increase in World temperature, the IPCC 1990 report[20] cited a B-a-U increase of 1°C above the 1990 temperature (2°C above the pre-industrial average global temperature) for the year 2025. The later 1992 IPCC report modified this to a B-a-U temperature increase of just 0.6°C above the year 1990 for the year 2025. However, there is such inertia within the biosphere's climatic system that much of the climatic benefit of lowering CO_2 emissions below the B-a-U scenario will not be felt until the end of the twenty-first century. Indeed, lowering emissions, as in Table 13.3, will only have a slight effect on climate, perhaps 0.1°C less than the B-a-U increase by the year 2025.

Those born now or at the beginning of the twenty-first century can in all probability expect to see the Earth warm by a degree or two over their life time. Indeed, just as the post Second World War generation has seen the Spring in mid-latitudes advance by about a week between 1960 and the mid-1990s, so those born now could tentatively expect to see a further advance of up to an extra fortnight over their lifetime, if we failed to employ greenhouse measures. The human impact of climate change for those born in the mid-latitudes now will be milder winters, more balmy summer nights, and changes in rainfall and climatic patterns. The milder winters will have a large human impact in terms of the physical range of diseases as well as a number of diseases affecting agriculture. The shifting of climatic belts will also have a direct impact on agriculture with regards to what species to grow where and an even greater impact on cities. Above all, for those born now, if we do make better use of fossil fuels, switch to more sustainable ways of living and begin to control human population, it is very likely that they will see a time around the middle of the next century when many human systems will be severely stressed.

Now the above paragraph may seem to some to be a 'doom and gloom' scenario. For instance a change in rainfall patterns in the UK might mean having to make major investments in upgrading water supplies which could mean high water bills, but nothing that the nation cannot afford. But this whole chapter underlines the opportunity that anthropogenic climate change offers us. That our species has tended to despoil rather than enhance our planet is not news. The principal problems *are* common knowledge and the subject of a number of international reports and conferences (see Chapter 2) and we have known what to do, but have not whole heartedly sought to get it done. But with climate change issues we have a common concern that unifies nearly all the principal World development difficulties: it does not matter that we cannot *exactly* forecast the effects of higher CO_2 when deciding whether or not to implement greenhouse measures, these other World development difficulties alone should provide us with ample motivation. We have, in anthropogenic climate change, an issue directly tied to key economic activities (energy supply and use), the redirection of part of that economic wealth could help to fund solutions. We have therefore in this issue both the vision, the relevance and the means. Little else, surely, provides such an opportunity to shift the World economy to a more sustainable mode.

GETTING THERE

In Chapter 12 we have reviewed some of the perceptual barriers and social traps that need to be overcome if we are to accept the challenge of opportunity that climate change concerns present. The bad news is that these social traps and perception barriers exist but the good news is that they can be overcome. Second, politicians will have to start to think longer term and the electorate will have to both become informed and also think the longer term. Getting the electorate to think in the longer term will be hard but not impossible. Politicians need to treat such 'electorate grooming' as a matter of priority.

Finally we have one advantage bestowed on us by Darwinian evolution and that is Darwinian evolution itself. For some three billion years nucleic acids (genes) have been replicated from one generation to the next, building in a core trait of self-survival. At the macro level this expresses itself in the way we think of our children – we want what is best for them. This then might be a useful tool in encouraging longer-term perspectives. Let us hope we can get away from short-termism for the stakes are high!

REFERENCES

Initials:

BBC	British Broadcasting Corporation
BMJ	British Medical Journal
CEGB	Central Electricity Generating Board
DoE	Department of the Environment
DoEn	Department of Energy
DTI	Department of Trade and Industry
ENDS	Environmental News Data Service
ITV	Independent Television
ITN	Independent Television News
ENDS Report	Environmental News & Data Service Report
Proc. R. Soc. London	Proceedings of the Royal Society of London
Pub Und Sci	Public Understanding of Science
REView	Renewable Energy View
WHO	World Health Organization

1. BP Economics Unit (1990). *BP Statistical Review of World Energy.* British Petroleum Corporate Communications Services: London
2. International Union for Nature Conservation, United Nations Environment Programme and World Wildlife Fund (1980). *The World Conservation Strategy.* IUCN, UNEP and WWF: Gland
3. Brandt Commission (1980). *North-South: A programme for survival.* Pan Books: London
4. World Bank (1988). *World Development Report.* Oxford University Press: Oxford
5. Science Summit (1993). *Population Summit of the World's Scientific Academics.* National Academic Press: Washington, DC
6. Clark, R. B. (1989). *Marine Pollution.* Clarendon Press: Oxford
7. France, R. (1992). Garbage in paradise. *Nature,* **355**, 504
8. Sachs, I. (1993). Transition strategies for the 21st century. *Nature & Resources,* 28(1), 4–7
9. International Union for the Conservation of Nature, WCSII draft (1989). *World Conservation Strategy for the 1950s – first draft.* IUCN: Gland
10. Ward, B. and Dubos, R. (1972). *Only One Earth.* Penguin: Harmondsworth
11. Clarke, R. and Timberlake, L. (1982). *Stockholm Plus Ten.* Earthscan (IIED): London
12. Meadows, D. H., Meadows, D. L., Randers J. and Behrens, W. W. III (1972). *The Limits to Growth,* Pan: London

13. Malthus, T. R. (1798) *Essay on the principle of population as it affects the future improvement of society with remarks on the speculation of Mr Godwin, Mr Condorcet and other writers.* In: Johnson, in St Paul's Churchyard
14. Brandt Commission (1983). *Common crisis – North–South: Co-operation for World Recovery*, Pan: London
15. Brandt, W. (1989). *World Armament and World Hunger: A call for action.* Gollancz: London
16. International Union for the Conservation of Nature, United Nations Environment Programme, World Wildlife Fund (1991). *Caring for the Earth.* Earthscan: London
17. World Commission on Environment and Development (1987). *Our Common Future.* Oxford University Press: Oxford
18. DoE (1988). *A perspective by the UK on the report of the world Commission on Environment and Development.* DoE: London
19. Cottrell, A. (1978). *Environmental Economics.* Edward Arnold: London
20. Intergovernmental Panel on Climate Change (1990). *Climate change: The IPCC Scientific Assessment.* Cambridge University Press: Cambridge
21. Hardin, G. (1968). The Tragedy of the Commons. *Science*, **162**, 1243–8
22. Wayne, R. P. (1988). Origins and evolution of the atmosphere. *Chemistry in Britain*, **24**(3), 225–30
23. Holland, H. D. (1990). Origins of breathable air. *Nature*, **347**, 17
24. Lovelock, J. E. (1979). *Gaia: A new look at life on Earth.* Oxford University Press: Oxford
25. Han, T. M. and Runnegar, B. (1992). Megascopic eukaryotic algae from the 2.1 billion year-old Neguanne iron formation. *Science*, **257**, 232–5
26. Berger, A., Imbrie, J., Hays, J., Kukla, G. and Saltsman, B., (1984). *Milankovitch and Climate.* Reidal Publishing: The Netherlands
27. Hays, J. D., Imbrie, J. and Shackleton, N. J. (1976). *Science*, **194**, 1112–32
28. Lorius, C., Jouzel, J., Raynard, D., Hansen, J. and Le Treut, H. (1990). The ice core record: Climate and sensitivity and future greenhouse warming. *Nature*, **347**, 139–45
29. Jouzel, J., Petit, S. R., Genthon, C., Barkov, N. I., Kotlyakove, V. M. and Petrov, V. M. (1987). Vostok ice core: A continuous isotope temperature record over the past climatic cycle (160 000 years). *Nature*, **329**, 403–8
30. Barnola, J. M., Raynaud, D., Korotkevich, Y. S. and Lorius, C. (1987). Vostok ice-core provides 160 000-year record of atmospheric CO_2. *Nature*, **329** 408–14
31. Jouzel, J., Barkov, N. I., Barnola, J. M., Bender, M., Chappellaz, J., Genthon, C., Kotlyakov, V. M., Li Perkov, V., Lorius, C., Petit, J. R., Raynaud, D., Raisbeck, G., Ritz, C., Sowers, T., Stievenard, M., Yiou, F. and Yiou, P. (1993). Extending the Vostok ice-core record of the palaeoclimate to the penultimate glacial period. *Nature*, **364**, 407–12
32. Bender, M., Sowers, D., Dickson, M-L, Orchards, J., Grootes, P., Mayewski, P. A. and Meese, D. A. (1994). Climate correlations between Greenland and Antarctica during the past 100 000 years. *Nature*, **372**, 663–6
33. Johnson, S. J., Clausen, H. B., Dansgaard, W., Fuher, K., Gundestrup, N., Hammer, C. O., Iversen, P., Jouzel, J., Stauffer, B. and Steffensen, J. B. (1992).

Irregular glacial interstadials recorded in new Greenland ice core. *Nature*, **359**, 311–3

34. Taylor, K. C., Lamory, G. W., Doyle, G. A., Alley, R. B., Groots, P. M., Mayenski, P. A., White, J. W. C. and Barlow, L. K. (1993). The flickering switch of the late Pleistocene climate change. *Nature*, **361**, 432–6
35. Bond, G., Broecker, W., Johnsen, S., McManus, J., Labyrie, L., Jouzel, J. and Bonani, G. (1993). Correlations between climate records from North Atlantic sediments and Greenland ice. *Nature*, **365**, 43–7
36. Lehman, S. (1993). Ice sheets, wayward winds and sea change. *Nature*, **365**, 108–10
37. Fronval, T., Jansen, E., Bloemendal, I. and Johnsen, S., (1995). Oceanic evidence for coherent fluctuations in Fennoscandian and Laurentide ice sheets on millennium timescales. *Nature*, **374**, 443–6
38. Jasper, J. P. and Hayes, J. M. (1990). A carbon isotope record of CO_2 levels during the late Quaternary. *Nature*, **347**, 462–4
39. Flint, R. F. (1959). *Glacial & Pleistocene Geology*. John Wiley: New York
40. Steine, S. (1994). Extreme and persistent drought in California and Patagonia during mediaeval time. *Nature*, **369**, 546–9
41. Lamb, H. H. (1965). Britain's changing climate in *The Biological significance of climate change in Britain*. Institute of Biology symposia No 14, IoB & Academic Press: London
42. Payette, S., Filion, L., Dewaide, A. and Began C. (1989). Reconstruction of tree line vegetation responses to long-term climatic change. *Nature*, **341**, 429–3
43. Cook, E., Bird, T., Peterson, M., Barbetti, B. B., A'arrigo, R., Francy, R. and Tans, P. (1991). Climatic change in Tasmania inferred from a 1089-year tree-ring chronology of Huon pine. *Science*, **235**, 1266–8
44. Karl, T. R., Liveze, R. E. and Epstein, E. S. (1984). *Bulletin of the American Meteorological Society*, **65**, 1302. Reported in Kerr, R. A. (1984). Wild string of winters confirmed. *Science*, **227**, 506
45. Kerr, R. A. (1989). 1988 ties for warmest year. *Science*, **243**, 891
46. Tyndall, J. (1863). On the influence of carbonic acid in the air upon the temperature of the ground of the absorbtion and radiation of heat by gases and vapours, and on the physical connection of radiation, absorbtion, and conduction. *Philosophical Magazine*, **41**, 237–76
47. Arrhenius, S. (1896). On the influence of carbonic acid in the air upon temperature of the ground. *Philosophical Magazine*, **41**, 237–76
48. Gribbin, J. (1988). Britain shivers in the Global greenhouse. *New Scientist*, 9th June, 42–3
49. Jones, P. D., Ya Grossman, P., Coughlan, M., Plummer, N., Wang, W. C. and Karl, T. R. (1990). Assessment of the urbanisation effects in time series of surface air temperature over land. *Nature*, **347**, 169–72
50. Maddox, J. (1990). Clouds and global warming. *Nature*, **347**, 329
51. Verdansky, V. I. (1926). *The Biosphere*. republished in 1986 by Synergistic Press: London & Arizona

52. Lovelock, J. E. (1979). *Gaia: A new look at life on Earth* (2nd edition 1987). Oxford University Press: Oxford
53. Lovelock, J. E. (1988). *The Ages of Gaia.* Oxford University Press: Oxford
54. Genthon, C., Barnola, G. M., Raynaud, D., Lorius, C., Jouzel, J., Barkov, N. I., Kortkevich, Y. S. and Kortlyakov, V. M., (1987). Vostok ice core: Climatic response to carbon dioxide and orbital forcing over the last climatic cycle. *Nature*, **329**, 414–8
55. Kuo, C., Lindberg, C. and Thomson, D. J. (1990). Coherence established between atmospheric carbon dioxide and global temperature. *Nature*, **343**, 709–13
56. Aldhous, P. (1991). 1990 warmest year on record. *Nature*, **349**, 186
57. Houghton, J. T., Callander, B. A. and Varney, S. K. (1992). *1992 IPCC Supplement*, in Intergovernmental Panel on Climate Change (1992) *Climate Change 1992: supplementary report to the IPCC scientific assessment.* Cambridge University Press: Cambridge
58. Intergovernmental Panel on Climate Change (1996). *Climate change 1995. Vol 1. The Science of Climate Change: The IPCC Second Assessment Report.* Cambridge University Press: Cambridge
59. Wigley, T. M. L. and Raper, S. C. B. (1992). The implications for climate and sea level of the revised IPCC emissions scenarios. *Nature*, **357**, 293–300
60. Alley, R. B. and Whillans, I. M. (1991). Changes in the West Antarctic ice sheet. *Science*, **254**, 959–63
61. MacAyeal, D. R. (1992). Irregular oscillations of the West Antarctic ice sheet. *Nature*, **359**, 29–32
62. Blankenship, D. D., Bell, R. E., Hodge, S. M., Brozena, J. M., Behrendt, J. C. and Finn, C. A. (1993). Active volcanism beneath the West Antarctic ice sheet. *Nature*, **361**, 526–9
63. Barrett, P. J., Adams, C. J., McIntosh, W. C., Swisher, III C. C. and Wilson, G. S. (1992). Geochronological evidence supporting Antarctic deglaciation three million years ago. *Nature*, **359**, 816–8
64. Parry, M. (1990). *Climatic Change and World Agriculture.* Earthscan: London
65. Adams, R. M., Rosenzweig, C., Peart, R. M., Ritchie, J. T., McCarl, B. A., Glyon, D., Curry, R. B., Jones, J. W., Boote, K. J. and Hartwell-Allen, L. Jr (1990). Global climate change and US agriculture. *Nature*, **345**, 219–23
66. Gribbin, J. (1989). Global granary in peril as rains desert America. *New Scientist*, 3rd June, 26
67. Blumental, C., Barlow, S. and Wrigley, C. (1990). Global warming and wheat. *Nature*, **347**, 235
68. Institute of Terrestrial Ecology (1990). *The greenhouse effect and terrestrial ecosystems of the UK.* ITE/NERC Research Publication No4. HMSO: London
69. Jones, R. J. A. and Thomasson, A. J. (1985). *An Agroclimatic Databank for England and Wales.* Technical monograph No 16. Soil Survey of England and Wales: Harpenden
70. Grabherr, G., Gottfried, M. and Paull, H. (1994). Climatic effects on mountain plants. *Nature*, **369**, 448
71. Holdgate, M. W. (1987). Changing habitats of the World. *Oryx*, **21**(3), 149–59

72. Bodmer, R. E., Mather, R. J. and Chivers, D. J. (1991). Rain forests of Central Borneo – threatened by modern development. *Oryx*, **25**(1), 21–6
73. Mittermeir, R. A., de Cusmao Camara, I., Padua M. T. J. and Blanck, J. (1990). Conservation in the Pantal of Brazil. *Oryx*, **24**(2), 103–12
74. Bonner, W. N. (1985). Conserving the Antarctic. *Biologist*, **32**(3), 103–12
75. Holt, W. V. and Moore, H. D. M. (1988). Semen banking – Is it now feasible for captive endangered species? *Oryx*, **22**(2), 172–8
76. Pielou, E. C. (1991). *After the Ice Age*. University of Chicago Press: Chicago
77. Gear, A. J. and Huntley, B. (1991). Rapid changes in range limits of Scot's pine 4000 years ago. *Science*, **251**, 544–6
78. Overpeck, J. T., Bartlein, P. J., Webb, T. III (1991). Potential magnitude of future vegetation change in Eastern North America: Comparison with the past. *Science*, **254**, 692–5
79. Institute of Terrestrial Ecology (1989). *Climate change, rising sea level and the British coast*. ITE/NERC Research Publication No1. HMSO: London
80. Prater, A. J. (1990). Environmental implications of utilising renewable energy barrages. In *Renewable energy: All power to the UK environment*. Institute of Biology, Environmental Sciences Division symposium proceedings. IoB: London
81. Smith, T. B. and Tirpack, D. A (1988). *The potential effects of global climate change in the US*. Draft report to Congress, Executive Summary
82. Daily, G. C. and Ehrlich, P. R. (1990). An exploratory model of the impact of rapid climatic change on the world food situation. *Proc. R. Soc. London: Biological Sciences*, **241**, 232–44
83. Broadus, J., Milliman, J., Edwards, S. *et al.* (1986). Rising sea level and damming of rivers: possible effects in Egypt and Bangladesh. In Titus, J. G. (ed). *Effects of changes in stratospheric ozone and climate. Sea Level Rise*. pp 165–89. US Environmental Protection Agency: Washington DC
84. Haines, A. (1991). Global warming and health. *BMJ*, **302**, 669–670
85. WHO (1992). *How does the World fare? The World Health Organization's Statistics Annual*. Press release WHO /24. WHO: Geneva
86. McMichael, A. J., Woodward, A. J. and van Leeuwen, R. E. (1994). The impact of energy use in industrialised countries on global pollution. *Medicine and Global Survival*, **1**(1), 23–32
87. Schneider, S. (1990). Prudent planning for a warmer planet. *New Scientist*, 17th November, 45–51
88. Spash, C. L. and d'Arge, R. C. (1989). The greenhouse effect and intergenerational transfers. *Energy Policy*, **17**, 88–95
89. Broecker, W. S. (1987). Unpleasant surprises in the greenhouse? *Nature*, **238**, 123–6
90. Nature (1991). *A drop to drink*. BBC 2, 2nd April. BBC TV: London
91. Anon (1991). Water shortage pits Man against nature. *Nature*, **350**, 180–1
92. Brown, L. R. and Kane, H. (1995). *Full house: Reassessing the Earth's population carrying capacity*. Earthscan: London
93. Intergovernmental Panel on Climate Change (1992). *Climate change 1992: The*

supplementary report to the IPCC scientific assessment. Cambridge University Press: Cambridge

94. NERC (1989). *Oceans and the Global Carbon Cycle.* Natural Environment Research Council: Swindon
95. Whitfield, M. and Williamson, P. (1989). The ocean carbon cycle and global climate change. *NERC News* (January), 25–7
96. Etheridge, D. M., Pearman, G. I. and de Silva, F. (1988). Atmospheric trace-gas variations as revealed by air trapped in an ice core from Law Dome, Antarctica. *Annals of Glaciology*, **10**, 28–33
97. Chappellaz, J., Barnola, J. M., Raynaud, D., Korotkevich, Y. S. and Lorius, C. (1990). Ice-core record of the atmospheric methane over the past 160 000 years. *Nature*, **345**, 127–31
98. Crutzen, P. J., Aselman, I. and Seiler, W. S. (1986). Methane production by domestic animals, wild ruminants, other herbivorous fauna and humans. *Tellus*, **38**(B), 271–84
99. Henderson, A., Churchill, S. P. and Luteyn, J. L., (1991). Neo-tropical plant diversity. *Nature*, **351**, 21–2
100. Mitchell, C., Sweet, J. and Jackson, T. (1990). A study of leakage from the UK natural gas distribution system. *Energy Policy*, **18**(9), 809–18
101. MacKenzie, D. (1990). Leaking gas mains help to warm the globe. *New Scientist*, 22nd November, 24
102. James, C. G. (1990). Natural gas in the greenhouse. *Nature*, **347**, 720. Also, Wallis, M. K. reply (1990) *Nature*, **347**, 720
103. Lelieveld, J. and Crutzen, P. J. (1995). Indirect chemical effects of methane on climate warming. *Nature*, **355**, 339–41
104. Skea, J. (1990). Energy scenarios – Demand and supplies. In *Renewable energy: All power to the UK environment.* Proceedings of an Environmental Division of the Institute of Biology. IoB: London
105. Staffelbach, T., Neftel, A., Stauffer, B. and Jacob, D. (1991). A record of atmospheric methane sink from formaldehyde in polar ice cores. *Nature*, **349**, 603–5
106. Mosier, A., Schimel, D., Valentine, D., Bronson, K. and Parton, W. (1991). Methane and nitrous oxide fluxes in native, fertilised and cultivated grasslands. *Nature*, **350**, 330–2
107. Stelle, L. P., Dlugokencky, E. J., Lang, P. M., Tans, P. P., Martin, R. C. and Masarie, K. A. (1992). Slowing down of global accumulation of atmospheric methane during 1980s. *Nature*, **358**, 313–6
108. Rudolph, J. (1990). Anomalous methane. *Nature*, **368**, 19–20
109. Whalen, M., Tanaka, N., Henry, R., Deck, B., Zeglen, J., Vogel, J. S., Southon, J., Shemesh, A., Fairbanks, R. and Broecker W. (1989). ^{14}Carbon in methane sources and in atmospheric methane: The contribution from fossil carbon. *Science*, **245**, 286–90
110. United Nations Environment Programme (1987). *Montreal Protocol on substances that deplete the ozone layer.* UNEP conference services number 87–6106

111. Pool, R. (1991). A global experiment in technology transfer. *Nature*, **351**, 6–7
112. MacKenzie, D. (1989). Substitute CFCs will stoke global warming. *New Scientist*, 13th May
113. Thiemens, M. H. and Trogler, W. C. (1991). Nylon production: An unknown source of atmospheric nitrous oxide. *Science*, **251**, 932–4
114. Ball, D. and Laxen, D. (1987). Ozone pollution over Britain. *London Environmental Bulletin*, **4**(2), 2
115. Caldereira, K. and Kasting, J. F. (1993). Insensitivity of global warming potentials to carbon dioxide emission scenarios. *Nature*, **366**, 251–3
116. Wang, W-C., Dudek, M. P., Liang, X-Z. and Kiegel J. T. (1991). Inadequacy of effective CO_2 as a proxy in simulating the greenhouse effect of other radioactively active gases. *Nature*, **350**, 537–77
117. Lashof, D. A. and Ahuja, D. R., (1990). Relative contributions of greenhouse emissions to global warming. *Nature*, **344**, 529–31
118. Hansen, J., Lacis, A. and Prather, M. (1989). Greenhouse effect of chlorofluorocarbons and other trace gases. *Journal of Geophysical Research*, **94**, 16417–21
119. Myers, N. (ed.) (1987). *The Gaia Atlas of Planet Management*. Pan: London
120. Northcott, J. (1991). *Britain in 2010*. PSI Publishing: London
121. WHO (1991). The urban crisis: Will megacities lead to mega crisis? *WHO Features*, **156**. WHO: Geneva
122. WHO (1991). *Health in a Changing World*. WHO Press Release No. 26. WHO: Geneva
123. WHO (1990). *Conference on the least developed countries – Don't forget health*. WHO Press Release No 44. WHO: Geneva
124. WHO (1990). World tuberculosis toll is rising. *WHO Features* No 148. WHO: Geneva
125. WHO (1991). *Malaria worsening in many areas*. WHO Press Release WHH/6. WHO: Geneva
126. WHO (1991). Cholera: Ancient scourge on the rise. *WHO Features* No 154. WHO: Geneva
127. WHO (1991). The global HIV/AIDS situation. In *Point of Fact* No 74. WHO: Geneva
128. WHO (1994). *World AIDS day 1st December*. WHO Press Release WHO/92. WHO: Geneva
129. Anon (1991). Anonymised screening for HIV: First results. *BMJ*, **302**, 1229
130. Evans, B. G., Gill, O. N. and Emslie, J. A. N. (1991). Completeness of reporting AIDS cases. *BMJ*, **302**, 1351–2
131. Minerva (1991). Views. *BMJ*, **302**, 918. (Reporting from *AIDS Analysis Africa*, March/April 1991
132. Dunn, J. (1990). Fuelling the population explosion. *The Engineer*, 14th June, 18
133. International Institute for Environment and Development and the World Resources Institute (1987). *World Resources 1987*. Basic Books: New York
134. BP Economics Unit (1992). *BP Statistical Review of World Energy*. Corporate Communications Services: London

135. BP Economics Unit (1996). *BP Statistical Review of World Energy*. Group Media & Publications: London
136. Brierley, D. (1991). BP is fuelled by oil find. *Sunday Times*, June 30th
137. Blaxter, K. (1988). Future problems. *Biologist*, **34**(4), 173–8
138. Blaxter, K. (1986). *People food and resources*. Cambridge University Press: Cambridge
139. Mackney, D. (1991). Resource depletion in a resource dependent economy. *Australian Biologist*, **4**(2), 110–21
140. Goldstein, W. (1989). From one crisis to the next: The fate of the oil markets in the 1990s. *Energy Policy*, **17**(1), 11–4
141. Cruver, P. C. (1989). Electricity's future. *Energy Policy*, **17**(6), 617–20
142. Anon (1991). In Parliament – Carbon dioxide's emissions. *Atom*, **413**, 29
143. Harding, C. (1991). Round the bend with nuclear power. *Atom*, **409**, 18–22
144. Eyre, B. (1991). The longer term direction for the nuclear power industry. *Atom*, **411**, 8–14
145. Carle, R. (1983). Super Phénix and beyond. *Atom*, **318**, 74–6
146. Monckton, N. (1994). Ready for UK rock lab. *Atom*, **436**, 22–7
147. Lemons, J. and Brown, D. A. (1990). The role of science in the decision to site a high-level nuclear waste repository at Yucca Mountain. *The Environmentalist*, **10**(1), 3–24
148. BBC (1990). The ten thousand year test. *Horizon*, 5th March. BBC2: London
149. Gavaghan, H. (1984). Healthy miners but fewer jobs. *New Scientist*, **101**, 22
150. Makins, R. (1982). Living with nuclear energy. *Atom*, **310**, 168–73
151. Fremlin, J. H. (1983). Power production at minimum risk. *Atom*, **316**, 28–30
152. Peirson, D. H., Cambray, R. S., Cawse, P. A., Eakins, J. D. and Pattender, N. J. (1982). Environmental radioactivity in Cumbria. *Nature*, **300**, 27–31
153. Black, D. (1984). *Investigation of the possible increased incidence of cancer in West Cumbria*. HMSO: London
154. Gardner, M. J., Snee, M. P., Hall, A.J., Powell, C. A., Downes, S. and Terrell, J. D. (1990). Results of a case–control study of leukaemia and lymphoma among young people near Sellafield nuclear plant in Cumbria. *BMJ*, **300**, 423–9
155. Urquart, J. D., Black, R. J., Muirhead, M. J., Sharp, L., Maxwell, M., Eden, O. B. and Jones, D. A. (1991). Case-control study of leukaemia and non-Hodgkin's lymphoma in children in Caithness near Dounray nuclear installation. *BMJ*, **302**, 687–92
156. Kinlen, L. J., Dickson, M. and Stiller, C. A. (1995). Childhood leukaemia and non-Hodgkin's lymphoma near large rural construction sites, with a comparison with the Sellafield nuclear site. *BMJ*, **310**, 736–68
157. English, D. R., Parkin, D. M. M., Kaldor, J. M. and Masuyer E. (1990). *European childhood leukaemia/lymphoma study*. (ECLIS) International Agency for Cancer Research, World Health Organization: Lyon
158. Watt Committee for Energy (1991). *Five years after Chernobyl*. The Watt Committee: London
159. Isles, C. G., Robertson, I., Macleod, J. A. J., Preston, T., East, B. W., Hole, D. J. and Lever, A. F. (1991). Body concentration of caesium^{137} in patients from the Western Isles of Scotland. *BMJ*, **302**, 1568–71

160. Kenward, M. (1990). Nuclear fusion: Pursuing the SOFT option. *Atom*, **408**, 6–7
161. Law, S. (1990). Fusion's future in Europe. *Physics World*, **3**(10), 13–14
162. Jackson, T. (1989). Is fusion feasible? *Energy Policy*, **17**(4), 407–12
163. McMullan, J. T., Morgan, R. and Murray, R. B. (1977). *Energy Resources*. Edward Arnold: London
164. Wright, J. K. and Rodliffe, R. S. (1989). The global impact of nuclear power. *Atom*, **398**, 10–18
165. Hafele, W. and Sassin, W. (1978). A future energy scenario. In Grainger, L. (ed.) *Energy resources: Availability and rational use*. IPC: London
166. Elkington, J. (1984). *Sun Traps*. Penguin: London
167. Dawson, J. (1981). The prospects for renewable energy resources in the UK. *Energy World*, (November), 3–9
168. DoEn (1991). *United Kingdom Energy Statistics 1991*. Government Statistical Service: London
169. Stoubough, R. and Yergin, D. (1979). *Energy future – Report of the energy project at the Harvard Business School*. Ballentine: New York
170. Flavin, C. and Lenssen, N. (1991). Power from the Sun. *Sun World*, **15**(2), 10–33
171. Hill, R. (1991). Solar power steps out of the shadows. *Physics World*, **14**(8), 30–34
172. Denton, J. D., Glanville, R., Gliddon, B. J., Harrison, P. L., Hughes, E. M., Swift-Hook, D. T. and Wright, J. K. (1975). The potential of natural energy resources. *CEGB Research*, **2**
173. Lewis, T. (1985). Wave energy: Evaluation for the commission of the European Community. Referred to in 'Europe misled over wave energy' *New Scientist*, 10th November, 26
174. Salter, S. H. (1974). Wave power. *Nature*, **249**, 720
175. DoEn (1989). *Taking power from water: Water energy technology in Britain*. DoEn: London
176. DoEn (1991). *UK leads in shoreline wave energy*. DoEn News Release **115**, 15th July. Department of Energy: London
177. Fell, B. (1974). *Life, space and time: A course in environmental biology*. Horer & Row: New York
178. DoEn (1981). Tidal power from the Severn Estuary. *Energy Paper*, **46**, DoEn/HMSO: London
179. DoEn (1989). The Severn Barrage Project: General Report. *Energy Paper*, **57**, DoEn/HMSO: London
180. DoEn (1991). *Energy trends: A statistical bulletin*. July DoEn: London
181. Wood, D. (1990). Mersey barrage project enters a critical phase. *REView*, **11**. DoEn: London
182. Grubb, M. J. (1988). The potential for wind energy in Britain. *Energy Policy*, **16**, 594–607
183. CEGB – Department of Information and Public Affairs (1983). *Sizewell 'B' PWR nuclear power station. CEGB: London*
184. Taylor, A. (1994). A question of attitude. *REView*, **22**. DTI: London
185. British Wind Energy Association (1987). *Wind for the UK*. BWEA: London

186. DoEn (1988). *Michael Spicer gives go-ahead to wind farms.* DoEn News Release, **38** 23rd March. DoEn: London
187. Hague, B. (1990). Another step forward. *REView*, **13** 15. DoEn: London
188. Moynihan, C. (1990/1). The renewables: Where are we now? *REView*, **13**, 3–4
189. DoEn (1991). *Government encouraging opportunities for wind power.* DoEn News Release, **159**, 11th August. DoEn: London
190. Anon (1993). Renewable energy's future blowing in the wind. *ENDS Report*, **216**, 19–21
191. Shillitoe, S. (1991). Alternative energy in Nicaragua. *The Chemical Engineer*, **491**, 17–19
192. Turtle, J., Nicol, D. A. C. and Klewe, R. C. (1986). Geothermal energy: Hot favourite for the future? *CEGB Research*, January, 3–17
193. Kerr, R. A. (1991). Geothermal tragedy of the commons. *Science*, **253**, 134–5
194. Wright, M. (1990). The geothermal programme – Where does it stand? *REView*, **13**, 8–9. DoEn: London
195. Wayman, M. and Parkh, S. R. (1990). *Biotechnology of biomass conversion.* Institute of Biology & Open University Press: Milton Keynes
196. Sinclair, J. (1991). Using today's technology to clean up the planet. *Our Planet*, **3**(3), 4–9
197. Anon (1994). Biomass power: The germination of a new industry. *ENDS Report*, **232**, 24–6
198. Mitchell, C. P. (1990). Energy forestry. In *Renewable energy: All power to the UK environment.* Proceedings of an Environment Division symposium of the Institute of Biology, 9th March. Institute of Biology: London
199. Laughton, M. A. (ed.) (1990). *Renewable energy sources.* Watt Report **22**. Watt Committee & Elsevier Applied Sciences: Barking
200. DoEn (1991). *Potential for straw as a renewable source of energy.* DoEn News Release, **153**, 6th August. DoEn: London
201. Hall, D. O., Mynick, H. E. and Williams, R. H., (1990). *Carbon sequestration* versus *fossil fuel subscription: Alternative roles for biomass in coping with greenhouse warming.* Centre for Energy and Environmental Studies. Report **255**. Princeton University
202. Greenhaugh, G. (1991). Global warming and the electricity supply industry. *Atom*, **416**, 24–7
203. Pimental, D., Rodrigues, G., Wang, T., Abrams, R., Goldberg, K., Staecker, H., Ma, E., Brueckner, L., Travato, L., Chow, C., Govindarajulu, U. and Boerke, S. (1994). Renewable energy: Economics and environmental issues. *BioScience*, **44**(8), 536–46
204. CEGB – Department of Information and Public Affairs (1988). *Statistical yearbook 1987/8.* Central Electricity Generating Board: London
205. Cambel, A. B. and Koomanoff, F. A. (1989). High temperature super conductors and CO_2 emissions. *Energy*, **14**(6), 309–522
206. Rose, D. J. (1974). Energy policy in the US. *Scientific American.* Reprinted in *Energy and environment – Readings from Scientific American.* Siever, (ed.). Freeman & Co: San Francisco

207. Leach, G., Lewis, C., Romig, F., van Buren, A. and Foley, G. (1979). *A low energy strategy for the UK*. Science Reviews: London
208. House of Commons Select Committee on Energy (1981). *The Government's Statement on the new nuclear power programme*. Vol 1 Reports and Minutes of Proceedings HC114-1/February
209. Chateau, B. and Lapillonne, B. (1978). Long term energy demand forecasting – A new approach. *Energy Policy*, **6**(2), 140–57
210. Anon (1980). Low energy strategies/ 14. *ENDS*, **58**, 11–13
211. Durning, A. (1991). Asking how much is enough? In Brown, L. R. *State of the World 1991*. Earthscan: London
212. WHO (1991). Use your head and seatbelt. *WHO Features*, **160**, WHO: Geneva
213. Anon (1991). Fuel economy worsens. *Acid News* (October), 4 (Reporting on an article by M. P. Walsh in the July 1991 edition of *Car Lines*)
214. Martin, D. J. and Shock, R. A. W. (1989). *Energy use and energy efficiency in the UK transport up to the year 2010*. Energy Support Unit. HMSO: London
215. Mills, E. (1991). Evaluation of European lighting programmes. *Energy Policy*, **19**(3), 266–78
216. Evans, R. D. and Herring, H. P. J. (1989). *Energy use and energy efficiency in the UK domestic sector up to the year 2010*. DoEn & HMSO: London
217. Schipper, L. and Ketoff, A. (1983). Home energy use in nine OECD countries 1960–1980. *Energy Policy*, **11**(2), 131–47
218. Anon (1992). Many households lack basic energy efficiency measures. *ENDS Report* **210**, 26
219. Schipper, L., and Hawk, D. V. (1991). More efficient household electricity use. *Energy Policy* **19**(3) 244–62
220. Energy Efficiency Office (1990). *Energy efficiency in domestic electric appliances*. DoEn & HMSO: London
221. *Guardian, The* (1991). Washing machines fail eco-test on water and electricity use. *The Guardian*, 25th November, 6. London & Manchester
222. Owens, J. and Wilhite, H. (1988). Household energy behaviour in Nordic countries – An unrealised energy saving potential. *Energy*, **13**(12), 853–9
223. Ross, M. (1989). Improving the efficiency of electricity use in manufacturing. *Science*, **244**, 311–7
224. Torvanger, A. (1991). Manufacturing sector carbon dioxide emissions in nine OECD countries 1973–87. *Energy Economics*, **13**(3), 168–83
225. Horowitz, M. J. (1989). Energy efficiency improvements and investment behaviour in small commercial buildings. *Energy*, **14**(11), 697–707
226. Cooter, M. (1991). NHS could save some £30m on power. *BMJ*, **302**, 805 (See also *Saving energy in the NHS*. HMSO: London)
227. Burton, R. (1990). St Mary's Hospital, Isle of Wight: A suitable background for caring. *BMJ*, **301**, 1423–5
228. DoEn (1983). *Investing in energy use as an alternative to an investment in energy supply*. DEN/S/3(NE)
229. Energy Policy Unit (1982). *Investment in energy supply and energy use*. DoEn: London. (Submitted to the Sizewell enquiry by the Council for the Protection of Rural England as document CPRE/S/21(NE) in 1983)

230. DoEn (1979). Energy conservation – Scope for new measures and long-term strategy. *Energy Paper*, **33**. DoEn & HMSO: London
231. Advisory Committee on Business and the Environment (1993). *ACBE Third Progress Report*. DoE: London
232. Committee on Science, Engineering & Public Policy (1991). *Policy implications of greenhouse warming*. National Academy Press. Reported in *Science*, **252**, 204–5
233. Ysohizawa, M. (1988). Electricity conservation in Brazil: Potential and progress. *Energy*, **13**(6), 469–83
234. Johanasson, T. B. and Williams, R. H. (1987). Energy conservation in the global context. *Energy*, **12**(10/11), 907–19
235. Goldemberg, J., Johansson, T. B., Reddy, A. K. N. and Williams, R. H. (1987). *Energy for a Sustainable World*. World Resources Institute: Washington, DC
236. Gillis, A. M. (1991). Why can't we balance the global carbon budget? *BioScience*, **41**(7), 442–8
237. Huntley, M. E., Lopez, M. D. and Karl, D. M. (1991). Top predators in the Southern Ocean: A major leak in the biological pump. *Science*, **253**, 64–66
238. Whittaker, R. H., (1975). *Communities and Ecosystems*, 2nd edition. MacMillan: New York
239. Nisbet, E. G. (1991). *Leaving Eden: To protect and manage the Earth*. Cambridge University Press. Cambridge
240. Whitmore, T. C. and Sayer, J. A. (1992). *Tropical deforestation and species extinction*. Chapman & Hall: London
241. Skole, D. and Tucker, C. (1993). Tropical deforestation and habitat fragmentation in the Amazon: Satellite data from 1978 to 1988. *Science*, **260**, 1905–9
242. Lanly, J-P., Singh, K. D. and Janz, K. (1991). FAO 1990 reassessment of tropical forest cover. *Nature & Resources*, **27**(2), 21–6
243. Laird, J. (1991). S. E. Asia's trembling rain forests. *Our Planet*, **2**(4), 4–11
244. Down to Earth (1991). *Pulping the Rain Forest*. International Campaign for Ecological Justice in Indonesia. Reported in Sattour, O. (1991) Indonesia pulps its rain forest. *New Scientist*, 24th August, 15
245. Carroll, J. B. and Thorpe, I. C. (1991). The conservation of the Livingstone fruit bat *Pteropus Livingstonii*: A report on an expedition to the Comares. *The Dodo – Journal of the Jersey Wildlife Preservation Trust*, 26–40
246. Rosencranz, A. and Scott, A. (1992). Siberia's threatened forests. *Nature*, **355**, 293–4
247. Cave, S. (1990). Tropical forests: Key to global climate. *Our Planet*, **2**(3), 8–11
248. Lean, J. and Warrilow, D. A. (1989). Simulation of the regional climatic impact of Amazon deforestation. *Nature*, **342**, 411–13
249. Dickson, R. E. (1989). Predicting climatic effects. *Nature*, **342**, 343–4
250. Gash, J. (1991). Tropical rainforest and climate. *NERC News*, Jan, 28–9
251. Cachier, H. and Ducret, J. (1991). Influence of biomass burning on equatorial African rains. *Nature*, **352**, 208–30
252. Hedges, B. S. and Woods, C. A. (1993). Caribbean hot spot. *Nature*, **364**, 375

253. Ashford, J. (1991). Nutty prices mean chop for Amazon trees. *New Scientist*, 21/26 December, 7
254. Pinedo-Vasques, M. (1990). Land use in the Amazon. *Nature*, **384**, 397
255. Josephson, J. (1991). Tropical forests: Effects of destruction *versus* preservation. *Environmental Science & Technology*, **25**(7), 1204–6
256. Vitousek, P. (1990). *Reforestation to offset carbon dioxide emissions*. EN-6910 Research Project 3041-1 prepared by Stanford University for the Electric Power Research Institute: California
257. Gaia (1990). Chopping down trees for books and journals: A 'green' practice? *Biologist*, **37**(5), 141
258. Cowie, J. (1991). The greening of SF book collecting. *Nexus*, **1**, 28–9
259. Marsh, C. (1993). Carbon dioxide offsets as potential funding for improved forest management. *Oryx*, **27**, 2–3
260. Flavin, C. (1990). Slowing global warming, in *State of the World 1990*. Unwin Paperback: London
261. Kulp, J. L. (1990). *The phytosystem as a sink for carbon dioxide*. Report EN-6786 for the Electric Power Research Institute: California
262. Finch, S. (1991). Are you being misled. *Publishing* (November), 48–9
263. Anon (1991). Newspaper recycling targets draws fire from environmentalists. *ENDS Report*, **200**, 13
264. Anon (1991). Paper group sets up one eco- labelling scheme. *ENDS Report*, **197**, 27–8
265. Allen, M. R. and Christensen, J. M. (1990). Climate change and the need for a new agenda. *Energy Policy*, **18**(1), 19–24
266. Greenhalgh, G. Energy conservation policies. *Energy Policy*, **18**(3), 293–9
267. BP Economics Unit (1991). *BP Statistical Review of World Energy*. British Petroleum Corporate Communications Services: London
268. Seiver R. (1980). *Energy and environment: Readings from Scientific American – Energy Policy in the US*. Commentary pp3940. Freeman & Co: San Francisco
269. Eikeland, P. O. (1993). US energy policy at a crossroads? *Energy Policy*, **21**, 987–98
270. Dickson, D. (1994). Discord over IPCC meeting re-opens climate dispute. *Nature*, **371**, 467
271. Ellington, A. and Burke, T. (1981). *Europe: Environment*. Eco Books: London
272. Nielson, J. H. (1990). Denmark's energy future. *Energy Policy*, **18**(1), 80–5
273. Surrey, J. (1990). Beyond 1992 – the single market and EC energy issues. *Energy Policy*, **18**(1), 42–54
274. Anon (1993). Fragile deal on EC carbon tax leaves UK out in the cold. *ENDS Report*, **219**, 37–8
275. Anon (1993). EC faces failure on CO_2 stabilization goal. *ENDS Report*, **217**, 36
276. DoEn (1977). Energy policy review. *Energy Paper*, **22**. DoEn and HMSO: London
277. DoEn (1978). *Energy policy a consultative document*. cmnd 7107. HMSO: London
278. Fells, I. (1991). What price energy policy? *New Scientist*, 26th October, 12

279. Energy Committee (1989). *Energy policy implications of the greenhouse effect* vol 1. HMSO: London
280. Cabinet Office (1992). *Climate change: The UK programme.* cmnd 2427. HMSO: London
281. Anon (1994). A short-term strategy on climate change. *ENDS Report*, **228**, 21–4
282. Anon (1995). NFFO-3/SRO-1 results announced. *REView* (Supplement to the February edition.) Department of Trade & Industry: London
283. Krause, F., Bach, W. and Kooney, J. (1990). *Energy Policy in the greenhouse: From warming fate to warming limit*, vol 1. Earthscan: London
284. DoE (1990). *This Common Inheritance.* cmnd 1200. HMSO: London
285. Department of Transport (1989). *Roads for Prosperity.* cmnd 693. HMSO: London
286. Motor Vehicle Manufacturer's Association (1989). *World motor vehicle data.* MVMA: Detroit
287. Bleviss, D. L. (1990). The role of the automobile. *Energy Policy*, **18**(2), 137–48
288. Rainbow, R. and Tan, H. (1993). *Meeting the demand for mobility.* Shell selected paper. Shell International: London
289. Ottewell, S. (1990). Paying the price of petrol. *The Chemical Engineer*, **484**, 15
290. Kula, E. (1992). *Economics of natural resources and the environment.* Chapman & Hall: London
291. Bator, M. F. (1958). The anatomy of market failure. *Quarterly Journal of Economics*, **72**, 351–79
292. Buchanan, J. and Stubblebine, (1962). Externality. *Econometrica*, **29**, 371–84
293. Ariadne (1991). Ariadne. *New Scientist*, 23rd November, 88
294. Elliot, L. and Milner, M. (1990). Surge in oil prices wipes out billions of shares. *The Guardian*, cols 1–4, 1, 3rd August
295. CEGB – Department of Information and Public Affairs (1989). *Statistical yearbook 1988/89.* Central Electricity Generating Board: London
296. Baker, J. (1989). Personal communication to 'all National Power staff' 9th November. Internal memo
297. Marshall, W. (1989). Personal communication to 'all CEGB staff'. 18th December. Internal memo
298. Marshall, W. (1989). *The future for nuclear power.* British Nuclear Energy Society Annual Lecture, 30th November, London
299. Collier, J. (1992). *Nuclear Electric.* BBC TV Select on BMH-The Business Channel
300. Thomas, D. and Pearson, C (1990). Electricity price shares soar. *The Financial Times* 12th December cols 2–4, 1. Also see Kinnock hits at electricity sell-off scandal. cols 1–4, 11
301. Skarbon, C. and Roberts, S. (1991). Cutting your fuel bills – Cutting heating costs. *Which* (October), 566–7
302. Samouilidis, J. E., Berhas, S. A. and Psarras, J. E., (1983). Energy conservation – Central *versus* decentralized decision making. *Energy Policy*, **11**(4), 302–11
303. DoEn (1982). Advisory Council on Energy Conservation report to the Secretary of State. *Energy Paper*, **49**. HMSO: London
304. Anon (1991). National Audit Office report. *Atom*, **417**, 29–30

305. Department of Trade & Industry (1992). *Energy Trends: A Statistical Bulletin.* September. DTI: London
306. Hanneberg, P. (1991). Economic instruments – Tomorrow's approach to protecting the environment. *Enviro*, **12**, 2–5
307. Jornstedt, O. (1991). Green taxes gaining ground in Sweden. *Enviro*, **11**, 27–8
308. Anon (1992). Growing with environmental care. *Nature*, **357**, 177
309. Meadows, D. H., Meadows, D. L. and Randers, J. (1992). *Beyond the Limits.* Earthscan: London
310. Anon (1992). Power offer on CO_2 gets short shrift. *ENDS Report*, **208**, 7
311. Schevega, J. D. and Leary N. A. (1991). Efficiency of climate policy. *Nature*, **354**, 193
312. Anon (1991). *The EC carbon/energy tax and the UK economy.* Cambridge Econometrics Press Release, 2nd December
313. Centre for Global Energy Studies (1992). *Carbon taxes: Levelling the playing field.* CGES: London
314. Department of Trade and Industry (1992). *Government seeks industry's view on carbon tax proposals.* DTI Press Release P/92/524, 7th August
315. Wilmhurst, P. (1994). Temperature and cardiovascular mortality. *BMJ*, **309**, 1029–30
316. Watt, G. C. (1994). Health implications of putting value added tax on fuel. *BMJ*, **309**, 3030–1
317. Organization for Economic Co-operation and Development (1992). *The OECD environment industry – Situation, prospects and government policies.* OECD: Paris
318. Feedback (1992). Feedback in Rio. *New Scientist*, 13th June, 88
319. Lovelock, J. E. (1988). Schumacher lecture – Stand up for Gaia. Reprinted from *Resurgance* in *Biologist* (1989), **36**(5), 241–7
320. Dew, M. A., Bromel, E. J., Schuberg, H. C., Leslie, O., Dunn, M. P. H and Parkinson, D. K. (1987). Mental health effects of the Three Mile Island nuclear reactor re-start. *American Journal of Psychiatry*, **144**(8), 1074–7
321. Edwards, R. (1995). Not yodelling but drowning. *New Scientist*, 11th November, 5
322. Narum, D. (1992). A troublesome legacy – The Reagan Administration's conservation and renewable energy policy. *Energy Policy*, **20**(1), 40–52
323. Boyes, E. and Stanisstreet, M. (1990). Misunderstandings of 'law' and 'conservation': a study of pupil's meanings for these terms. *School Science Review*, **72**(258), 51–7
324. Levey-Leblond, J-M. (1992). About misunderstandings about misunderstandings. *Public Understanding of Science*, **1**(1), 17–21
325. Platt, J. (1973). Social traps. *American Psychologist*, **28**, 641–51
326. Block, J. (1989). The role of efo-control, ego-resiliency and IQ in delaying gratification in adolescence. *Journal of Personality and Social Psychology*, **57**(6), 1041–50
327. Wilson, J. Q. and Hernstein, R. J. (1985). *Crime and Human Nature.* Touchstone Books: New York

328. Southern, D. A. and Read, M. S. (1994). Overdosage of opiate from patient controlled analgesia devices. *BMJ*, **309**, 1002
329. Wilkinson, T. G. and Wilkinson, P. (1985). Environmental groups and energy priorities. *Science and Public Policy*, **12**(5). 273–8
330. Nowak, M. and May, R. M. (1992). Evolutionary games and spatial chaos. *Nature*, **359**, 826–9
331. Nowak, M. and Sigmund, K. (1993). A strategy of win–stay, lose–shift that out performs tit-for-tat in the Prisoner's Dilemma game. *Nature*, **364**, 56–8
332. Vine, E., Barnes, B. K. and Ritschard, R. (1988). Implementing home energy systems. *Energy*, **13**(5), 401–11
333. Sassone, P. G. and Martucci, M. V. (1984). Industrial energy conservation: The reasons behind the decisions. *Energy*, **9**(5), 427–37
334. Gruber, E. and Brand, M. (1991). Promoting energy conservation in small and medium-sized companies. *Energy Policy*, **19**(3), 279–87
335. US Congress, Office of Technology Assessment (1991). *Energy efficiency in the Federal Government: Government by good example?* OTA-E-492. US Government Printing Office: Washington, DC
336. Lenssen, N. (1993). Providing energy in developing countries. In Brown, L. (1993) *State of the World.* Earthscan: London
337. Imperial Chemical Industries (1990). *ICI and the environment.* ICI External Relations Department
338. Hoskins, B. J., Hall, N. M. and Valdes, P. J. (1994). Storm tracks in a warmer, moister World. European Conference on Energy and the Water Cycle 16th June. In *Abstracts of oral & poster sessions.* Royal Society: London
339. Berz, G. A (1991). Global warming and the insurance industry. *Nature & Resources*, **27**(1), 19–28
340. Jennings, J. S. (1995). *Future sustainable energy supply.* Public Affairs – Shell Centre: London
341. Pearce, D. (1993). *Economic Values and the Natural World.* Earthscan: London
342. Cohen, J. (1995). *How many people can the Earth support.* Norton: New York
343. Nordhaus, W. (1990). *To slow or not to slow: The economics of the greenhouse effect.* Dept Economics: Yale University
344. Pearce, D. (1991). *Blueprint 2.* Earthscan: London
345. Pigou, A. (1929). *The Economics of Welfare.* Macmillan: London
346. Galbraith, J. K. (1992). *The Culture of Contentment.* Sinclair-Stevenson: London
347. Grubb, M. (1992). Climate change. *Nature*, **358**, 448
348. MacKenzie, D (1990). Scientists clash with politicians over CO_2 emissions. *New Scientist*, 10th November, 13
349. Anon (1990). Global warming gas forecast inflated to aid Thatcher image. *The Independent*, 7th June, 3, cols 1–3
350. Minerva (1991). Views. *BMJ*, **303**, 322, reporting from the *British Journal of Psychiatry* (1991), **159**, 19–32
351. Anon (1993). Gummer slips up on energy consumption. *ENDS Report*, **224**, 7
352. Middttun, A. and Baumgartner, T. (1986). Negotiating energy futures: The politics of energy forecasting. *Energy Policy*, **14**, 219–41

353. Godlee, F. and Walker, A (1991). Importance of a healthy environment. *BMJ*, **303**, 1124–6
354. Godlee, F. and Walker, A. (1992). *Health and the Environment*. BMJ Books: London
355. Dunwoody, S. and Peters, H. P. (1992). Mass media coverage of technological and environmental risks: A survey of research in the US and Germany. *Pub. Und. Sci.*, **1**(2), 199–230
356. Anon (1992). Papers featured in the New York Times cited more often. *BioScience*, **42**(2), 150
357. Bell, A. (1994). Media (mis)communication on the science of climate change. *Pub. Und. Sci.*, **3**(3), 259–75
358. Jones, D (ed.) (1990). The greenhouse conspiracy. *Equinox*. TVF for Channel 4, August 1990
359. Jones, D (ed.) (1990). The greenhouse conspiracy. Programme script of Jones, D., TVF. *ibid*
360. Schneider, S. (1989). *Global Warming*. Lutterworth Press: Cambridge
361. International Institute for Environment and Development & World Resources Institute (1987). *World Resources 1987*. Basic Books: New York
362. Tyme, J. (1978). *Motorways versus Democracy*. Macmillan Press: London
363. Pooley, D. (1982). Energy options for the long-term future. *Energy World*, **18** 12–15
364. Sioshansi, F. P. (1991). The myths and facts of energy efficiency: Survey of implementation issues. *Energy Policy*, **19**(3), 231–41
365. Lenssen, N. (1993). Providing energy in developing countries. In Brown, L. (ed.) (1993) *State of the World*. Earthscan: London
366. Greenhalgh, G. (1989). The comforting illusion of energy conservation. *Atom*, **393**, 6–7
367. Boehmer-Christiansen, S. A. (1990). Energy policy and public opinion. *Energy Policy*, **18**(9), 828–37
368. Jeffery, J. W. (1982). The nuclear economic fraud. *The Ecologist*, **12** (2), 80–6
369. Electricity Council (1986). *Handbook of electricity supply statistics*. Electricity Council: London
370. Anon (1991). Can PR cool the greenhouse? *Science*, **252**, 1784–5
371. Anon (1994). More US people favour nuclear energy. *Atom*, **435**, 10
372. Anderson, C. (1992). How much green in the greenhouse? *Nature*, **356**, 369
373. Macilwain, C. (1994). Earmark lobbyists paid $60 million a year. *Nature*, **371**, 367
374. Government Statistical Service (1992). *The UK Environment*. HMSO: London
375. Anon (1992). Recession fails to dent public concern on the environment. *ENDS Report*, **212**, 3
376. Doble, J (1995). Public opinion about issues characterized by technological complexity and scientific uncertainty. *Pub.Und.Sci.*, **4**(2), 95–118
377. Marean, C. W. and Gifford-Gonzalez, D. (1991). Late quaternary extinct ungulates of East Africa and palaeoenvironmental implications *Nature*, **350**, 418–20
378. Abate, T. (1994). Climate and the collapse of civilization. *BioScience*, **44**(8), 516–9
379. Hodell, D. A., Curtis, J. H. and Benner, M. (1995). Possible role of climate in the collapse of Classic Maya civilization. *Nature*, **375**, 391–4

380. Knox, J. C. (1993). Large increases in flood magnitude in response to modest changes in climate. *Nature*, **361**, 430–2
381. deMenocal, P. B. (1995). Plio-Pleistocene African climate. *Science*, **270**, 53–8
382. Dayton, L. (1994). The fat, hairy women of Pompeii. *New Scientist*, 24th September, 10
383. Kerr, R. A. (1995). A volcanic crisis for ancient life. *Science*, **270**, 27–8
384. Rampino, M. R. and Self, S. (1992). Volcanic winter and accelerated glaciation following the Toba super-eruption. *Nature*, **359**, 50–2
385. Taylor, K. E. and Penner, J. E. (1994). Response of the climate system to atmospheric aerosols and greenhouse gases. *Nature*, **369**, 734–7
386. Jones, A., Roberts, D. L. and Slingo, A. (1994). A climate model study of indirect radiative forcing by anthropogenic aerosols. *Nature*, **370**, 450–3
387. Mitchell, J. F. B., Johns, T. C., Gregory, J. M. and Tett, S. F. B. (1995). Climate response to increasing levels of greenhouse gases and aerosols. *Nature*, **376**, 501–4
388. Li, X., Maring, H., Savoie, D., Voss, K. and Prospero, J. M. (1996). Dominance of mineral dust in aerosol light scattering in the North Atlantic trade winds. *Nature*, **380**, 416–22
389. Kerr, R. A. (1995). Scientists see greenhouse, semioffically. *Science*, **269**, 1667
390. Intergovernmental Panel of Climate Change (1995). *Summary for policymakers of the Contribution of working Group I to the IPCC 2nd Assessment Report*, 1995. UNEP & WMO: Bracknell
391. Vaughan, D. G. and Doake, C. S. M. (1996). Recent atmospheric warming and retreat of ice shelves on the Antarctic Peninsula. *Nature*, **379**, 328–30
392. Charatan, F. B. (1995). Hundreds die in US as temperatures reach 41°C. *BMJ*, **311**, 277
393. Bloom, D. E. (1995). International public opinion on the environment. *Science*, **296**. 354–7
394. Myers, N. (1993). Environmental refugees in a globally warmed world. *BioScience*, **43**(11), 752–60
395. Anon (1996). Climate debate must not overheat. *Nature*, **381** 539
396. Geller, H., DeCicco, J., Laitner, S. and Dyson, C. (1994). Twenty years after the embargo: US oil import dependence and how it can be reduced. *Energy Policy*, **22**(6), 471–84
397. Greene, D. L. and Leiby, P. N. (1993). *The Social Costs to the US of Monopolization of the World Oil Market 1972–1991*. ORNL-6744. Oak Ridge National Laboratory, Oak Ridge, TN
398. Office of Technology Assessment (1991). *Changing by degrees: Steps to reduce greenhouse gases*. Reported in *Science*, **251**, 621
399. Reichhardt, T. (1997). No net cost in cutting carbon emissions. *Nature*, **389**, 429
400. Masood, E. (1996). Sparks fly over climate report. *Nature*, **381**, 639
401. Keeling, C. D., Chin, J. F. S. and Whorf, T. P. (1996). Increased activity of northern vegetation inferred from atmospheric CO_2 measurements. *Nature*, **382**, 146–9
402. Intergovernmental Panel on Climate Change (1996). *Climate change 1995: The science of climate change – Summary for policymakers and Technical Summary of the Working Group I report*. UNEP & WMO: Bracknell

APPENDIX I

ENERGY AND UNITS

Powers of 10

10^{-1}	deci	d	10^{3}	kilo	k
10^{-2}	centi	c	10^{6}	mega	M
10^{-3}	milli	m	10^{9}	giga	G
10^{-6}	micro	μ	10^{12}	tera	T
10^{-9}	nano	n	10^{15}	peta	P
10^{-12}	pico	p	10^{18}	exa	E

Billion = 1000 million = 10^{9}

Energy, work and power

Work done at the rate of a Joule (J) a second has a power rating of 1 Watt (W)

1 kilowatt (kW) of power consumed over one hour = 1 kilowatt hour (1 kWh) = 1000 (Watts) × 60 (seconds) × 60 (minutes) = 3600 Joules of energy

TWh = tera Watt hour = $10^{12} \times 60 \times 60 = 3.6 \times 10^{16}$

1 kilojoules = 1000 Joules

1 EV = 1 electron volt = 1.6×10^{-19} J

Energy in terms of million tonnes of oil equivalents (mtoe)

1 ton = 1.016 tonnes 1 tonne = 1000 kilograms (kg)

1 mtoe = 40×10^{12} Btu (British thermal units)
= 397×10^{6} therms
= 0.805×10^{6} tonnes of natural gas
= 12×10^{9} kWh
= 4×10^{9} kWh of electricity if generated in a power station with a (Carnot) efficiency of 33.3% (one third)

btoe = billion tonnes of oil equivalent = mtoe × 10^{3}
1 EJ = 1 Exa Joule = 23.15 mtoe (million tonnes of oil equivalent).

Quantities

1 Barrel = 0.136 tonnes = 35 Imperial gallons = 42 gallons (US)

APPENDIX II

BIOLOGICAL, GEOLOGICAL AND CHEMICAL NOMENCLATURE

The lay reader can safely ignore this section. But students (especially social science students) should be aware that scientific terms are not immune from the vagaries of language.

Biological nomenclature

Where applicable the biological nomenclature in this text is in line with guidelines from the Institute of Biology (IoB), which should be in line with the major academic (book) publishers. (However the IoB numerical and graphical recommended format or presentation of data, is not used as international standards are insufficiently developed). It should though be noted that there are common discrepancies between science dictionaries, and even between specialist science dictionaries.

Geological nomenclature

Geological nomenclature is as fraught, due to the complex history of the quaternary ice age deposits, and regional variations worldwide. For instance what are known as 'Wisconsinian' deposits in the United States correlate with 'Devensian' deposits in the UK, and 'Weichselian' in northern Europe. I have tried to get around this by hyphenating two or more synonymous terms (mainly UK and US) so that students on either side of the Atlantic stand a chance of understanding what is being discussed. So the term 'emm-sangion' glacial is used.

Chemical nomenclature

The chemical nomenclature used in this text is not the latest internationally approved standard. For our purposes this primarily affects the dicussions of CFCs and the nomenclature used in this text is that employed by the Intergovernmental Panel on Climate Change (IPCC), on the basis that most climatological literature draws upon IPCC documentation.

APPENDIX III

ACRONYMS AND PRINCIPAL CHEMICALS

ABB	Asea Brown Boveri
ABC	American Broadcasting Company (US)
ABRACOS	Anglo-Brazilian Climatic Observations Study
ACBE	Advisory Committee on Business and the Environment (UK)
AIBS	American Institute for Biological Sciences (US)
AIDS	Acquired Immune Deficiency Syndrome
B-a-U	Business-as-Usual
BOFS	Biogeochemical Ocean Flux Study
CAP	Common Agricultural Policy (EEC)
CEGB	Central Electricity Generating Board (UK)
CFC	Chlorofluorocarbon
CO_2	Carbon dioxide
CH_4	Methane
DEn	Department of Energy (UK)
DoE	Department of the Environment (UK)
ECM	Electrical conductivity measurement
EC	European Community
EEC	European Economic Community
EFR	European Fast (breeder) Reactor
EPRI	Electric Power Research Institute (US)
ESI	Electricity supply industries
FACE	Forest Absorbing CO_2 Emissions Foundation (ND)
FAO	UN Food and Agricultural Organization
FBR	Fast Breeder Reactor
FCCC	UNCED Framework Convention on Climate Change
FDR	Federal Republic of Germany
GDP	Gross Domestic Product
GISP	Greenland ice sheet project
GNP	Gross National Product
GRIP	Greenland ice core project
GWP	Global warming potential
HEP	Hydroelectric Power
HERS	Home Energy Rating System (US)
HIV	Human Immunodeficiency Virus

ICI	(former) Imperial Chemicals Industry (UK)
IoB	Institute of Biology (UK)
IPCC	Intergovernmental Panel on Climate Change
IR	Infra-red (light/radiation)
ITER	International Thermonuclear Experimental Reactor
IUCN	International Union for the Conservation of Nature
JET	Joint European Torus
MP	Members of Parliament (UK)
NCCR	National Center for Climatic Research (US)
NET	Next European Torus
NFFO	Non-Fossil Fuel Obligation (UK)
NOAA	National Oceanic and Atmospheric Administration (US)
N_2O	Nitrous oxide
OECD	Organization for Economic Co-operation & Development
OPEC	Organization of Petroleum Exporting Countries
OTA	Office of Technology Assessment (US)
ppb	Parts per billion
ppm	Parts per million
ppmv	Parts per million by volume
PSMD	Potential Soil Moisture Deficit
PWR	Pressurized Water Reactor
R&D	Research and development
R/P	Reserve to production ratio
TGS	Thermal Growing Season
TOE	Tonnes of oil equivalent
UK	United Kingdom of Great Britain & Northern Ireland
UKAEA	UK Atomic Energy Authority
UNCED	UN Conference on the Environment & Development
UNEP	United Nations Environment Programme
UNESCO	UN Educational, Scientific & Cultural Organization
US	United States (of America)
UV	Ultra-violet (light/radiation)
VAT	Value Added Tax (UK)
WCS	World Conservation Strategy
WHO	UN World Health Organization
WMO	World Meterological Organization
WOCE	World Ocean Circulation Experiment (EC)
WWF	(formerly World Wildlife Fund) World-Wide Fund for Nature

APPENDIX IV

GLOBAL WARMING POTENTIALS

Chemical	*Global Warming Potential (Time horizon)*		
	20 years	*100 years*	*500 years*
Carbon dioxide	1	1	1
Methane	56	21	6.5
Nitrous oxide	280	310	170
HFC-134a	3400	1300	420
HFC-152a	460	140	42
HFC-143a	5000	3800	1400

(Summarised from the IPCC, 1995[390])

INDEX